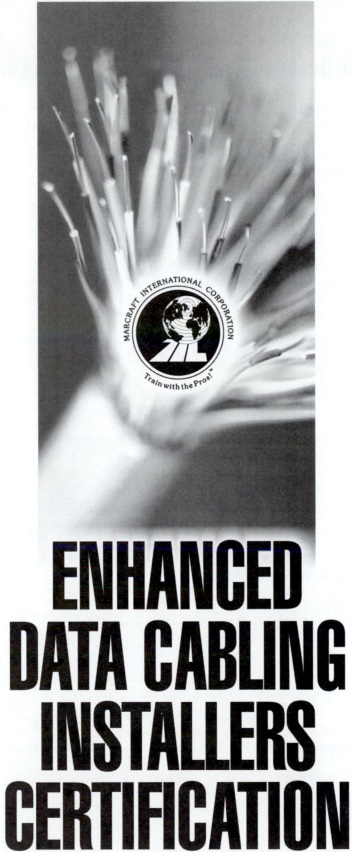

MARCRAFT INTERNATIONAL CORPORATION

Train with the Pros!™

D0164582

ENHANCED DATA CABLING INSTALLERS CERTIFICATION

Trademark Acknowledgments

Written by Brent L. Wright and Don Ritchey
Edited by Whitney G. Freeman
Original graphics created by Michael R. Hall and Cathy J. Boulay

10 9 8 7 6 5 4 3 2
ISBN: 0-13-091617-X

P/N DC-200
Printed in the United States of America
1-11/00

Preface

The advent of the Data Cabling Installer Certification from ETAI (Evolving Technologies Association International), has provided the IT industry with an introductory, vendor-neutral certification for personnel that install Category 5 copper data cabling. Marcraft International has been producing IT certification courses for various facets of computer technology and networking since 1988. This course represents an important element in that group of IT industry certifications. It expands the knowledge base of people involved in the commercial construction industry, as well as computer service and network administration personnel, since data cabling ties computers together between rooms, buildings, and continents.

The *Enhanced Data Cabling Installers Certification* course from Marcraft contains a textbook featuring integrated, hands-on lab procedures that prepare students to pass the certification examination, and to accomplish the actual work required. It provides all of the theory necessary to understand the concepts behind cabling standards and to pass the theory examination. In addition, it fully prepares the student to actually perform the tasks required of a certified cable installer.

Data Cabling Installer Certification

ETAI is an organization that establishes certification criteria for entry-level data cable installers in the cabling industry. This organization has created and sponsors the Data Cabling Installer Certification (DCIC) Exam, which is designed to certify cable installers in the installation, testing, and troubleshooting of copper data cable and category rated devices and components. Certification is accomplished by satisfactorily passing the DCIC exam. For more information on ETAI and the Data Cabling Installer Certification exam, visit http://www.etainternational.org.

The Marcraft *Enhanced Data Cabling Installers Certification* training guide provides students with the knowledge and skills required to pass the Data Cabling Installer Certification exam and become a certified cable installer. The DCIC is recognized nation-wide and is the hiring criterion used by major communication companies. Therefore, becoming a certified data cable installer will enhance your job opportunities and career advancement potential.

Marcraft has embraced the DCIC because it sets a standard for excellence that we have been preparing students to achieve from our beginning. Therefore, we are offering this course to prepare students to successfully challenge the DCIC examination. From its size, one should gather that this course is not simply a cram course for the test. Instead, it is a complete training course designed to prepare for the exam, as well as to provide the fundamental knowledge base required to establish a career in this rapidly changing industry.

The textbook, with integrated lab procedures, and accompanying test-prep materials are intended for anyone interested in pursuing the Data Cabling Installer Certification. While it contains all of the pedagogical support materials required for use in a classroom environment, it can also be used by experienced technicians to prepare for the exam in a self-study mode.

This first edition of the *Enhanced Data Cabling Installers Certification* training guide includes information about all of the current DCI objectives needed to pass the certification exam. The latest information has been provided in every chapter as appropriate for the new TIA/EIA and ANSI standards.

Key Features

The pedagogical features of this book were carefully developed to provide readers with key content information, as well as review and testing opportunities. Each chapter begins with a list of learning objectives that establishes a foundation and systematic preview of the chapter and concludes with a summary of the points and objectives to be accomplished. Key terms are presented in bold throughout the text.

Appendices

Appendix A contains Manufacturer Instruction Sheets, which provide supplemental information used in the lab procedures. Appendix B provides complete information on the ETAI Certification exams. Appendix C is a comprehensive glossary of words and meanings to provide quick, easy access to key term definitions that appear in each chapter. Appendix D provides a list of common acronyms used in the cabling industry.

Evaluation and Test Material

An abundance of test materials is available with this course. At the end of each chapter, there is a question section with open-ended review questions which are designed to test critical thinking. Additional test material in the form of multiple-choice questions can be found on the interactive CD-ROM.

Lab Exercises

Applying the concepts of the chapter to hands-on exercises is crucial to preparing for a successful career as a data cable installer. Each chapter includes integrated lab sections that call upon the students to perform exercises that reinforce the chapter content via hands-on exploration. Questions appear at the end of each lab enabling students to assess their understanding of the lab.

Interactive CD-ROM

The *Enhanced Data Cabling Installers Certification* training guide is accompanied by a comprehensive, electronic practice test bank on a CD-ROM, which is sealed on the back cover of the book. This CD testing material was developed to simulate the Data Cabling Installer Certification Exam process and to allow students to complete mock tests, determine their weak points, and study more strategically.

During the question review, the correct answer is presented on the screen, along with the reference heading where the material can be found in the text. A single mouse click will take you directly to the corresponding section of the embedded electronic textbook.

ORGANIZATION

In general, it is not necessary to move through this text in the same order that it is presented. Also, it is not necessary to teach any specific portion of the material to its extreme. Instead, the material can be adjusted to fit the length of your course.

Chapter 1 - *Basic Standards and Practices* introduces the student to how the industry is organized and how installation of cabling is standardized. Some insight is provided into why standards are so important. There is also an introduction into the different types and categories of industry-wide cabling. Labs offered in this chapter provide hands-on activities that introduce the tools, cable identification, cable examinations, and terminations used in the industry.

Chapter 2 - *Cable Ratings and Performance* builds on the introductory material from Chapter 1 to show the different ratings applied to cabling used in the industry and to show how the performance of the different types of cable affects installations. The series of hands-on labs included in this chapter examine the different types of cabling and termination methods involved in the installation of cabling in the industry today.

Chapter 3 - *Cable Installation and Management* examines the different types of buildings and cabling facilities involved in cable installations. It also addresses the different styles of cable terminations used. The hands-on lab procedures included in this chapter assemble and install different types of connections on cabling. They also introduce the testing of cable connections.

Chapter 4 - *Testing and Troubleshooting* takes an in-depth look at the testing and troubleshooting unique to cable installation, offering techniques and tips as the testing tools are examined. The hands-on labs in this chapter examine the various pieces of test equipment used to test cable and their calibration procedures. These lab activities also cover proper procedures for using the equipment to test and troubleshoot sample cable installations and terminations.

Chapter 5 - *Industry Standards* provides comprehensive exploration of the importance and variety of standards used by the cabling industry. The student will gain a thorough understanding of specifications and requirements of various standards and TSBs observed by the industry for different cable installation tasks.

Chapter 6 - *Pulling Cable* deals with how to get cables through conduits that contain massive amounts of installed cabling. The unique problems associated with pulling data cable are examined along with proper techniques for acceptable cable installations. The hands-on procedures associated with this chapter allow students to experiment with pulling of different types of cable and terminations techniques.

Chapter 7 - *Understanding Blueprints* offers a basic knowledge of blueprint reading. This final chapter is essential. Without blueprints, cable installations would be haphazard and unruly. To understand blueprints is to have clarity for the cable installer. The hands-on procedure in this chapter provides an activity in examining and identifying items associated with blueprints.

Teacher Support

A full-featured Instructor's Guide is available for the course. Answers for all of the end-of-chapter Quiz Questions are included along with a paragraph reference in the chapter where that item is covered. Sample schedules are included as guidelines for possible course implementations. Answers to all Lab Questions are provided so that there is an indication of what the expected outcomes should be.

AUTHOR

Brent L. Wright is the founder, President/Dean, and Chairman of the Board of Wrightco Technologies Inc. He is also the Executive Director of the Wrightco Educational Foundation, a non-profit educational trust, providing educational opportunities to students entering the Communications, Information Technology, and Security Alarm industries.

As Executive Director of the Evolving Technologies Association International (ETAI), and Chairman of its Certifications Committee, Brent L. Wright has been active in the development of meaningful certifications, as well as in promoting and upgrading technology industry standards.

WRIGHTCO TECHNOLOGIES

Wrightco Technologies is a nationally recognized leader in technology training, specializing in fiber optics, data communications, telecommunications, video, and security alarm systems. Wrightco is a post-secondary institute of higher education, operating sixteen multi-state campus locations serving the local and international markets.

Wrightco Technologies also provides and administers the ETA International certifications discussed in this book.

ACKNOWLEDGMENTS

To my wife, Joan B. Wright, who provided support, encouragement, love, and understanding while I was taking time from the family.

To my experienced technical staff of professionals at Wrightco Technologies Incorporated and Technical Training Institute for the research and review of the material contained in this book. The staff effort was led by my son-in-law Donald W. Ritchey, Fredrick D. Carlson, Richard G. Ardini, Thomas R. Lewis, and Crissy A. Stiffler. To all of these employees, a well-deserved thank you!

To Whitney G. Freeman, Michael R. Hall, and Cathy J. Boulay at Marcraft International Corporation, who provided the layout, graphics, and some additional lab/text materials for this book. Thanks also to Stuart Palmer for all his efforts in perfecting the hands-on portions of the book.

Table of Contents

Chapter 1 - Basic Standards and Practices

Chapter 2 - Cable Ratings and Performance

Chapter 3 - Cable Installation and Management

Chapter 4 - Testing and Troubleshooting

Chapter 5 - Industry Standards

Chapter 6 - Pulling Cable

Chapter 7 - Understanding Blueprints

Appendix A - Manufacturer Instruction Sheets

Appendix B - ETA International Certifications

Appendix C - Glossary

Appendix D - Acronyms

Index

CHAPTER

1

BASIC STANDARDS AND PRACTICES

OBJECTIVES

Upon completion of this chapter and its related lab procedures, you should be able to perform these tasks:

1. State the advantages that come from establishing industrial cabling standards.

2. Identify the standards that specifically apply to cable installation.

3. Describe the basic operation of the telephone system in the United States.

4. Identify the components of the telephone system.

5. Identify the wire types used in telephone systems.

6. Describe the accepted color coding strategies commonly used in telephone installations.

7. Identify typical physical topologies, and state the advantages and disadvantages for each.

8. List the accepted logical topologies.

9. State the general instructions for running copper cabling.

10. Given a specific application, identify the Grade of cabling that would be appropriate.

11. Identify Type-66 and Type-110 insulation displacement connectors.

12. Identify the various types of modular jacks commonly used with telecommunications systems.

Basic Standards and Practices

INTRODUCTION

From the earliest days of Man, our ability to communicate has been what has separated us from the animal kingdom. As man evolved, so did the need to communicate, not just in a local area, but in an ever-expanding world.

In the early stages, smoke signals, lantern signals, or even drums accomplished short distance communication, while long distance communication was limited to messenger service, either on foot, or by horseback. These methods were very time consuming, and, as in the case of a messenger, it could take weeks to reach certain destinations.

Our means of communication were very primitive to say the least, until February 14, 1874, when Alexander Graham Bell filed for U.S. patent No. 174,465. Though the patent was initially filed on February 14, 1874, it would not be until March 10, 1874 that Mr. Bell, and his assistant Thomas Watson, produced two telephones that were capable of transmitting a recognizable signal.

The Bell Telephone Company was formed on July 9, 1877. Chief investor, Gardinar Hubbard, was named trustee, and had complete control of the business and its policies. This is significant in the fact that Hubbard initiated the idea that telephones would be leased to the customers.

Until the late 1970's, customers could not install their own telephone equipment. Telephone jacks, provided by the Telephone Company, had to be installed by employees of the Telephone Company.

By the end of 1977, the Supreme Court ruled that the consumer had the right to connect any FCC-approved equipment to the telephone line. This ruling is listed under Part 68 of docket 19528.

The consumer now had a choice, and the era of **Customer Provided Equipment (CPE)** was born. While the acronym refers to the equipment itself, the Interconnect companies' primary service is to supply and/or install the customer provided equipment.

Customer Provided Equipment (CPE)

EVOLUTION OF CABLING STANDARDS

The **Electronic Industries Association (EIA)** was founded in 1924 as the Radio Manufacturers Association. The EIA is a trade organization of manufacturers, which sets standards for use by its member companies.

Electronic Industries Association (EIA)

Telephony

Computer and
Communications
Industry Association
(CCIA)

Telecommunications
Industry Association
(TIA)

United States
Telephone
Association (USTA)

United States
Telecommunications
Suppliers Association
(USTSA)

MultiMedia
Telecommunications
Association (MMTA)

TIA/EIA T568A

Technical Systems
Bulletin (TSB)

As already mentioned, before the late 1970s, it was illegal for customers to do the wiring of their own telephones, or the hooking up of other equipment to their telephone lines. In addition, vendors of computers always maintained the computers and the equipment that the customer purchased.

As **Telephony** (the technology and manufacture of telephone equipment capable of electrically transmitting sound between distant stations) and computer technology matured, consumers had to buy new wiring, connectors, and systems to accommodate the changes. They began to complain about the costs to update their systems, because every system had its own type of connectors and cables that could not be interchanged with any other system. The **Computer and Communications Industry Association (CCIA)** wanted to gain and maintain market confidence. The CCIA went to the EIA about developing cabling standards for commercial buildings.

The **Telecommunications Industry Association (TIA)** began as a group of equipment suppliers in the form of a committee of the **United States Telephone Association (USTA)**. In 1979, the TIA split off of the USTA to form the **United States Telecommunications Suppliers Association (USTSA)**. In 1988, TIA merged with EIA to form TIA/EIA. TIA represents the telecommunications industry (i.e., the companies that provide communications materials, products, systems, distribution services, and professional services). In 1997, the **MultiMedia Telecommunications Association (MMTA)** combined with TIA.

Until 1985, there was no standardization of cabling systems. There was no need for standardization prior to the late 1970s because the local Telephone Companies maintained the cabling needs of the customer.

In 1985, EIA agreed with CCIA that standards were necessary and agreed to design the standards. EIA assigned the enormous task to the TR-41 committee. The TR-41 assigned subcommittees to assist in the task. It took EIA many years to design the building cabling standards. In Oct. 1990, the TIA/EIA 569 Commercial Building Standard for Telecommunications Pathways and Spaces was released. In May 1991, TIA/EIA 570 Residential and Light Commercial Telecommunication Wiring Standard was released. In July 1991, the TIA/EIA 568 Commercial Building Telecommunications Wiring Standard was released. In Oct. 1995, TIA/EIA released a revision to the TIA/EIA 568 calling it the **TIA/EIA T568A** Commercial Building Telecommunications Cabling Standard.

The TIA/EIA is constantly doing revisions to the standards. As technology continues to change and push the limits, the standards also will be updated to accommodate the changes in performance. When such an update is ready for release, it is called a **Technical Systems Bulletin (TSB)**.

Benefits of Standardization

There are many benefits in the standardization of building cabling designs. One such benefit is consistency of design and installation. The system itself may be complex and unique; however, the design and installation of the system is relatively the same for all such systems. The overall goal is to ensure that all telephone equipment is compatible, and normally interchangeable.

Another benefit of standardization is conformance to physical and transmission line requirements. All equipment will be able to be installed on any existing system, or demarcation, without any major system design changes being undertaken.

Standardization provides a basis for examining a proposed system expansion, or any other change. You can update or upgrade the current system without having to purchase a totally new system to accommodate the changes.

Standardization provides for uniform documentation as well as troubleshooting. This allows competition, as well as lower prices for services, because more than one company can do the repairs or installations.

BASIC TELCO OPERATIONS

Central Offices (COs)

Switching Centers

Trunk Lines

Private Branch
Exchanges (PBXs)

The basic Telco organization is depicted in Figure 1-1. Notice how the various **Central Offices (COs)** connect directly to the individual subscriber lines. The COs, in turn, connect to large **Switching Centers** through specially designed **Trunk Lines**. For large office buildings and corporate layouts, **Private Branch Exchanges (PBXs)** take over some of the same responsibilities as the residential COs.

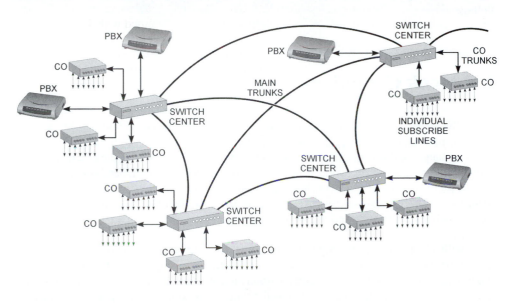

Figure 1-1: Basic Telco Organization

These trunk lines carry telephone signals that have been enhanced and combined (multiplexed) to make the transport of vast quantities of information (both voice and data) more economical. **Main Trunks** that stretch between the various switching centers are capable of carrying even larger amounts of multiplexed information, and they don't necessarily need to be physical in nature. It is true that before the advent of microwave and satellite transmission capabilities, trunk lines were actually physical wires that require many stages of signal restoration, and conditioning, to ensure that a relatively successful transfer of information would normally occur.

Main Trunks

However, modern telephone transmission equipment has made much of the older land-line technology obsolete, although fiberglass transmission lines are becoming more and more common in wide-band applications involving video.

On-Hook Conditions

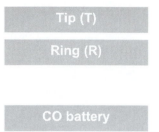

The Telco Central Office provides the dial tone to the consumer using a pair of wires called CO Tip and CO Ring. **Tip (T)** is usually connected to the positive side of the battery at the Central Office. **Ring (R)** is usually connected to the negative side of the battery at the Central Office. You must have both a tip and a ring in order to receive the dial tone.

When the telephone is not in use (the handset is sitting on the cradle or the phone is on-hook), there are 48 VDC continuously applied across the tip, ring, and a series impedance. This is called the **CO battery**. The CO battery voltage is supplied by the Central Office, from a large bank of batteries that are connected in parallel, and is highly filtered. The CO battery allows continued telephone service, even when power from the electric company fails. However, as long as an on-hook condition prevails, the telephone looks like an open circuit, and no current will flow through the wires.

TIP: When working on telephone lines in the premises, if unable to disconnect the phone service, take the phone off the hook to avoid electrical shock by reducing the continuous voltage.

Off-Hook Conditions

When the telephone set is off-hook, or in use, the series impedance (600-900 ohm load) is placed across the tip and ring, reducing the voltage at the CO battery to 6-12 VDC. This reduced voltage condition is called the **talk battery**, during which a **loop current** of 20-90 milliamps is produced. The **range** is broad because the loop current depends on the distance that the premise is from the CO, the **loop resistance**, and how many phone sets are connected to the loop. The loop resistance itself could get as high as 1,500 ohms depending on how long the line is between the CO and the off-hook telephone.

Even with a loop resistance as high as 1,500 ohms, the loop current should still be at least 20 milliamps. For shorter distances, the loop current can be as high as 120 milliamps. Even though modern telephones no longer use carbon microphones, the 20mA current that was required for their satisfactory operation is still a standard used in both telephone and teletype line connections.

Ringing Voltage

In order to ring a telephone that is in the on-hook condition, the CO applies a **ringing voltage** of about 80 to 130 VDC using a sine wave of 20-30 Hz sent down the tip and ring lines. A capacitor prevents the telephone from ringing when alternating current (ac) is not present. The ringing voltage is usually around 100 V (280 V peak-to-peak), but the telephone should be able to ring correctly even if the ringing voltage drops as low as 40 V. This could very well be the situation on telephone lines that develop impedances up to 1,500 ohms.

Unless the telephone being called is on-hook, and the CO is able to detect a high-impedance condition, the ringing voltage will not be applied to its line. The ringing itself is normally cycled repeatedly every 6 seconds, being on for 2 to 2.5 seconds, and off for 3.5 to 4 seconds. In applications where a party-line (one line connecting several customers to a central office) exists, frequencies other than the standard 20Hz signal can be used. The telephone ringers can be adjusted to prevent more than one telephone from ringing at any one time.

Placing a Call

There are two ways that a calling or dialing signal is sent to the Central Office to place a call. The first and oldest method is by using a **rotary dial** to send pulses, which are then translated by the Central Office equipment and used to identify the telephone number of the party being called. When the number is dialed, a set of contacts is opened and closed for a number of times corresponding to the number displayed behind the finger opening in the dialer. This opening and closing action makes and breaks the loop current circuit, as shown in Figure 1-2.

rotary dial

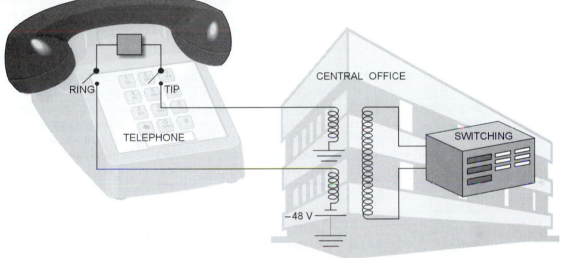

Figure 1-2: Loop Current Circuit

As shown, the circuit which connects the telephone to the CO is in the break position and no current is flowing. The switches at the tip and ring will close to make the connection to the CO, and current will flow from the CO, through the series impedance in the telephone, back to the CO.

There are also pulse-dial telephones that use push buttons instead of the rotary dialer. Telephones that use this dialing method employ a switching circuit to produce the required loop current pulses. Regardless of the method used to achieve pulse dialing, the amount of time that is required to place a call makes it unsuitable for high-speed data transfers.

The second method is called **tone dialing**. This method uses oscillators to generate a combination of two tones, instead of pulses, in transmitting the selected digits to the Central Office. One tone is designated a high frequency, and the other is designated a low frequency.

tone dialing

The selected combination of tones is determined by using a 3 x 4 matrix layout of the common telephone keypad, as shown in Figure 1-3.

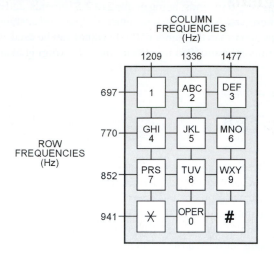

Figure 1-3:
Keypad for DTMF

Notice that there are four low-frequency oscillators (row frequencies) and three high-frequency oscillators (column frequencies) used in this method of dialing, called **Dual-Tone, Multi-Frequency** (**DTMF**). The various combinations produce a distinct tone for each number on the keypad. The tones are processed by the Central Office equipment to determine which digits are dialed.

Keypad frequencies are selected to eliminate the possibility of **intertone distortion** (the misinterpretation by the receiving circuitry as to which key is being pressed). As the tones propagate through the telephone system, the system circuitry acts upon the frequencies to produce harmonics. Also, sum and difference frequencies may be created when the tones **heterodyne** (create new frequencies by mixing). The selected keypad frequencies are not harmonically related to one another, either fundamentally, or by their sum or difference frequencies. This gives telephone systems using DTMF keypads a high degree of reliability.

Voice Vs. Data Signals

Network cabling falls prey to several physical limitations, the most obvious being the signal degradation that occurs across long lengths. As the cabling distance between computers increases, it soon becomes impractical to use dedicated wiring to carry the digital data. Fortunately, the public telephone network helps to solve this otherwise perplexing problem.

The telephone system was originally designed to accommodate a range of frequencies between 300 Hz and 3,300 Hz, or a **bandwidth** of 3,000 Hz, because this was quite adequate for transmitting analog voice. However, because of the fact that computer networks deal primarily with digital data, this same telephone system had a major, built-in drawback; it severely distorted digital data. In order to use the audio characteristics of the phone lines, a device called a **modem** (named for the **mo**dulation and **dem**odulation of its signals) had to be used to encode the computer's digital 1's and 0's into analog signals within this bandwidth.

Although the design of the public phone system limited the frequency at which data could be transmitted over these lines, the modem did enable computers to use them to conduct a data exchange just as if the computers were separated by only a few feet of cable.

Voice Signals

Voice signals are analog in nature, and are therefore sinusoidal, and of constant variation in size and frequency. The simple analog sine wave shown in Figure 1-4 has a peak value of 1 volt, a frequency of 1 kHz, and a **period** of 1 ms.

period

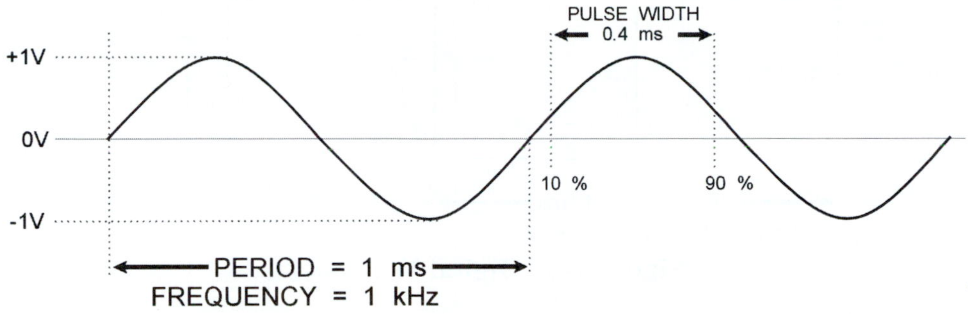

PULSE WIDTH
0.4 ms

+1V

0V

-1V

10 %   90 %

PERIOD = 1 ms
FREQUENCY = 1 kHz

**Figure 1-4:
Analog Waveform**

As an ac waveform, it has positive and negative alternations, and it recurs in time in a predictable manner. Measuring from the 10% point at the leading edge of the signal, and the 90% point at the trailing edge provides a conventional method of measuring its pulse width, or the time of one alternation. As shown in Figure 1-4, the pulse width at these points is 0.4 ms.

It is extremely difficult to determine the voltage level because the amplitude varies for each moment of time through the alternation. This is directly due to the sine structure of the ac wave. If the point at which the pulse width is measured is taken with respect to voltage levels around the alternations, then the pulse width becomes as relative as the amplitude measurements, and the period, pulse width, or time of an alternation becomes equally difficult to determine. Because digital equipment requires a predictable waveform pattern, the extreme variety of change in an analog signal makes it totally unsuitable.

To illustrate the problem of using analog signals for data communications, consider a 1kHz ac sine wave with peak amplitude of 6 volts. A considerable amount of electronic confusion would occur were this signal to be transmitted through a TIA/EIA-232 interface. If the interface were set so as to ignore any signal level less than 3 volts, the amplitude could be cut in half. Under those circumstances, any synchronizing pulse that had been sent with the signal would be lost. Parts of the alternations not cut off would be transmitted as bits. The time between bits would not equal the time of a bit, causing the receiver to incorrectly identify them.

Although both analog and digital ac waveforms are susceptible to **noise distortion** en route to the receiver, the problem is particularly difficult with analog. This is because the noise is difficult to remove from the signal without causing even further distortion. In some types of channels, the frequency of the noise may be such that it actually passes through the line more easily than the desired signal! The receiver would then reject the signal and accept only noise, which would be futilely processed.

noise distortion

A desirable quality of analog signals is that a 1kHz sine wave has a **bandwidth** of 1 kHz. In a digital system, a 1kHz square wave has a bandwidth of 11 kHz. An analog system with a 10kHz bandwidth could accommodate ten analog signals, yet barely one digital signal. Clearly, an analog system would be the best choice to use in a narrow-bandwidth channel.

bandwidth

Data Signals

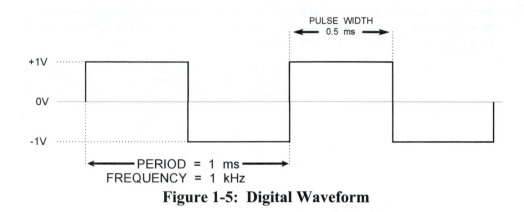

Figure 1-5: Digital Waveform

Computer data is digital in nature, and because of the monumental growth of data communications, and computer networks, an interesting development has been taking place among the various Telcos. Even the look of data outside the local network environment is becoming digital, as the local and long distance telephone companies convert analog systems to digital. Digital data has the advantage of being much more resistant to noise than analog data. It is also in the same format as the square waves found in computers and other types of data processing equipment, as shown in Figure 1-5. Therefore, the pulse width is measured across the entire half cycle, rather than at the 10% and 90% amplitude points

Binary Numbers

binary numbering system

The reason why square waves are used in digital communications is because they're representative of the base two, **binary numbering system**, which uses 1s and 0s. At this point, let's review how the binary numbering system works.

As already stated, the binary numbering system consists of 1's and 0's, and various combinations of these two digits can be used to represent any of the decimal numbers that we use every day. Binary numbers are broken down into groups or columns, in a similar fashion to decimal numbers. The big difference is in the values that are represented by each of the columns. For the familiar decimal numbering system, we use increasing/decreasing powers of 10, as shown in Table 1-1 (fractions not included).

Table 1-1: Decimal Column Values

Millions	Hundred Thousands	Ten Thousands	Thousands	Hundreds	Tens	Ones
1000000	100000	10000	1000	100	10	1
1	1	1	1	1	1	1

We understand that any digit (0 to 9) receives its value by being multiplied times the value of the column position it assumes. Each digit to the left is ten times the value of the digit to its immediate right, etc. For the binary numbering system, we use increasing/decreasing powers of 2, as shown in Table 1-2 (fractions not included).

Table 1-2: Binary Column Values

One-Hundred Twenty-Eights	Sixty-Fours	Thirty-Twos	Sixteens	Eights	Fours	Twos	Ones
128	64	32	16	8	4	2	1
1	1	1	1	1	1	1	1

Here, the digit (0 or 1) also receives its value by being multiplied times the value of the column position it assumes. In this case, however, each digit to the left is two times the value of the digit to its immediate right, etc. The easiest way to understand the binary numbering system is to break it down into its individual column values. Think of it in terms of 1 egg, a dozen eggs, etc. If you wanted to represent the decimal number 7, using an 8-digit binary number, you would write 00000111. If you were not required to include at least eight digits for the binary number, you could simply write it as 111.

To determine how this number would represent an equivalent decimal 7, look closely at Table 1-2 directly underneath the binary number listings of 1, 2, 4, 8, 16, 32, 64, and 128. The binary numbering system uses the digit 1 in the column representing the power of two needed to represent the required number. In the example 00000111, notice that there is a 1 under the ones column, the twos column, and the fours column. By adding the values of these columns together, 1 + 2 + 4, binary 111 would equal decimal 7.

Another example would be the situation in which the decimal number 28 needed to be written as a binary number. Using the same method as previously described, you would write 00011100 as an 8-digit binary equivalent for decimal 28. Locating the required column values from Table 1-2 would result in 16 + 8 + 4 = 28. If necessary, you should take the time to do a few practice conversions, and become more familiar with the binary numbering system, until you get the hang of it. On the following page, Table 1-3 shows the 8-digit binary equivalents for decimal numbers 0 to 255.

Square Waves

A square wave is either on or off, high or low, and it's this equivalency to the binary system that makes square waves suitable for data communications. In representing alphanumeric systems, square waves are relatively easy to generate and to code. Although analog signals can also be coded, the first telegraph communication systems—and later, the Teletype—were digital systems. So it shouldn't come as a big surprise that much of the development in data processing can be traced back to efforts at improving telegraph signal transmission and reception. This was digital data, and the machines that were developed to automate telegraphy have pioneered modern digital systems.

Table 1-3: Decimal to Binary Conversion Table

000	00000000	064	01000000	128	10000000	192	11000000
001	00000001	065	01000001	129	10000001	193	11000001
002	00000010	066	01000010	130	10000010	194	11000010
003	00000011	067	01000011	131	10000011	195	11000011
004	00000100	068	01000100	132	10000100	196	11000100
005	00000101	069	01000101	133	10000101	197	11000101
006	00000110	070	01000110	134	10000110	198	11000110
007	00000111	071	01000111	135	10000111	199	11000111
008	00001000	072	01001000	136	10001000	200	11001000
009	00001001	073	01001001	137	10001001	201	11001001
010	00001010	074	01001010	138	10001010	202	11001010
011	00001011	075	01001011	139	10001011	203	11001011
012	00001100	076	01001100	140	10001100	204	11001100
013	00001101	077	01001101	141	10001101	205	11001101
014	00001110	078	01001110	142	10001110	206	11001110
015	00001111	079	01001111	143	10001111	207	11001111
016	00010000	080	01010000	144	10010000	208	11010000
017	00010001	081	01010001	145	10010001	209	11010001
018	00010010	082	01010010	146	10010010	210	11010010
019	00010011	083	01010011	147	10010011	211	11010011
020	00010100	084	01010100	148	10010100	212	11010100
021	00010101	085	01010101	149	10010101	213	11010101
022	00010110	086	01010110	150	10010110	214	11010110
023	00010111	087	01010111	151	10010111	215	11010111
024	00011000	088	01011000	152	10011000	216	11011000
025	00011001	089	01011001	153	10011001	217	11011001
026	00011010	090	01011010	154	10011010	218	11011010
027	00011011	091	01011011	155	10011011	219	11011011
028	00011100	092	01011100	156	10011100	220	11011100
029	00011101	093	01011101	157	10011101	221	11011101
030	00011110	094	01011110	158	10011110	222	11011110
031	00011111	095	01011111	159	10011111	223	11011111
032	00100000	096	01100000	160	10100000	224	11100000
033	00100001	097	01100001	161	10100001	225	11100001
034	00100010	098	01100010	162	10100010	226	11100010
035	00100011	099	01100011	163	10100011	227	11100011
036	00100100	100	01100100	164	10100100	228	11100100
037	00100101	101	01100101	165	10100101	229	11100101
038	00100110	102	01100110	166	10100110	230	11100110
039	00100111	103	01100111	167	10100111	231	11100111
040	00101000	104	01101000	168	10101000	232	11101000
041	00101001	105	01101001	169	10101001	233	11101001
042	00101010	106	01101010	170	10101010	234	11101010
043	00101011	107	01101011	171	10101011	235	11101011
044	00101100	108	01101100	172	10101100	236	11101100
045	00101101	109	01101101	173	10101101	237	11101101
046	00101110	110	01101110	174	10101110	238	11101110
047	00101111	111	01101111	175	10101111	239	11101111
048	00110000	112	01110000	176	10110000	240	11110000
049	00110001	113	01110001	177	10110001	241	11110001
050	00110010	114	01110010	178	10110010	242	11110010
051	00110011	115	01110011	179	10110011	243	11110011
052	00110100	116	01110100	180	10110100	244	11110100
053	00110101	117	01110101	181	10110101	245	11110101
054	00110110	118	01110110	182	10110110	246	11110110
055	00110111	119	01110111	183	10110111	247	11110111
056	00111000	120	01111000	184	10111000	248	11111000
057	00111001	121	01111001	185	10111001	249	11111001
058	00111010	122	01111010	186	10111010	250	11111010
059	00111011	123	01111011	187	10111011	251	11111011
060	00111100	124	01111100	188	10111100	252	11111100
061	00111101	125	01111101	189	10111101	253	11111101
062	00111110	126	01111110	190	10111110	254	11111110
063	00111111	127	01111111	191	10111111	255	11111111

Another look at the square wave of Figure 1-5 indicates that it has a peak amplitude of 1 volt, a period of 1 ms, and a frequency of 1 kHz. Its basic specifications are similar to the sine wave discussed earlier, but the similarities between the two waveforms end there. The time required for a single data bit is equal to the square wave's pulse width, and it is measured in the same way as that of a sine wave.

The difference is that it doesn't matter if the measurement is taken on the leading or trailing edge, because the assumption is that both edges of a square wave are identical. The pulse width measures the same between 10% and 90%, or 40% and 60%. The leading and trailing edges of the wave provide the transmitting and receiving equipment with very predictable vertical slopes. Sine wave pulse-width measurements depend on where the measurement is taken, and therefore are open to interpretation.

Of course, if data bits are switched through wide area networking machines, such as routers, and clocked at GHz rates, the bit edges are all that matters. Once the bits have been coded and modulated, the 10%/90% measuring points become critical. Figure 1-6 illustrates how undesired noise may be superimposed on a square wave.

**Figure 1-6:
Square Wave Noise**

astable multivibrator

repeaters

Notice how the square wave has been heavily distorted. An in-line amplifier (regenerator or repeater) boosts the signal level and restores the square wave's original shape. Many modern local area network hubs normally incorporate repeaters, which may contain an **astable multivibrator** (Schmitt Trigger) using threshold voltages equal to the signal's original amplitude. The Schmitt Trigger is a specially designed flip-flop circuit that provides a means of restoring the squareness to a wave that contains a substantial noise component. Therefore, a nonstandard input transition will not be passed on to the output. The square wave is clean of noise and distortion at the output of the regenerator. There may be many **repeaters** along the way if the distance from the transmitter to the receiver is long. However, the overall performance of the system is improved, and the quality of the data is much higher over that of analog systems, because the repeaters regenerate a new signal at each point.

Repeaters can be used with analog systems, and frequently they are. However, the corrupting noise will be amplified along with the signal, which is a big disadvantage. Analog repeaters best serve to compensate for signal losses resulting over long cable lengths, but are not very useful at eliminating or decreasing noise levels.

The analog repeater can't simply restore the signal to its original shape in the same manner as a digital repeater, because an analog signal contains an infinite number of states, whereas a digital signal is composed of only two. Even if occurring multiple times within a single bit, only one of two state changes will ever be used to represent some part of the data. This remains true no matter how complex a digital waveform becomes. The analog signal, however, contains an infinite number of possibilities. The repeater cannot restore the originally transmitted data without knowing exactly what it is beforehand.

harmonics

As mentioned previously, a disadvantage in transmitting digital signals is the wide bandwidth required. This is because square waves are composed of an infinite number of odd **harmonics**, or exact multiples, of a fundamental frequency. For example, the frequency components that are combined to create a 1kHz square wave are depicted in Figure 1-7.

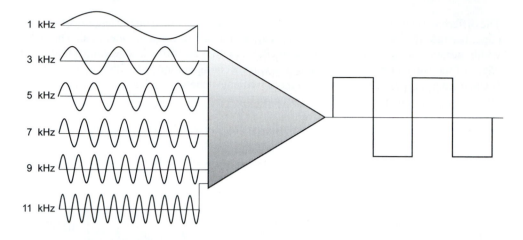

Figure 1-7: Frequency Components of a 1kHz Square Wave

The fundamental frequency is 1 kHz, and the odd harmonics of 3 kHz, 5 kHz, 7 kHz, and so forth, are summed together to create the square wave. Because of the fact that harmonics beyond the fifth order (11 kHz, here) have a negligible effect on the square wave's structure, they can be ignored. Figure 1-8 depicts various degrees of **harmonic distortion** that might occur at various channel bandwidths for a 1kHz square wave, as evidenced by the appearance, in the output, of harmonics other than the fundamental input component.

harmonic distortion

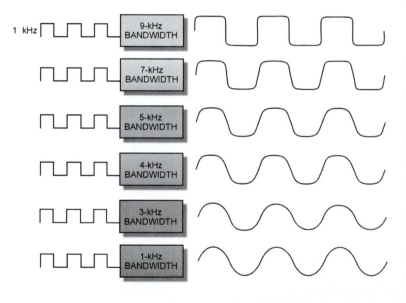

Figure 1-8: Harmonic Distortion with Inadequate Bandwidth

communications channel

The square wave is composed of a horizontal peak for each alternation, and two vertical sides, which are the leading and trailing edges. The higher frequencies are primarily responsible for the vertical slope of the leading and trailing edges. This wave shape indicates the presence of rapid signal changes, and is descriptive of the higher-order harmonics. The lower-order harmonics are represented by the flat top, or the horizontal peak of the square wave. To avoid the harmonic distortion illustrated in Figure 1-8, the bandwidth of the transmit channel must be at least equal to the fifth harmonic of the square wave's fundamental frequency. Otherwise, the square wave will become distorted as one or more of the higher-order harmonics are attenuated.

Notice how the corners of the square wave are rounded slightly, but the square wave remains flat on top, when the 1kHz square wave is passed through a **communications channel** having a 9kHz bandwidth. The 11kHz fifth harmonic has been lost, causing the rounding of the corners, and an increase of rise and fall time of the square wave. The rounding of the corners is more severe, and the rise and fall times increase, if the same 1kHz square wave is transmitted through a channel with a 7kHz bandwidth. When it's applied to a channel with a 1kHz bandwidth, the square wave takes on the characteristics of the first-order harmonic, at which point a digital signal with sharply defined edges has become an analog signal, with amplitude and pulse-width characteristics that will be open to receiver interpretation.

Although the wide bandwidth requirements of digital signals are a disadvantage, this can actually be turned into an opportunity to improve the system's performance. Because of the fact that channel capacity (data volume) is proportional to bandwidth, the greater the bandwidth, the greater is the volume of data that can be transmitted. Many channels of data can be multiplexed to flow through a single, wide-bandwidth channel, utilizing as much of the available bandwidth as possible.

It should be easy to see why the terms "data communication" and "digital communication" have become synonymous. Because **machine language** itself is digital in nature, digital methods have become the choice data structure for computer networks. Modern digital systems have proven to be of superior quality and performance, and although the telephone system still retains a small segment of analog networks, these are being gradually replaced with digital networks.

Remember that when information from computers is transmitted through an analog system, it must first be converted from digital to analog. At the receiving end, it must be converted back to digital. However, with a fully digital system, this conversion need never take place.

TELEPHONE WIRE

Although older types of telephone cables were flat, 28AWG, 4-, 6-, or 8-conductor non-twisted variations, as shown in Figure 1-9, cable suitable for modern **telephone wire** must be able to carry both voice and data information, and because of this, special design measures are required. The wire must be 22 or 24 **American Wire Gauge** (**AWG**), always used in **Tip/Ring** (**T/R**) pairs, and the pairs must be twisted together. Gauge is actually the measure of the diameter of the conductor. The numbering system works in reverse, in that the higher the number, the smaller the wire's diameter.

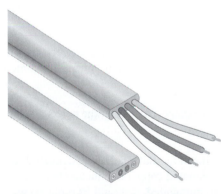

For each wire pair, the Tip and Ring must always maintain the proper polarity, with the Tip usually wired or connected to the positive side of the battery, and the Ring usually wired or connected to the negative side.

Look again at Figure 1-2 to see how this is accomplished, and always remember to check the polarity before undertaking any installation work. In older installations, don't assume that the green Tip wire and the red Ring wire were connected correctly until you have checked them with a voltmeter.

Figure 1-9: Flat 4-Conductor Telephone Cable

Just because standard telephone systems used to use only two wires (one pair) to accomplish bi-directional voice audio, dc power for circuit operation, dial tone service, dialing signals, and ringing, it didn't mean that you wouldn't find four, eight, or more wires being used in the phone cords or outlets. A second pair (black Tip and yellow Ring) might have been used to provide power to a dial light, but more than likely it was unused.

Before the mid 1980s when telephone wiring was installed by the telephone company, what is called **Plain Old Telephone System** (**POTS**) wiring was used. It was usually quad wire that often picked up interference or cross talk. The fact is that the **quad wire** color coding just described is considered obsolete, and it is not used anymore for new telephone installations.

Properly installed TIA/EIA T568A compliant wiring eliminates those and many other problems. The new wire will handle much higher frequencies that are often needed for computer network applications.

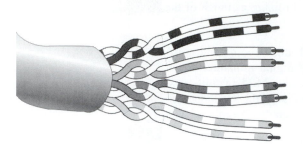

Figure 1-10: 24AWG UTP Telephone Cable

The standard for new residential phone systems is CAT5 wiring with 4 pairs. This will handle new wide-bandwidth services such as ISDN. In modern installations, for example, a common wiring scheme would use 4-pair (solid 8-conductor) 24AWG UTP telephone cable, as shown in Figure 1-10. Such wire would be suitable for running up to four voice telephone circuits, up to two RS232 circuits, ISDN, 10baseT Ethernet, 100baseT, 100baseVG, 16Mbps Token Ring, and ATM at 155 Mbps. This cable would also be CAT5 compliant to 100 MHz, use the industry standard color codes, and be available by the foot or in 1000-foot bulk packs. These installations would also take into account future expansion possibilities, in which case the additional pairs would remain unused, but available when needed.

There is also 4-pair (solid 8-conductor) 24AWG UTP telephone cable, similar to that mentioned above, that is compliant to the proposed CAT5E standard. This rating is good for data transmission frequencies up to 350 MHz.

Twisted Pair Cable

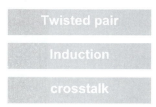

A **Twisted pair** consists of two insulated copper conductors twisted around each other to reduce induction from one wire to the other. **Induction** is simply the electromagnetic transfer of signal energy from one conductor to another conductor. The twists in the pair prevent this interference, or **crosstalk**, from reaching other pairs within the same sheathings, as well as cables running beside each other.

The twists are varied in length to reduce the potential for signal interference between pairs in the same cable. Twisted pair cable can have two or more twisted pairs under a common jacket. Hence we have 4-pair, 25-pair, 600-pair, or even higher.

The manufacturer determines the amount of twist in the pairs. The tighter the twists, the less likely that distortion or noise will occur.

During an installation or retrofit, you may encounter the older, flat quad wire mentioned earlier. Quad wire is simply a cable that contains four insulated copper conductors that are not twisted together. Because quad wire is no longer recommended, it should be replaced with 100-ohm UTP whenever possible. Quad's lack of pair twisting obviously makes it susceptible to the types of noise and interference previously discussed, and shown in Figure 1-11. Even in residential applications, the trend is toward increasing use of modems, fax machines, and other multi-line installations. Multi-line operations cannot tolerate crosstalk or electrical noise.

Figure 1-11: Noise Distortion of a Data Signal

S/N = 3.1 dB 1V

Shielded Twisted Pair Vs. Unshielded Twisted Pair

Shielded Twisted Pair (**STP**) is a cable medium with one or more pairs of twisted insulated copper conductors, within a metal foil or shield wrapped around the conductors, then bound in a single, plastic sheath.

Shielded Twisted Pair (STP)

Shielded Twisted Pair greatly reduces unwanted electrical signals (radiation) from entering or exiting the cable. One downfall to STP is that if the foil or shield is not properly grounded, it will attract the electrical signals it is attempting to direct away from the twisted pairs within its jacket.

Unshielded Twisted Pair (**UTP**) is a cable medium with one or more pairs of twisted insulated copper conductors bound in a single plastic sheath. There is a strong belief that twisted pair cables carrying data over 16 megabits of data per second should be carried on UTP, because at such speeds these cables emanate radiation that should be allowed to escape. If this radiation were not allowed to escape, the signal would interfere with itself.

Unshielded Twisted Pair (UTP)

Unwanted electrical radiation is known as **Electromagnetic Interference** (**EMI**), or Electromagnetic Interface. EMI can be any electrical or electromagnetic phenomenon, man made or natural, either radiated or conducted, that results in unintentional or undesirable transmission.

Electromagnetic Interference (EMI)

The unwanted energy that may be conducted into a cable is not limited to EMI. RFI also plays a role in the conduction of unwanted energy onto a communication circuit. **Radio-Frequency Interference** (**RFI**) is the disruption of signal caused by any source that generates radio waves such as a television, radio, or computer.

Radio-Frequency Interference (RFI)

As already mentioned, copper wires are twisted together to maintain excellent performance. The twists in the pair prevent interference such as crosstalk from other pairs within the same sheathings, as well as cables running beside each other. As you already know, crosstalk is the unwanted signal transfer between adjacent cable pairs. However, there are actually two kinds of crosstalk, Near-End Crosstalk (NEXT) and Far-End Crosstalk (FEXT). **Near-End Crosstalk** is the interference of two signals traveling in opposite directions within the same cable sheath. The stronger transmitted signal drowns out the weaker one, being received from the far end of the circuit. **Far-End Crosstalk** is the interference of two signals traveling in the same direction in a cable, where the stronger signal drowns out the weaker one.

Near-End Crosstalk

Far-End Crosstalk

Color Coding

Telecommunication cables are color-coded to enable cable pair identification. The coding schemes have been established by the TIA/EIA. As mentioned earlier in this section, each cable pair has one wire known as the "Tip", and the other wire that is known as the "Ring". Tip refers to the positive wire in the circuit, and Ring refers to the negative wire in the circuit.

For 3-pair cable, the solid-color twisted pair marking system is used. This method uses one color to identify each of six wires.

- Pair #1: Green/Red (tip/ring).

- Pair #2: Black/Yellow (tip/ring).

- Pair #3: White/Blue (tip/ring).

For 4-pair UTP horizontal cable, the standard color-coding method uses band-striped twisted pair markings. This method identifies the wire with the base color, and a smaller band of color repeated along the length of the wire. The tip wire will have mostly the base color with patches of the band color, while the ring color will have mostly the band color with patches of the base color.

- Pair #1: White/Blue. The tip wire is predominately white with blue bands. The ring wire is predominately blue with white bands.

- Pair #2: White/Orange. The tip wire is primarily white with orange bands. The ring wire is primarily orange with white bands.

- Pair #3: White/Green. The tip wire is mostly white with green bands. The ring wire is mostly green with white bands.

- Pair #4: White/Brown. The tip wire is predominantly white with brown bands. The ring wire is predominantly brown with white bands.

During any installation, you must follow the **color codes**, depicted in Figure 1-12. Failure to do so will result in needless frustration, and a good deal of wasted time. The amount of twist in the pairs is determined by the manufacturer. The tighter the twists are maintained, the less likely that distortion will occur.

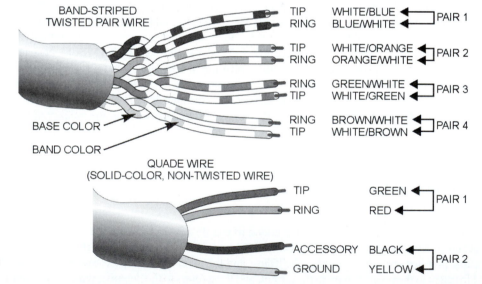

**Figure 1-12:
Color Coded Wire**

Color-coding schemes have been developed for cables containing greater numbers of pairs, such as the one that is used for 25-pair cables, shown in Table 1-4.

Table 1-4: Color Coding for 25-Pair Cables

Pair	Tip Wire	Ring Wire
1	white/blue	blue/white
2	white/orange	orange/white
3	white/green	green/white
4	white/brown	brown/white
5	white/slate	slate/white
6	red/blue	blue/red
7	red/orange	orange/red
8	red/green	green/red
9	red/brown	brown/red
10	red/slate	slate/red
11	black/blue	blue/black
12	black/orange	orange/black
13	black/green	green/black
14	black/brown	brown/black
15	black/slate	slate/black
16	yellow/blue	blue/yellow
17	yellow/orange	orange/yellow
18	yellow/green	green/yellow
19	yellow/brown	brown/yellow
20	yellow/slate	slate/yellow
21	violet/blue	blue/violet
22	violet/orange	orange/violet
23	violet/green	green/violet
24	violet/brown	brown/violet
25	violet/slate	slate/violet

Twisted Pair Connectors

With the majority of new network installations favoring the use of UTP cable from the wiring closets at each floor (horizontal distribution), workstations would normally connect to the network through a UTP interface following the T568B standard for data. Telephones would follow the T568A standard. Figure 1-13 illustrates the wiring differences between these two connectors, which would run from the equipment at the node, and plug into the corresponding modular wall jacks mounted approximately 12 inches from the floor.

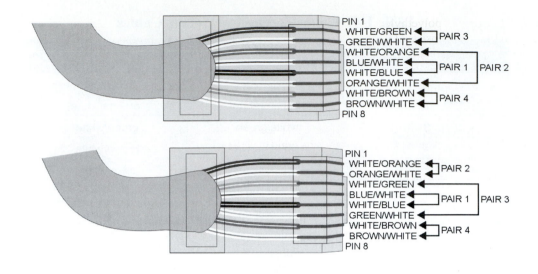

**Figure 1-13: UTP
Connectors for
Telephone and Data
Wiring**

ALTERNATE CABLING

Although many new telephone and computer networking installations now use similar, if not identical, STP and UTP cabling in their communications networks, some networking applications do not rely on telephone wire at all. Cabling installation technicians need to be familiar with the networking systems that utilize various other types of communication medium, such as coax and fiber optic cable. Undoubtedly, installation jobs will come along that require a mixture of several different communication media to complete the overall layout.

Coaxial Cable

coaxial cable

polyethylene

Another commonly used copper wire medium is **coaxial cable**. Enclosed within a protective copper braid, is a single insulated copper conductor, with a protective outer insulation surrounding the braid. It contrasts sharply with the UTP cable mentioned earlier, which has become today's common telephone cabling with four pairs of twisted wires inside.

Historically, coaxial cable has been used in wide-bandwidth applications, particularly in the cable television industry, and is familiar to most people as the conductor that carries cable TV into their homes. It was also used extensively in many earlier computer networks. Coaxial cable used for networking purposes has a maximum data rate of 100 Mbps, along with decent noise immunity. For other applications, it can provide data rates of up to 4,000 Mbps, and bandwidths of up to 1,000 MHz, but is more expensive to install than twisted pair wire.

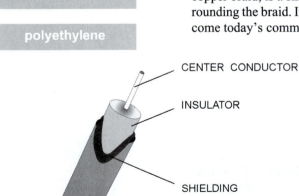

Coaxial cable used to offer better noise immunity than UTP, and still does in some situations. Enclosed in a **polyethylene** jacket are two conductors, insulated from one another. Figure 1-14 shows a cutaway view of coaxial cable.

Figure 1-14: Coaxial Cable

A flexible polyethylene insulator surrounds the center conductor, while the second conductor consists of braided wire and/or foil, wrapped around the insulator, and connected to ground potential. External noise, such as crosstalk, is effectively shorted to ground by the braid encircling the center conductor. An outer jacket of polyethylene insulation covers the braided conductor. Polyethylene retains a high degree of flexibility over broad temperature ranges, and is very resistant to salt, water, and oil.

At frequencies over 100 kHz, coaxial cable becomes more effective, while at lower frequencies, external noise can leak through the braid to cause signal distortion on the center conductor. Although data transmission rates with coaxial cable can exceed 400 Mbps, resistive losses increase rapidly at higher frequencies. For long-distance transmission, repeaters must be used.

As stated earlier, the cable TV industry has extensively used coaxial cable. In fact, many LANs owe their high data rates to advancements pioneered by the cable TV industry. While LANs typically transmit on coax from 1 Mbps up to about 100 Mbps, cable TV networks can utilize coaxial cable providing a bandwidth of 400 MHz. Using proper multiplexing techniques, that can allow a single cable to carry 52 TV channels.

Coaxial cable is more expensive than UTP, and is not compatible with it. Although this cost is only slightly more per foot than twisted wire, when thousands of feet of cable are needed, this can become significant. Because twisted pair cabling now competes equally with coax in a network environment, there's little justification for using coaxial cable for most Ethernet LAN installations. With the availability of wide-bandwidth, CAT5 UTP cabling (100MHz bandwidth), you should use CAT5 twisted pair for the local network cabling infrastructure whenever possible.

Thicknet

All Ethernet LANs once used coaxial cable as their transmission medium. Because of its superior strength, and because it would not bend easily, the cable was called **Thicknet**. Thicknet is also referred to as 10base5, and is used primarily as a **backbone** for interconnecting network devices. The cable, known as RG8/U, is 0.4 inches in diameter, yellow or orange-brown (plenum-rated) in color, and displays dark colored rings every 2.5 meters. **Plenum** ratings are often required for installation in air handling spaces (also called plenums) to meet certain fire regulations. The rings represent the minimum spacing of in-line transceiver taps to the backbone. The transceivers, in turn, connect to the computer nodes via a DB15-pin connector and cable (no longer than 50 feet).

Thicknet

backbone

Plenum

Medium Attachment Unit (MAU)

Up to 100 transceivers are supported on each segment, with the number of connections being limited to prevent signal attenuation and interference. If a segment end is unused, it must be terminated with a 50-ohm resistance to prevent signal reflections. Figure 1-15 shows a typical Thicknet connection.

A **Medium Attachment Unit (MAU)** contains the transceiver circuitry, along with various digital devices used to format data bits into Ethernet frames. Coaxial cable transceivers simply transmit and receive signals. They do not manage or massage the signal, nor do they amplify it, or remove noise from it. Because they are physically attached, they cannot be assigned to another physical device using software configuration techniques.

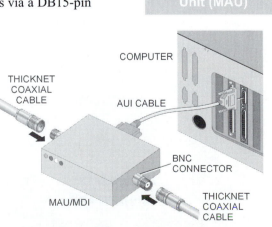

Figure 1-15: Thicknet Connection

Either a male or female DB15 connector is used to connect a 15-wire AUI cable between a MAU and the computer interface. It is called an **Attachment Unit Interface (AUI)**, and is depicted in Figure 1-16. In the pinout diagram, notice the reversal of pin number designations between the male and female connectors on the AUI cable.

PIN FUNCTION	PIN NUMBER
LOGIC REF	1
COLLISION +	2
TRANSMIT +	3
LOGIC REF	4
RECEIVE +	5
POWER RETURN	6
NOT USED	7
LOGIC REF	8
COLLISION –	9
TRANSMIT –	10
LOGIC REF	11
RECEIVE –	12
POWER (+12 Vdc)	13
LOGIC REF	14
NOT USED	15

MALE   FEMALE

PIN #1   PIN #1

PIN #15   PIN #15

Figure 1-16: Attachment Unit Interface

In the MAU, the formatted frames are uploaded and downloaded to the Thicknet coaxial cable using a mechanical connection, or interface, from the MAU to the coax. This connection is called a **Medium Dependent Interface (MDI)**, and there are two types, **intrusive taps** and **non-intrusive taps**.

When using an intrusive tap, the coax backbone must be physically cut, and a barrel-type BNC connector installed on each end, to allow the insertion of the transceiver between the backbone Thicknet cable segments. A female DB15 MDI interface attaches the MAU transceiver to the computer's **Network Interface Card (NIC)** via a 15-pin AUI cable from the node, as shown in Figure 1-17.

Medium Dependent
Interface (MDI)

Intrusive taps

non-intrusive taps

Network Interface
Card (NIC)

The cable does not need to be broken in a non-intrusive tap, so the network doesn't have to be brought down each time another user is added. A bayonet spear is inserted into a hole piercing the coax, which contacts the center conductor of the cable. Because the network continues to run while the procedure takes place, the

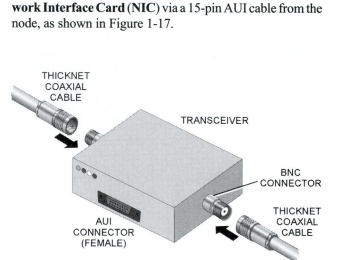

Figure 1-17: Intrusive Tap

non-intrusive method is preferred. A non-intrusive tap, also called a "vampire tap", is shown in Figure 1-18.

**Figure 1-18:
Vampire Tap**

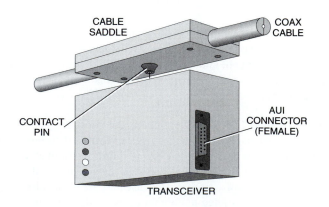

Thinnet

Thinnet, or 10base2, is also called "Cheapernet", or thin coax. Its RG58A/U coaxial cable has less shielding than Thicknet, and consequently, the LAN segment lengths must be shorter. The maximum length of a Thinnet segment is 185 meters (compared to 500 meters for Thicknet). In order to prevent undesired signal reflections, each unused segment end requires a 50-ohm termination. Each segment can support up to 30 transceivers, and the transceivers need be only 0.5 meters apart.

Transceivers are connected by cutting the cable and installing BNC connectors on the cable ends. T-connectors are used to attach the cable ends together to form the Ethernet bus, and to connect the bus to the computer through the network interface card. The NIC containing the transceiver circuitry is shown in Figure 1-19, in a typical arrangement. Two short RG-58 cables can be joined using **barrel connectors**, if necessary.

barrel connectors

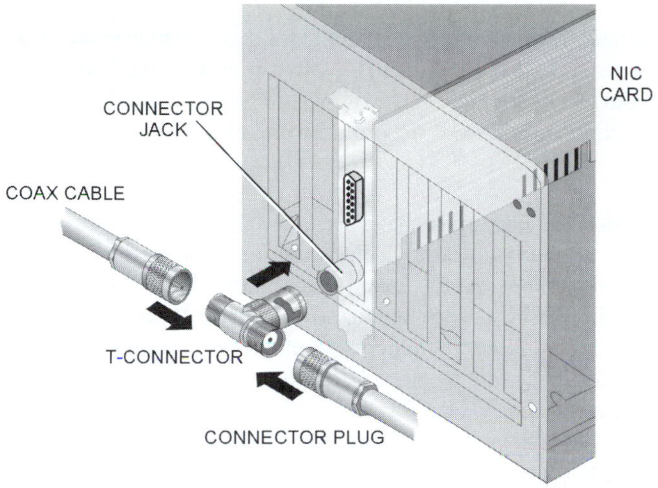

Figure 1-19: Thinnet Arrangement

In modern networks, coaxial cable is seldom used at the LAN level of networking. In terms of bandwidth and channel capacity, UTP and fiber optics have overtaken it. In a small peer-to-peer LAN, however, coax may be easier to configure than UTP, and will work just as well.

Fiber Optic Cable

Another type of data communications cable is a **fiber optic cable**, which consists of a thin strand of glass or plastic, surrounded by a jacket. This type of media transfers data by sending light waves through the glass or plastic conductor. The data to be transmitted is converted from electrical energy to light energy and transmitted through the cable. Fiber optic cable has significantly lower losses than copper wire, and is immune from external electrical interference (lightning, motors, generators, microwaves), which creates much of the distortion found in copper-based systems. Although it is the most expensive type of cable to install, it offers the best in noise immunity.

fiber optic cable

Within the last ten years, fiber has bubbled down to the local network level due to a drop in prices—although it remains the most expensive media—and broader standardization. The actual cost of fiber cable isn't expensive. The connectors, special splicing tools needed, and hardware to convert signals from electrical to light and vice versa all combine to drive up the cost of an optical system. Aside from its expense, another major disadvantage of fiber optics is the need for very careful alignment of splices and connections.

For LAN implementation, fiber optic cable provides a very high-performance link between various LANs. It can operate over longer distances between stations, before signal deterioration becomes a problem.

Even though it's an expensive choice, there are several reasons to install fiber in a network. Fiber optic cable does not radiate electromagnetic interference containing signal information that can be detected outside of the conductor. Because no copper is used in the cable, data bits won't be corrupted by EMI, crosstalk, or other external noise that may garble bits in a UTP or coaxial system. Also, it is fairly difficult to tap into, revealing a decided signal loss when an attempt is made to do so. It therefore offers a high degree of data security.

The optical cable isolates devices connected to either end of it, which makes it a good choice where completely separate systems are linked together, such as two LANs in different buildings. Connecting them can be tricky because of ground loops caused by common electrical planes. A fiber optic connection avoids the problem and eliminates the danger to personnel and equipment.

Fiber Distributed Data Interface (FDDI)

Laboratory experiments with fiber optics have produced data rates as high as 200,000 Mbps, utilizing bandwidths as high as 1,000 GHz. Telecommunication carrier frequencies hover in the 40Mbps to 8,000Mbps range. At the local level, data rates are standardized at 10Mbps and 100Mbps, using the IEEE 802 protocols and the **Fiber Distributed Data Interface (FDDI)** protocol. When compared to copper systems, losses are quite small. At the long-distance carrier level, losses are not to exceed 2dB/km (over 1 kilometer). Compare this to 11.5db/km over 100 meters using UTP. Obviously, a fiber cable will carry more data farther, and with fewer losses, than its copper-based cousin.

Fiber systems are currently being used in point-to-point configurations with conventional star, ring, and bus topologies. Due to the large data-carrying capacity of a fiber system, it's installed where it will be expected to handle large volumes of data. This is the main reason why telecommunications and cable television industries have used it for many years.

Although fiber optic cable is now widely used in telephone systems, and although there are standards in place for implementing fiber optic cabling on LANs, few LAN hardware manufacturers have switched over to fiber optic networking as of yet. This data cabling course utilizes labs that specifically target the skills necessary for CAT5 copper cabling installation certification, and to a lesser extent, includes activities that improve skills using coaxial cabling and associated connectors. The subject of fiber optic cable installation is important enough to deserve an entire course being dedicated specifically to it. For more information, see Marcraft's Fiber Optic Cabling Installation Certification course.

Lab 1 – Cable Identification

<u>Lab Objectives</u>

To identify the cable types provided with this course

To identify the parts that make up the Thicknet active tap kit

<u>Materials needed</u>

- Lab 1 work sheet
- Cable, CAT5 UTP, 100-ohm, non-plenum, 12-inch (with one end stripped)
- Cable, CAT5 STP, 100-ohm, non-plenum, 12-inch (with one end stripped)
- Cable, Type-1A STP, 150-ohm, non-plenum, 12-inch (with one end stripped)
- Cable, RG6, 75-ohm, coax, 12-inch (with one end stripped)
- Cable, RG58, 50-ohm, coax, 12-inch (with one end stripped)
- Cable, RG8, 50-ohm, coax, 12-inch (with one end stripped)
- Thicknet tap kit
- Pencil

<u>Instructions</u>

Before beginning this lab, read through all of the following steps at least once, thoroughly.

1. Examine the pictures shown in Figure 1-20.

2. From the collection of 12-inch cable samples, select the CAT5 UTP, 100-ohm, non-plenum segment.

TIP: The cable's jacket should indicate that it is a Category 5 cable, and it may be either light gray, or white, in color.

3. Examine the CAT5 UTP, 100-ohm, non-plenum segment carefully, and once you are sure you have properly identified it, use the pencil to place a checkmark (√) in its corresponding caption box for Figure 1-20.

4. From among the 12-inch cable samples select the CAT5 STP, 100-ohm, non-plenum segment.

TIP: The cable's jacket should indicate that it is a shielded Category 5 cable, and it may be blue in color.

5. Examine the CAT5 STP, 100-ohm, non-plenum segment carefully, and once you are sure you have properly identified it, use the pencil to place a checkmark (√) in its corresponding caption box for Figure 1-20.

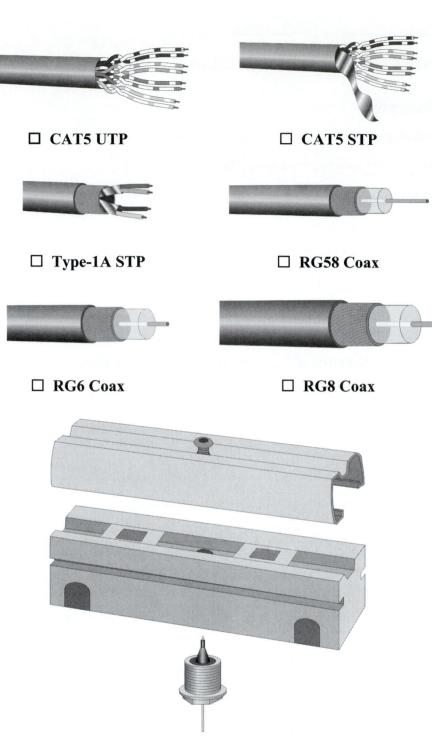

☐ CAT5 UTP   ☐ CAT5 STP

☐ Type-1A STP   ☐ RG58 Coax

☐ RG6 Coax   ☐ RG8 Coax

☐ Thicknet Active Tap Kit

**Figure 1-20: Cable Types
and Thicknet Tap**

6. From among the 12-inch cable samples select the Type-1A STP, 150-ohm, non-plenum segment.

TIP: The cable's jacket should indicate that it is a shielded cable, specified for IBM, and it will probably be black in color. The cable itself may appear somewhat flattened.

7. Examine the Type 1A STP, 150-ohm, non-plenum segment carefully, and once you are sure you have properly identified it, use the pencil to place a checkmark (√) in its corresponding caption box for Figure 1-20.

8. From the 12-inch cable samples select the RG6, 75-ohm, coax segment.

TIP: This cable's jacket should indicate that it is designed for cable television transmission (CATV designation), and will probably display its 75-ohm impedance as well. It is normally black in color.

9. Examine the RG6, 75-ohm, coax segment carefully, and once you are sure you have properly identified it, use the pencil to place a checkmark (√) in the corresponding caption box shown in Figure 1-20.

10. From the 12-inch cable samples select the RG58, 50-ohm, coax segment.

TIP: RG58 is probably the smallest diameter cable in the sample set. Its jacket should be black, and will be marked with the RG58 designation.

11. Examine the RG58, 50-ohm, coax segment carefully, and once you are sure you have properly identified it, use the pencil to place a checkmark (√) in the corresponding caption box shown in Figure 1-20.

12. From the 12-inch cable samples select the RG8, 50-ohm, coax segment.

TIP: RG8 is probably the largest diameter cable in the sample set. Its jacket should be yellow or orange, and may be marked with the Ethernet designation. It will also indicate an impedance of 50 ohms.

13. Examine the RG8, 50-ohm, coax segment carefully, and once you are sure you have properly identified it, use the pencil to place a checkmark (√) in the corresponding caption box shown in Figure 1-20.

14. From the parts supplied for this course, locate the AMP Thicknet tap kit.

TIP: The kit is basically composed of the clamp assembly, pressure block, button-head socket screw (tightened with the included hex wrench), braid terminators (2), tap body, and a probe assembly. An application tool (shown in the Tools Identification lab) is used to drill through the cable to the center conductor and to thread the probe assembly.

15. Examine the Thicknet tap kit carefully, and once you are sure you have properly identified its various parts, use the pencil to place a checkmark (√) in the corresponding caption box for Figure 1-20.

16. Read the instruction sheets on pages 360 and 361 for more information about the Thicknet active tap.

17. Once all of the cable types have been recognized and the various parts of the Thicknet tap kit have been identified, return the parts to their storage area.

Lab 1 Questions

1. Which one of the cable types included in this lab procedure is used for cable television transmission?

2. Concerning the installation of the Thicknet active tap, what is the purpose for the hex wrench being included in the kit?

3. According to the instruction sheet for the installation of the Thicknet active tap, what are the two purposes for which the application tool is used?

4. What is the diameter of the RG8 coaxial cable around which the Thicknet active tap must clamp?

5. What was the impedance of the following cable types that you examined during this lab?
   (a) CAT5 UTP cable
   (b) CAT5 STP cable
   (c) Type-1A STP cable
   (d) RG6 coaxial cable
   (e) RG58 coaxial cable
   (f) RG8 coaxial cable

6. What would you expect the color of the outer jacket for RG8 coaxial cable to be?

7. Which of the cable types from the samples you examined possessed the smallest diameter?

TOPOLOGIES

Topology is the configuration of a communication network. The **physical topology** refers to the way that the network *looks*. The **logical topology** refers to the way the network *works*.

There are three main types of topologies in use today. All additional topologies are versions or combinations of these three main types: star, bus, and ring.

The first type, known as the **star topology**, is one in which all of the telephones or workstations are wired directly to a central service unit, or workstation, which establishes, maintains, and breaks the connections between them.

Some advantages of the star topology include:

1. If one wire run is damaged, it will not affect the other devices in that system.

2. It is easy to locate a problem in the system.

3. Additions and changes can be made very easily, without interfering with the rest of the system.

The main disadvantage of the star topology is the amount of wire used to make the runs. The more cable that is required, the more expensive it is to set up the system.

The star topology is shown in Figure 1-21. Although a computer networking arrangement is depicted, the networked devices could just as easily be telephones.

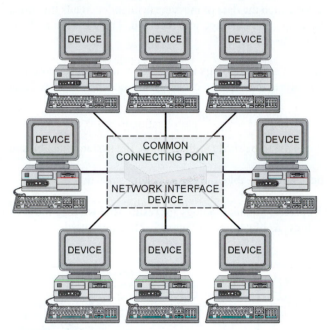

**Figure 1-21:
Star Topology**

The second type of topology is called a **bus topology**. A bus topology connects all of the devices to a single cable, with terminators on each end. The information, or data, is sent to all of the devices along the run, but only the device possessing an address matching the one encoded in the message will accept the data.

The terminators are placed on the ends of the cable run to prevent the signal from bouncing. A bouncing signal would travel from one end of the cable to the other, preventing the transmission of any other information.

The advantages of a bus topology, shown in Figure 1-22, are:

1. The bus topology is **passive** (it does not actively move the data signal). Therefore, if one device fails, it does not affect the others.

2. The system does not have to be shut down in order to add other devices.

The disadvantages of a bus topology are:

1. If a cut in the wire should occur, the entire network will be brought down. The connected devices will work in stand-alone mode, but will not be able to communicate with each other.

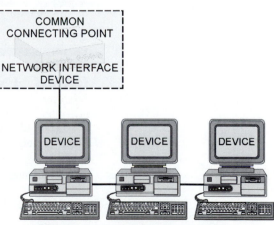

Figure 1-22: Bus Topology

2. Only one device can transmit at any one time. The more devices on a cable run, the slower the network.

The third type of topology is called a **ring topology**. In a ring topology, all of the devices are connected in a circle. For a three-device network, device A would be connected to device B, device B would be connected to device C, and device C would be connected to device A. Each device would read each message and act as a repeater, boosting the signal level, and passing it to the next device in line. Each device does the same procedure as the preceding device. The device for which the data or message is intended reads the message, and changes its status to indicate that the message was received. The message is then forwarded to the original transmitting device.

The main advantage of a ring topology is that there is not a large amount of cable needed for this topology. Therefore, the expense of the wiring system will be reduced.

Disadvantages of the ring topology are:

1. If any additions are to be made to the system, the system must be shut down.

2. If a break occurs in the cable, the entire system shuts down.

Although the TIA prefers the star topology for **residential wiring**, the use of a combination star and bus topology, as illustrated in Figure 1-23, is permitted.

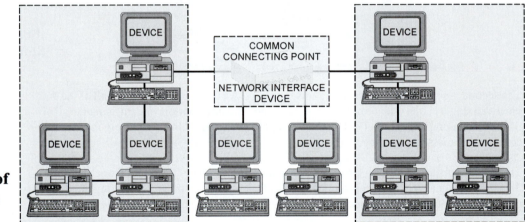

**Figure 1-23:
Combination of
Star and Bus
Topology**

BASIC INSTALLATION PRACTICES

It's a good idea at this time to review the following general tips for running network cable. This will help you to become acquainted with basic data cabling concepts:

- Never put a splice behind a wall or in an area where it cannot be accessed.

- Always test the cable runs after the installation is complete.

- If possible, avoid placing splices on cable runs altogether. If there is a real problem, pull a new cable.

- Do not pull 4-pair cable with more than 25 pounds of pulling force.

- Do not run data cables in parallel with any type of electrical cables.

- Maintain the following separation between telecommunications wiring and other types of wiring:

    - Electrical bare light, or power of any voltage   5 feet
    - Electrical open wiring under 300 volts   2 inches
    - Electrical wires in conduit, or in armored or nonmetallic   none
      sheath cable/power ground wires
    - Radio and television antenna lead and ground wires   4 inches
    - Open signal/control wiring under 300 volts   none
    - Community television systems coaxial cables, with   none
      grounded shield
    - Telephone service drop wire, aerial or buried   2 inches
    - Neon signs, and associated wiring from transformer   6 inches
    - Fluorescent lighting   5 inches
    - Lightning wire   6 feet

- Minimum separations between residential and light commercial telecommunications wiring are specified in TIA/EIA 570. When in doubt, use the rule of sixes:

    - Six feet of separation between telephone wiring and open high-voltage wiring, lightning grounding wire, or grounding rods
    - Six inches of separation from all other high-voltage wire, unless in conduit

- Never bend the cable sharply, or cut the sheath intended to protect the conductors.

- Always maintain correct polarity.

- Always maintain the line number with respect to the pair number.

- Always use plastic, nonmetallic staples.

- Always leave a pull cord in the conduit for future cable pulls.

- Never run voltage and telecommunication cables in the same conduit.

- If possible, avoid running telecommunication cables under carpeting to avoid damage, or slow degradation, to the cable.

- Try to run cabling through the inner walls rather than the outer walls if possible.

- Do not run Telco wiring parallel to electrical wiring. Instead, always cross electrical wiring at 90-degree angles. Also, never share bore holes with any electrical wiring.

- Avoid running wire near known sources of heat.

- Always conceal wire runs. Do not leave wire runs exposed.

- Always leave a one- to three-foot service loop of wire slack at outlets, or termination points, for later additions, changes, or repairs.

The following are general tips for the installation of telecommunication outlets:

- Never share bore holes with any type of electrical wiring.

- All data cabling outlets should be set up to handle either two 100-ohm UTP cables, or one 100-ohm UTP cable, and one cable of the following:

  - 2-pair, 150-ohm STP

  - 2-strand, 62.5/125-micron optical fiber

- Place Telco outlets at the same wall height as electrical outlets, but no closer than 12 inches, keeping in mind that they should not share the same stud space.

- Always consider the current layout of the furniture, and how it could be rearranged when you are installing telecommunication outlets.

Proposed Cabling Grades

There is a proposed modification to the TIA 570 standard called 3490A. It proposes different grades of cabling in residential installations according to the types of services it provides. A number of services are available for residential installation. These services include: telephone, cable television, data, security, paging, etc. The two proposed grades are **Grade 1** and **Grade 2**. Grade 1 covers residential installations of telephone, CATV, and data. The minimum cable requirements are a 4-pair, 100-ohm UTP cable, that meets or exceeds CAT3 transmission requirements, and 75-ohm coaxial cables. Grade 2 covers residential installations of all Grade 1 installations, plus **fiber optic wiring** and **multimedia wiring**. The minimum cable requirements consist of 4-pair, 100-ohm UTP cables, that meet or exceed CAT5 transmission requirements, and 75-ohm coaxial cables. The addition of 2-strand 62.5/125 fiber-optic cable is optional. In order to meet full compliance, all of the accompanying modular jacks must be wired according to T568A or T568B.

CATV stands for **Community Antenna TeleVision**, or **CAble TeleVision**. Generally using a 75-ohm coaxial cable, CATV systems retransmit distant TV signals to individual subscribers. The original broadcast frequencies are converted in order to conveniently transport them over the cable. Frequency spectrum normally allocated to other communication services can be used within a CATV system to provide many additional channels of television programming. CATV systems can also provide other services such as specifically tailored music channels, and Internet connections (through the use of cable modems).

Grade 1

Grade 2

fiber optic wiring

multimedia wiring

Community Antenna TeleVision

CAble TeleVision

Lab 2 – Cable Examinations

Lab Objective

To examine the construction and makeup of the various cable types provided with this course

Materials needed

- Lab 2 work sheet

- Cable, CAT5 UTP, 100-ohm, non-plenum (spool)

- Cable, CAT5 STP, 100-ohm, non-plenum (spool)

- Cable, Type-1A STP, 150-ohm, non-plenum (spool)

- Cable, RG6, 75-ohm, coax (spool)

- Cable, RG58, 50-ohm, coax (spool)

- Cable, RG8, 50-ohm, coax (spool)

- Utility knife

- Wire stripper, CAT5-type

- Pencils, colored (set)

- Tension scale/Tape measure

Instructions

Before beginning this lab, read through all of the following steps at least once, thoroughly.

1. Using the 6-inch ruler to measure, cut a 2-inch segment of 100-ohm, CAT5 UTP cable from the spool with the cutting blade on the wire stripper.

2. With the 2-inch segment of 100-ohm, CAT5 UTP cable laying flat, use the utility knife to carefully cut a slit through the cable's outer jacket along its entire length.

TIP: Try not to damage any of the inside wires as you make the slit.

3. Spread the outer jacket open and remove the contents. Lay the contents out on the work surface for observation, as shown in Figure 1-24.

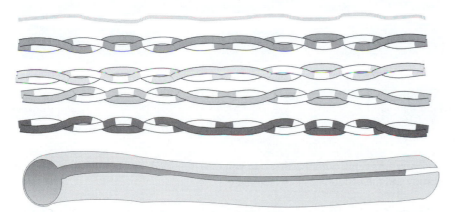

Figure 1-24: Examining the Contents of CAT5 UTP

4. Count the number of wire pairs, and use the ruler to determine if there are any differences in the number of twists per inch between the pairs.

TIP: Recall your reading about twisted-pair cable, and the purpose for the twisting.

5. Examine the orange wire pair and determine the identity of the tip and of the ring.

6. After untwisting the orange wire pair, use the wire strippers to remove ½ inch of insulation from the tip wire.

TIP: You will want to use the 24-gauge marking position on the wire stripper.

7. Examine the tip wire closely, and determine whether it is stranded or solid wire.

8. On the Lab 2 work sheet, use the colored pencils to draw the color schemes for the four twisted pairs of the 100-ohm, CAT5 UTP cable segment you disassembled.

TIP: Try to depict the twist, as well as the color markings, for each of the pairs. Identify the tip and ring, and don't worry if your drawing doesn't win any art awards.

9. Using the 6-inch ruler to measure, cut a 2-inch segment of 100-ohm, CAT5 STP cable from the spool with the cutting blade on the wire stripper.

10. With the 2-inch segment of 100-ohm, CAT5 STP cable laying flat against your work surface, use the utility knife to carefully cut a slit through the cable's outer jacket along its entire length just as you did before.

TIP: Again, try not to damage any of the inside wires as you make the slit.

11. Spread the outer jacket open and remove the contents. Lay the contents out on the work surface for observation, as shown in Figure 1-25.

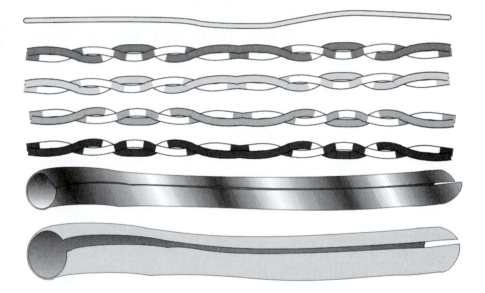

Figure 1-25: Examining the Contents of CAT5 STP

12. Compare this cable with the first one you disassembled, and identify both the similarities and the differences between them.

13. Using the 6-inch ruler to measure, cut a 2-inch segment of 150-ohm, Type-1A STP cable from the spool with the cutting blade on the wire stripper.

14. With the 2-inch segment of 150-ohm, Type-1A STP cable laying flat against your work surface, use the utility knife to carefully cut a slit through the cable's outer jacket along its entire length just as you did before.

15. Spread the outer jacket open and remove the contents, and pull the foil and the wire pairs from inside of the braided shield.

16. Lay the contents out on the work surface for observation, as shown in Figure 1-26.

Figure 1-26: Examining the
Contents of Type-1A STP

17. Compare this cable with the previous STP cable you disassembled, and identify both the similarities and the differences between them.

18. Using the 6-inch ruler to measure, cut a 2-inch segment of RG6 coaxial cable from the spool with the cutting blade on the wire stripper.

19. With the 2-inch segment of RG6 cable laying flat against your work surface, use the utility knife to carefully cut a slit through the cable's outer jacket along its entire length just as you did before.

20. After removing everything from the cable jacket, locate the insulated section of solid center conductor.

21. Use the ruler and the utility knife to carefully remove ½ inch of insulation from the center conductor on one end of the segment.

TIP: See if you can do this without putting any nicks in the center conductor. The wire stripper will help to remove the insulation once the cut has been made.

22. Once again, lay the contents out for observation, as shown in Figure 1-27.

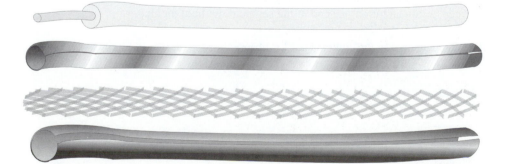

Figure 1-27: Examining
the Contents of RG6

23. Note the braided shield, and whether the center conductor is solid or stranded.

TIP: Although the center conductor for RG6 is obviously solid, there are types of coaxial cable that use stranded conductor. These coaxial types are generally more flexible than those with solid center conductors.

24. Using the 6-inch ruler to measure, cut a 2-inch segment of RG58 coaxial cable from the spool with the cutting blade on the wire stripper.

25. With the 2-inch segment of RG58 cable laying flat against your work surface, use the utility knife to carefully cut a slit through the cable's outer jacket along its entire length just as you did before.

26. After removing everything from the cable jacket, locate the insulated section of solid center conductor.

27. Use the ruler and the utility knife to carefully remove ½ inch of insulation from the center conductor on one end of the segment, without nicking it.

28. Once again, lay the contents out for observation, as shown in Figure 1-28.

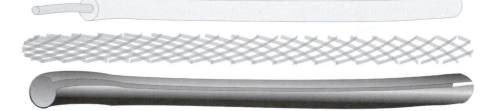

Figure 1-28: Examining the Contents of RG58

29. Compare the disassembled RG58 segment with the RG6 segment you disassembled earlier, and note the differences.

30. Using the 6-inch ruler to measure, cut a 2-inch segment of RG8 coaxial cable from the spool with the cutting blade on the wire stripper.

31. With the 2-inch segment of RG8 cable laying flat against your work surface, use the utility knife to carefully cut a slit through the cable's outer jacket along its entire length just as you did before.

32. After removing everything from the cable jacket, locate the insulated section of solid center conductor.

33. Use the ruler and the utility knife to carefully remove ½ inch of insulation from the center conductor on one end of the segment, without nicking it.

34. Once again, lay the contents out for observation, as shown in Figure 1-29.

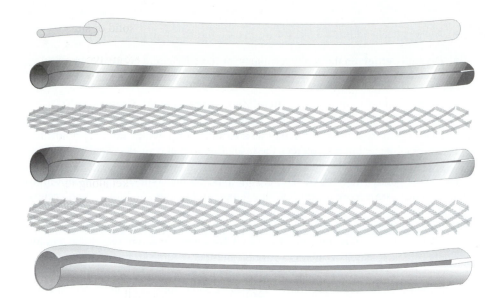

Figure 1-29: Examining
the Contents of RG8

35. Note the differences between the contents of the RG8 segment, and those of both the RG6 and RG58 segments you disassembled earlier.

36. Return all cable spools to their designated areas, dispose of the disassembled cable segments, put away all tools, and clean your work surface.

Lab 2 Questions

1. What is the significance of any differences in the number of twists per inch between the pairs in UTP/STP cable?

2. In 4-pair UTP/STP which wire, tip or ring, is mostly white?

3. When comparing RG6, RG58, and RG8, which type of coaxial cable contains the most shielding?

Termination Methods

There are two basic **termination methods** used in telecommunications. The first method is the **binding post connection**. This method uses screw terminals for the connections, as shown in Figure 1-30.

The general tips on binding post connections are as follows:

- Never nick the conductors when removing the sheath.

- Conductors always wind clockwise, between the two washers.

- Insure that the insulation does not get caught under the screw terminals.

- Always trim off the excess bare wire to avoid shorting out the connection.

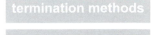
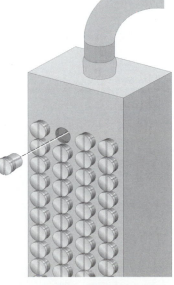

Figure 1-30: Binding Post
Termination

- Always leave additional slack wire in case future repairs are required.

- Do not overtighten the screw terminals, or the wire could break off. In addition, the threads will strip out.

Insulation Displacement Connector (IDC)

The other method of termination is **Insulation Displacement Connector (IDC)**. This method is faster and gives a more reliable connection. An IDC does not require stripping of the conductors. The conductor is forced into a terminal strip, which not only holds the conductor firmly in place, but also pierces the insulation to make an electrical connection.

The two most common styles of IDC are the type-66 and type-110 termination blocks. Type-110 blocks are a more recent IDC development than type-66 blocks, and occupy less space than type-66 blocks for an equal number of connections.

A type-110 block is an IDC invented by AT&T for interconnecting telephone and higher-speed data cables. It usually contains a variable number of slots arranged in rows. Patch cables are used to configure the interconnection once the source and destination cables have been attached. These blocks are more likely to be rated for CAT5 wire than for type-66 blocks, and often have RJ11 or RJ45 connectors already attached to them. Both block types

impact or punch-down tool

require the use of the **impact or punch-down tool**, as shown in Figure 1-31, to terminate the wires. However, the 110-type block requires the use of a different blade than the older type-66 blocks.

Figure 1-31: Impact or Punch-down Tool

An older type-66 block is an insulation displacement system, originally designed to wire up telephones and similar communications systems. It usually contains 200 metal slots (for interfacing 50 pairs from two sources), and comes in various formats for those pins. It is designed for solid wire (22- to 26-gauge), which is forced or "punched" into each slot, displacing the insulation and making the connection. It was not originally intended as a data, or audio connector, although some newer versions are used for those applications. These versions are rated for CAT5 wire, and also have RJ11 or RJ45 connectors pre-attached to them.

Newer type-66 blocks used for CAT5 are distinctly different than the type-66 blocks used for previous category standards. The clips on the type-66 block used for CAT5 are designed to reduce reactive coupling, improving near-end crosstalk (NEXT) performance. The type-66 block used for CAT5 has the tip and ring spacing closer together while the spacing between the pairs remains the same. Therefore, the clips can be smaller than those found on the type-66 block used for CAT4 terminations. Although the spacing is closer together between tip and ring clips, there is ample room for the termination tool as well as testing equipment to fit between them.

The type-66 blocks used for CAT4 have the clip pairs evenly spaced. These differences in spacing are shown in Figure 1-32. Notice that the termination block itself is fitted with 66 rows of terminals, instead of 50. This custom configuration allows the block to accommodate eight RJ45 modular connectors, or at least 64 connection pairs. A normal 50-row block would only be able to outfit six RJ45 connectors.

TYPE 66 IDC
BLOCK

STANDARD PAIR SPACING

10.2 mm
(0.40")

5.1 mm
(0.20")

CATEGORY 5 PAIR SPACING

10.2 mm
(0.40")

4.8 mm
(0.19")

**Figure 1-32: Standard and
CAT5 Pair Spacing**

Basic Hand Tools

There are several hand tools that are normally found in every cable installer's toolbox. These include a two-way punch-down or impact tool, electricians scissors or snips, diagonal cutting pliers, common and Phillips screw drivers, wire strippers, tone generating probes, and Modapt test adapters.

Siemon's S814 Punch-down Tool

The Siemon punch-down tool is also known as an **impact tool**. It is usually supplied with two types of blades. One is for the 66-type block, and the other is for the 110-type block. Both of these blades have two sides to them. One side is for terminating and cutting, and the other side terminates without cutting the wire to allow the wire to be looped to another terminal.

impact tool

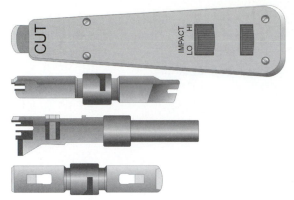

Figure 1-33: Impact or Punch-down Tool with Blades

The impact tool itself has two settings on it. They are Hi and Lo. The Hi setting is used to terminate wire on a 66-type block. The Lo setting is used to terminate wire for the 110-type block. You need to be sure to check the settings on the impact tool before you terminate the wires on any block, because an improper setting can damage the block you are working with.

The blades themselves are labeled. The blade that you use for the 66-type block has a "66" written directly on it for identification. Similarly, the blade that you use for a 110-type block has "110" written on it. You'll want to be sure that your cutting blades are kept sharp. A dull blade can cause improper termination. The punch-down tool is shown in Figure 1-33.

Electricians' Scissors

Electricians' scissors

Electricians' scissors, or snips, are designed to cut and remove insulation from small-diameter wires and cables. The back side of the scissors is notched for stripping insulation from 22- and 24-gauge wire.

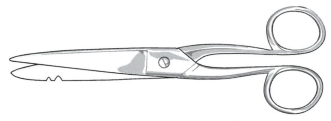

Figure 1-34: Electricians' Scissors

Safety, when using your snips, is very important. When cutting wire, the small end may become airborne, possibly causing injury to the eyes, or a puncture of the skin. One should cover the cutting area with the hand, thus shielding the eyes from possible injury. Because electricians' scissors are very sharp, you must use extreme caution when using them to avoid cutting your fingers, or injuring someone else. Figure 1-34 depicts a pair of electricians' scissors.

Diagonal Cutting Pliers

Diagonal cutting pliers

Diagonal cutting pliers, as shown in Figure 1-35 (also known as "dykes"), are primarily used for cutting or removing insulation from various diameters of wire and cable. Consider the size of the dykes you are using, and do not try to use them to cut off more than they were designed for. They come in various sizes, so be aware of the correct size required for the diameter of wire or cable that you are cutting.

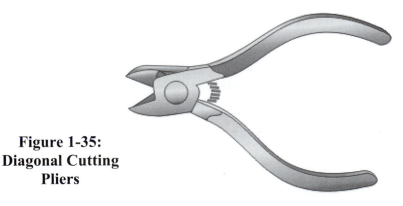

Figure 1-35: Diagonal Cutting Pliers

Once again, as with the electricians' scissors, safety is important. Small sections of wire, or cable, can become a projectile causing injury to the eyes or puncture injuries to the skin. Protective eye gear can be very helpful in these situations.

Screwdrivers

There are two major types of **screwdrivers** contained in your tool package, of differing head sizes and lengths. One type is called the **slotted-head**, and it is depicted in Figure 1-36.

Figure 1-36: Slotted-head Screwdriver

screwdrivers

slotted-head

Because the slotted-head screwdriver presents more difficulty when attempting to maintain contact between the driver and the slotted screw, you should use caution when driving screws from an awkward position. If, while under pressure, the screwdriver slips from its contact with a screw's slotted head, you can be seriously injured, or you could cause severe damage to the surrounding surfaces. The slotted-head screwdriver works best when the slot in the head of a screw or fastener fits snugly around the driver's blade. Therefore, you should do your best to select the size of the screwdriver so that its head matches the slot embedded in the screw or fastener.

Phillips-head

The second type of screwdriver that is most commonly found in the cable installer's toolbox is the **Phillips-head**. This type of driver is easy to recognize because of its cross-blade pattern, as shown in Figure 1-37. Using this type of driver/screw combination, it is much less likely that the head of the driver will slip out of position while under pressure. However, when a Phillips screwdriver is being used with the improper size of screw or fastener, damage to the head of the screw can occur, making it impossible to drive it down to its maximum depth. Damage to the screw or fastener can be so severe that it becomes impossible to either drive the screw down, or to remove it from the workpiece. For this reason, always try to match the head of a Phillips-type screw with the correct driver. Phillips screwdrivers vary in both the pitch and depth of the cross-blades.

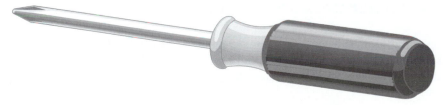

Figure 1-37: Phillips-head Screwdriver

Machine screws must be selected to correctly match the size of pre-threaded metal holes to prevent stripping of the grooves from the screw and/or hole.

If you have trouble getting a screw or fastener started, remember to drill a pilot hole before beginning the driving process. As always, remember that safety is extremely important when using all types of screwdrivers. Be sure to pay close attention to where you are driving a screw or fastener, and be aware of what may be hidden from view behind the surface into which you are driving.

Wire Strippers

Wire strippers are designed to make the job of stripping insulation from wire or cable a less daunting chore. Most wire strippers are also outfitted with cutters, and are suitable for stripping STP and UTP cable, speaker cable, flat wire, and computer cable of various types. An adjusting bolt is normally employed in order to match different cable sizes to the stripper guides. Figure 1-38 depicts a basic wire stripper.

When adjusted properly, wire strippers should score and remove a layer of insulation only, and not be permitted to score, damage, or cut the actual wire in any way. If any conductor is nicked by a cutting blade during the insulation removal process, it could be rendered useless.

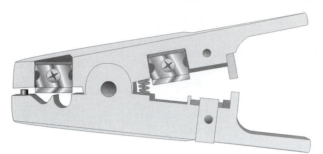

Figure 1-38: Wire Stripper

Take care in stripping wire and cables not to cut too deeply into the insulation. Another type of wire stripper is illustrated in Figure 1-39. Notice that its design is more basic than the first example, and it is designed to accommodate the thicker RG8 cable only. Insertion of the cable is accomplished by using the thumb slide to retract the blade.

Figure 1-39: RG8 Wire Stripper

Tone Generators and Signal Probes

Figure 1-40 illustrates a **tone generator** and **signal probe** combo. Tone generators allow the introduction of audio tones across a specific wire pair in order to test for the continuity of a given conductor, the signal attenuation along the conductor, or crosstalk between adjacent conductors.

Tone generators suitable for the cable installer's tool box should be battery powered, and contain circuitry that allows for selective filtering of the tone frequencies. Accordingly, you should be aware of the fact that changes in both temperature and battery power can alter the frequency of the tones produced by all tone generators.

Always operate a tone generator with a fresh battery for best results. Operating the device with a weak battery can produce inaccurate or misleading results.

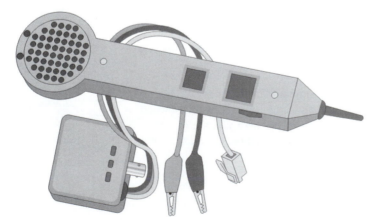

Figure 1-40: Tone Generator and Signal Probe

When tone generators are used in conjunction with signal probes, bad wire pairs can be identified and isolated. The signal probe is used to detect the tone being generated on a specific wire pair. The signal probe depicted in Figure 1-40 contains a built-in amplifier and speaker. The probe also contains a filter circuit in order to remove any power-related noise from the wires being tested.

Siemon's MODAPT® Modular Adapter

The Siemon Company's **MODAPT®**, shown in Figure 1-41, is a test adapter that is outfitted with a special, 8-wire modular plug that is capable of connecting to any 2-,4-,6-, or 8-wire modular jack. This adapter enables the efficient attachment of various pieces of test equipment to the modular jacks by using alligator clips.

The MODAPT® can plug directly into modular station jacks, or can be used in conjunction with other types of test equipment, to test 66-type terminal blocks. The MODAPT® shows which pins to clip to for tip and ring, according to USOC, T568A, and T568B. It will allow alligator clips from butt sets, tone generators, or other test equipment, to connect to specific terminals within the modular jacks. The MODAPT® shown comes with a universal test cord, and two built-in jacks for in-line testing.

Figure 1-41: MODAPT® Test Adapter

Modular Crimpers

Modular crimpers, such as those shown in Figure 1-42, are designed to cut wire, strip insulation, and attach medium "RJ11" and large "RJ45" modular plugs to 2-, 4-, 6-, or 8- wire cables. The model shown is extremely well made, sporting a durable steel frame and cushioned hand grips.

Figure 1-42: Modular Crimpers

Lab 3 – Tool Identification

Lab Objective

To properly identify common tools used for CAT5 and coaxial cable installations

Materials needed

- Lab 3 work sheet
- Cable stripper, RG58/RG6

- Cable stripper, RG8-type
- Wire stripper, CAT5-type
- Punch-down (impact) tool, steel
- Punch-down (impact) tool, plastic
- Punch-down blades, 110 and 66 types
- Electrical tape
- Twine
- Electricians' scissors
- Utility knife
- Pliers, diagonal cutting
- Screwdriver, slotted-head, small
- Screwdriver, Phillips-head, small
- Insertion/extraction tool
- Pliers, needle-nose
- Pliers, groove-joint
- Wrench, adjustable
- Dental pick
- Crimping tool, RJ45
- Crimping tool, RG58/RG6
- Crimping tool, pin
- Nippers, end-cutting
- Fish tape, steel
- Thicknet application tool
- Hex wrench (from cable stripper, RG58/RG6)
- Pencils, colored (set)
- Tension scale/Tape measure
- Label markers/dispenser
- Label markers, replacement roll
- Cable tester, continuity-type
- Level II testers, w/case & cables
- Multimeter, digital

<u>Instructions</u>

Before beginning this lab, read through all of the following steps at least once, thoroughly.

1. Examine the pictures shown in Figure 1-43.

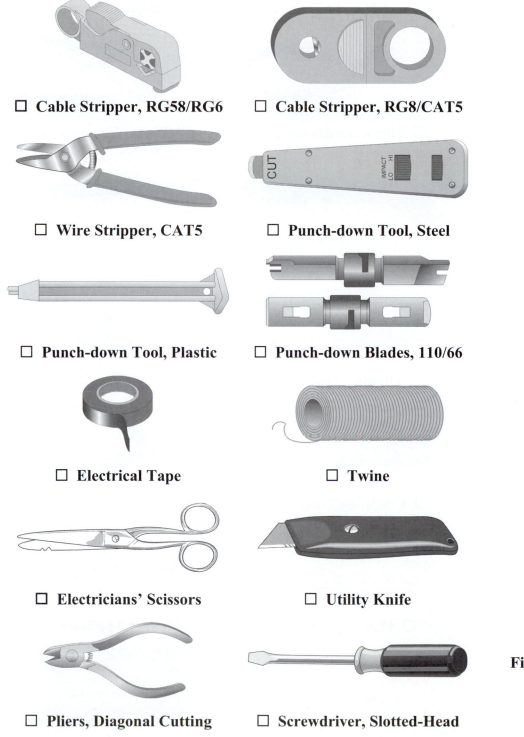

☐ **Cable Stripper, RG58/RG6**   ☐ **Cable Stripper, RG8/CAT5**

☐ **Wire Stripper, CAT5**   ☐ **Punch-down Tool, Steel**

☐ **Punch-down Tool, Plastic**   ☐ **Punch-down Blades, 110/66**

☐ **Electrical Tape**   ☐ **Twine**

☐ **Electricians' Scissors**   ☐ **Utility Knife**

☐ **Pliers, Diagonal Cutting**   ☐ **Screwdriver, Slotted-Head**

**Figure 1-43: Common
Tools for Cable
Installations**

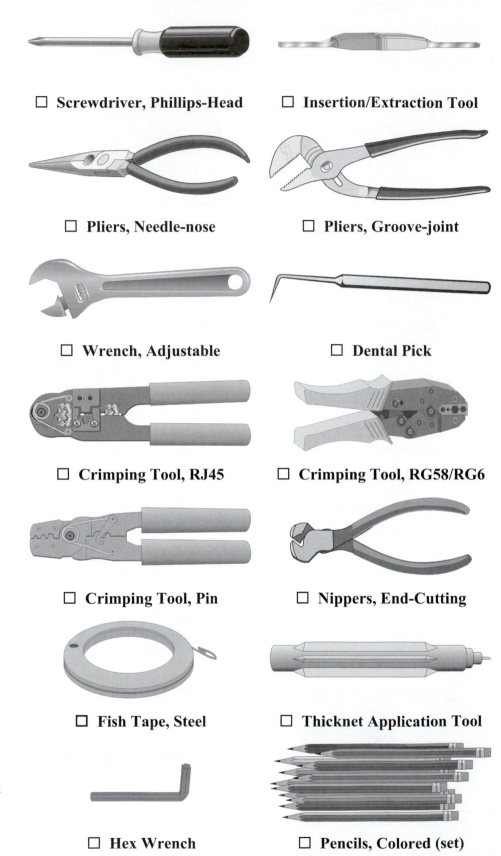

☐ **Screwdriver, Phillips-Head**

☐ **Insertion/Extraction Tool**

☐ **Pliers, Needle-nose**

☐ **Pliers, Groove-joint**

☐ **Wrench, Adjustable**

☐ **Dental Pick**

☐ **Crimping Tool, RJ45**

☐ **Crimping Tool, RG58/RG6**

☐ **Crimping Tool, Pin**

☐ **Nippers, End-Cutting**

☐ **Fish Tape, Steel**

☐ **Thicknet Application Tool**

Figure 1-43: Common Tools for Cable Installations (cont.)

☐ **Hex Wrench**

☐ **Pencils, Colored (set)**

□ **Tension Scale/Tape Measure**

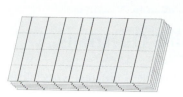
□ **Label Markers**

□ **Cable Tester, Continuity-type**

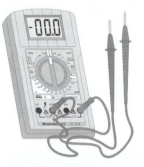
□ **Multimeter, Digital**

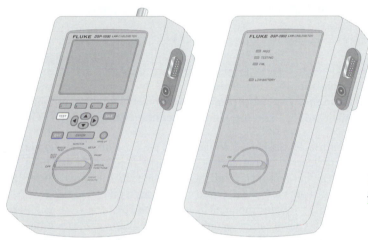

**Figure 1-43: Common Tools
for Cable Installations (cont.)**

□ **Level II Testers**

2. Beginning at the top-left corner of Figure 1-43, locate the tool shown in the block, examine it carefully, and place a check in the box corresponding to its diagram.

3. Return the tool you just examined to its storage area, or previous location.

4. From left to right, repeat steps 1 through 3 for each tool depicted in Figure 1-43.

5. If any tools are missing, notify your instructor immediately.

6. When you have completed checking and identifying all of the tools, check to be sure that they have been returned to their storage area, or previous location.

7. Be sure that you are leaving your work area in a clean and orderly condition.

Lab 3 Questions

1. How many different types of cable or wire strippers are included in the tool list?

2. What are the two types of punch-down blades provided with this course?

3. Identify a tool reviewed in the text that is not supplied for this course.

4. How many different types of crimpers are included in the tool list?

Modular Plugs and Jacks

Modular plugs

modular jacks

Modular plugs are telecommunications male connectors, while their female counterparts are **modular jacks**. They may be keyed or unkeyed, and may have a variable number of contact positions, although not all positions need to be connected in every cabling scenario. The newer-style modular insulation displacement jacks replace the older, screw-terminal modular jacks in residential applications that are wired with one or two jacks per room.

registered jack

Federal Communications Commission (FCC)

Universal Service Order Code (USOC)

jack

When referring to specific modular plugs and jacks, take care to be more precise, because terms such as RJ11, RJ12, RJ45, keyed RJ45, etc., are often used incorrectly to describe the types of available modular jacks and plugs that are widely used in telecommunications. The acronym "RJ" stands for the term **registered jack**, and is most often referred to as RJXX, where XX could be one of a number of telephone connection interfaces (plugs and jacks) that are registered with the U.S. **Federal Communications Commission** (**FCC**).

These various interfaces originated as part of AT&T's **Universal Service Order Code** (**USOC**), and were later adopted as Part 68, Subpart F, Section 68.502 of FCC regulations. The term **jack** can sometimes mean both the plug and jack, but usually refers only to the female jack. Plugs must attach to the required cabling for a given application. Although jacks are normally thought of as being mounted in various wall plate or patch panel configurations, they can also appear on patch cables, or on work area cabling, where they may provide for cabling extensions. Such use is to be discouraged, however, unless no other solution exists. This is because of the additional signal loss that can occur at each plug/jack connection.

telecommunication jacks

Modular **telecommunication jacks** may be configured as a 6- or 8-pin configuration, and all jack and plug mating connections must satisfy the specified mechanical tests detailed in the FCC USOC already mentioned. Certain of these test requirements address the mating of:

- 6-position plugs with 6-position jacks

- 4-position plugs with 6-position jacks

- 8-position plugs with 8-position jacks

- 6-position plugs with 8-position jacks

RJ11

The most common telecommunication jacks are the RJ11 types, which can have up to six conductors, but are usually implemented with only four. RJ11 jacks are the ones that your

household or office phones are likely plugged into, using ordinary "untwisted" telephone wire, which is sometimes called "gray satin" or "flat wire". The RJ11 jacks then connect to the "outside" wires, known as twisted pair, that connect to the telephone company Central Office (CO), or to a Private Branch Exchange (PBX). An RJ11 plug and jack are shown in Figure 1-44.

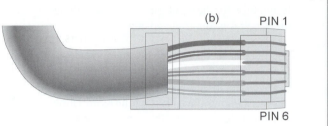
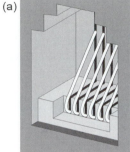

Figure 1-44: RJ11 Jack (a) and Plug (b)

The four conductors are usually characterized as a red and green pair and a black and white pair. The red and green pair typically carry voice or data. The phone company connection may use the black and white pair for low-voltage signals, such as phone lights, or they may be used for other kinds of signaling on a PBX system. Internal and external computer modems are usually connected to a RJ11 jack. From the rear view, as shown in Figure 1-45, the USOC specifies one pair of wires (pair 1) in a 4-, 6-, or 8- position modular jack. The design of these modular jacks is such that the 4-position modular plug will fit into the 6-position and 8-position modular jacks, and the 6-position modular plug will fit into the 8-position modular jack.

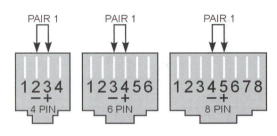

Figure 1-45: USOC RJ11

RJ14

The RJ14 registered jack is similar to the RJ11, with the exception that the four wires are used for two separate phone lines. Such a configuration will typically use one set of wires for each line. One line will utilize the red and green wire set, while the second line will utilize the yellow and black wire set. Each wire set can carry one analog "conversation" (voice or data).

From the rear view in Figure 1-46, the USOC specifies two pairs of wires in a 4-, 6-, or 8-position modular jack. Pair 1 would occupy the two center pins, and pair 2 would occupy the next two outward pins. Once again, the 4-position modular plug will fit into the 6-position and 8-position modular jacks, and the 6-position modular plug will fit into the 8-position modular jack.

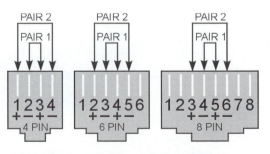

Figure 1-46: USOC RJ14

Telco 8P8C

The eight-pin modular connector is subjected to excessive name abuse. The USOC specialized RJ-45 wiring pattern is so completely different from that of the T568A and T568B configurations that it should not be called RJ-45 at all. RJ-45 is really a Telco designation for the eight-pin connector with an interface programming resistor.

Installers have been referring to T568A and T568B plugs and jacks as RJ-45s for so long that even manufactures and vendors advertise them as RJ-45s. When used for the termination of Data/Telco or just Data applications the male connector on the end of a patchcord is a "plug" and the female or receptacle is a "jack". When used for normal Telco applications the "Generic and the "RJ61X" are used, as shown in Figure 1-47. The RJ61X is the configuration is the most commonly used.

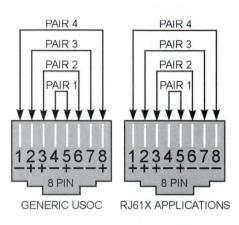

Figure 1-47: Telco 8P8C

T568A and T568B

The RJ45 is a single-line registered jack designed for digital transmission over ordinary phone wire, either untwisted (usually a flat wire such as common household phone extension cord) or twisted (often round in shape). The interface has eight pins or positions.

Untwisted wire can be used for connecting a modem, printer, or a data PBX at rates up to 19.2 kbps. However, for faster transmission rates (required in applications such as Ethernet 10baseT network connections), RJ45 jacks will need to be installed with twisted pair wire.

There are two varieties of RJ45 modular jacks: keyed and unkeyed. A keyed plug has a small bump on its end, and the female jack is fitted with a complementing groove to match it. Both jack and plug must match in order to effect a reliable connection. An RJ45 jack and plug are shown in Figure 1-48.

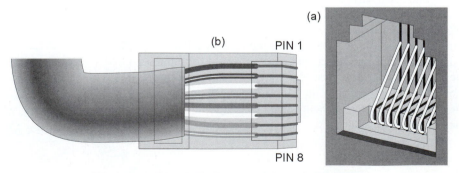

Figure 1-48: RJ45 Jack (a) and Plug (b)

From the rear view, as shown in Figure 1-49, the TIA/EIA T568A plug specifies an RJ45 pin configuration that is often used in Ethernet (10baseT) on pairs 3 and 2. In order to use this configuration in Fast Ethernet (100baseT) systems, CAT5-compliant jacks, plugs, patch panels, and cables must be used. This configuration can also be used in Token Ring networking on pairs 1 and 2.

The Tip and Ring assignments for TIA/EIA T568A are:

- Pin 1 = T3
- Pin 2 = R3
- Pin 3 = T2
- Pin 4 = R1
- Pin 5 = T1
- Pin 6 = R2
- Pin 7 = T4
- Pin 8 = R4

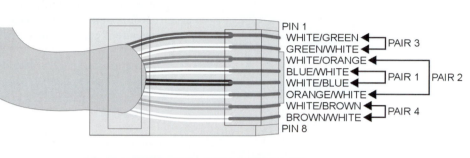

Figure 1-49: TIA/EIA T568A Plug

As seen from the rear view in Figure 1-50, the TIA/EIA T568B specifies an RJ45 pin configuration that is essentially the same as the AT&T 258A wiring scheme. Often, the TIA/EIA T568B specification is used in Ethernet (10baseT) on pairs 2 and 3. Just as with TIA/EIA T568A, CAT5 jacks, plugs, patch panels, and cables must be used for Fast Ethernet (100baseT) systems. This configuration can also be used in Token Ring networking on pairs 1 and 3. The Tip and Ring assignments for TIA/EIA T568B are:

- Pin 1 = T2
- Pin 2 = R2
- Pin 3 = T3
- Pin 4 = R1
- Pin 5 = T1
- Pin 6 = R3
- Pin 7 = T4
- Pin 8 = R4

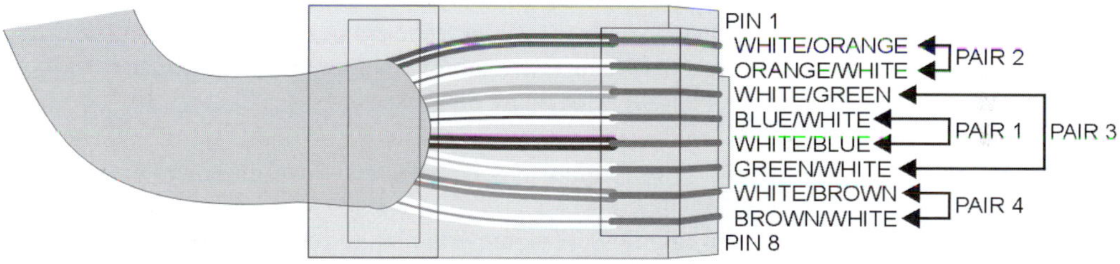

Figure 1-50: TIA/EIA T568B Plug

The TIA/EIA T568A standard specifies that all light commercial jacks should be wired according to TIA/EIA T568A or TIA/EIA T568B, unless otherwise specified. The TIA/EIA T568A is usually used for two-line telephone operations, and the TIA/EIA T568B is usually used for data applications.

Modern business networking requirements now include specifications for the **telecommunications outlets** that are installed at, or near, workstations. Each individual workstation must support at least two recognized cable types, which would allow for both data and voice applications.

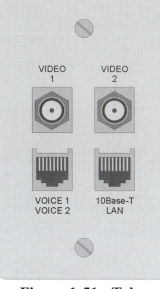

Figure 1-51: Telco Outlet

Therefore, outlets for telecommunications services should be distributed throughout a given facility, with each outlet containing voice, data, and video jacks as determined by the master design. As much as possible, each outlet should be mounted using a common face plate. Voice and data jacks must meet CAT5 specifications, while video jacks should use F-style coaxial connectors. An example of a telecommunications outlet is depicted in Figure 1-51.

Exact locations for each outlet shall be determined during the design process. At a minimum, two horizontal cable runs should extend from the telecommunications closet to the outlets in a star configuration.

Telecommunications outlets should meet the following criteria:

- At least two station cable pairs must be connected at every installed outlet.

A minimum of two horizontal cable runs should be made from the telecommunications closet to the outlets in a star configuration. At least one of the cables should connect to an 8-pin, modular RJ45 wall plate, with a configuration conforming to the TIA/EIA T568B interface. This outlet shall be clearly designated as a data communications interface.

- The minimum number usable termination points at the wiring closet is 3 per user.

A good plan is to keep telephone voice circuits and data circuits connected to separate outlets if this is physically possible, and financially feasible. This will help to avoid the excess handling of data cabling during the inevitable manipulations associated with normal telephone line maintenance. The choice as to which types of cables are to be installed (coax, UTP, fiber optic, etc.) must be determined during the planning stage. Many new installations specify ISDN, fiber optic, and video connections even though these services may not yet actually be used. Having the capability to upgrade the system at some point in the future is a critical consideration for many businesses.

- Outlets may be recessed, flush-mounted, or surface-mounted.

Regardless of where you plan to install the telecommunications outlets, or how many you will install, you will want to keep them all at least 12 inches from any power outlets.

Where possible, the telecommunications outlets should be located with centerlines at least 18 inches above floor level (or 12 inches above permanent bench surfaces). Do not locate a telecommunications outlet lower than 12 inches above floor level at any time. Surface-mounted outlets must be well-secured, if used. Recessed outlets which are mounted in any wall that has been designated as a fire barrier will require metal mounting boxes.

- When installed on building exteriors, outlets must be weather resistant.

Outlets must have spring-loaded, hinged covers that protect either the individual jacks, or the entire face plate area, from the intrusive effects of moisture, dust, and vibration. In many cases, these covers should possess the added feature of being lockable.

Installation Tips for Modular Cables and Jacks

When installing any 6-position modular cables from the outlet jack to the telecommunications equipment, the equipment is designed to use corresponding, **straight-through** pin-to-pin numbering (pin 1 at the jack to pin 1 on the cable, etc.). Typically, the 4-position plug and jack, found on the telephone handset, is smaller than the jack and plug that goes into the wall. The cable used from the wall to the telephones is usually stranded wire, giving it improved flexibility.

straight-through

Quality Installations

For quality installations, keep the following suggestions in mind:

- Always use the minimum number of connections possible, due to the fact that each additional connection to the system will degrade its overall performance.

- During installations, always install additional modular jacks, as well as **service loops**, to prepare the system for future upgrades, changes, moves, and repairs.

service loops

- All installations must be performed in a neat and orderly fashion. Documentation should include information as to where all outlets and cable runs are located, and should use a clearly understandable method of labeling the cable runs.

- Always install cables that are rated at the highest data speeds currently available, and never connect cables to any equipment or connector that is rated lower than that specified for the category of wire being used by the system.

- Once all of the telecommunications components have been installed, the entire system (cabling runs and associated equipment) must be tested for proper performance.

Lab 4 – Terminating CAT5 with RJ45/T568A Jacks

<u>Lab Objective</u>

To properly terminate CAT5 UTP cable onto CAT5-compliant RJ45 termination jacks using the T568A configuration

<u>Materials Needed</u>

- Lab 4 work sheet

- Cable, CAT5 UTP cable, 100-ohm, non-plenum (spool)

- Electricians' scissors

- Pliers, diagonal cutters

- Termination jacks, CAT5/RJ45, T568A (2)

- Tension scale/Tape measure

Instructions

Before beginning this lab, read through all of the following steps at least once, thoroughly.

1. Observing the proper safety procedures, use the diagonal cutters, or their equivalent, to cut approximately 5 feet of CAT5 cable from the spool.

2. Being careful not to nick or cut any of the individual conductors, strip approximately 2.5 inches of cable sheath (not to exceed 3 inches) from one end of the 5-foot section, and cut off the strain-relief threads as well, using the electricians' scissors.

3. Untwist each of the cable pairs for a length of approximately 2 inches, as shown in Figure 1-52.

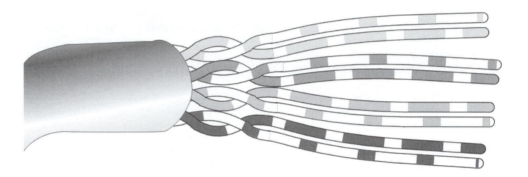

Figure 1-52: Untwisting the Cable Pairs

TIP: Keep the individual pairs grouped together for ease of identification, because some tip wires may not have any visible trace of color and may appear to be solid white.

4. Using the TIA/EIA T568A configuration, insert the untwisted CAT5 cable conductors into a CAT5-compliant termination jack cap.

TIP: Look on the jack cap first for a manufacturer's color code. If you find one, use it, because some jacks are manufactured with internal cross connections. For stop-end jacks, make sure that the conductors are inserted so that the cable jacket resides inside of the jack cap, with the tips of the wires inserted to the very end of the cap.

5. If you are using straight-through jack caps, insert the CAT5 conductors until they protrude past the front of the cap, and the cable jacket is inside of the cap, as shown in Figure 1-53. Then, using the electrician scissors, trim the excess cable flush with the jack cap's head.

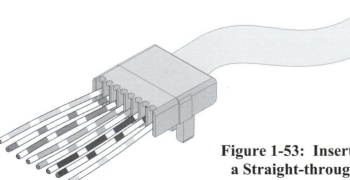

Figure 1-53: Inserting Wires in a Straight-through Jack Cap

6. For snap jacks, press the jack cap onto the housing, until you hear a click or snap.

TIP: For stop-end jacks, insert the jack cap/housing combination into the crimping tool, apply pressure to the cable to make sure that the conductors remain inserted to the very ends of the jack, and apply firm, even pressure to the crimping tool to attach the jack.

7. Repeat steps 2 through 6 on the opposite end of the CAT5 cable with the remaining termination jack.

8. If necessary, repeat this lab exercise until you have gained the desired level of proficiency.

TIP: Snap jacks can be pried apart, using a slotted-head screwdriver, and reused for practice. Stop-end jacks must be clipped off, and thrown away, and replaced in order to practice on the same piece of cable. If you must clip the cable ends to remove stop-end jacks for practice purposes, under no circumstances allow the cable to become shorter than four feet (48 inches) long. This cable will be used for later labs. If necessary, practice using pieces of scrap cable.

9. When you are satisfied, use the tension scale/tape measure to check the length of the cable. Make sure that it is at least 48 inches in length.

TIP: The tension scale/tape measure is exactly one meter long. Measure the cable to the 3-foot marker, place the start of the measure at that point, and measure at least 12 more inches beyond it. If the total length of the cable is less than four feet, you will have to discard it (or add it to the practice cable scrap pile) and try again using a new piece of cable. You may reuse the jacks if they are the snap type.

10. Save the T568A-configured cable (at least 48 inches long), on which you have just mounted CAT5-compliant termination jacks, for later use.

11. Check to be sure that all of the tools, and parts, have been returned to their storage area, or previous locations.

12. Before leaving your lab work area, check to be sure that you are leaving it in a clean and orderly condition.

Lab 4 Questions

1. What is the purpose of twisting the cable pairs in CAT5 cable?

2. List the pin/pair configuration for the TIA/EIA T568A standard.

3. What is the minimum length required for the completed T568A terminated CAT5 cable in this lab?

Lab 5 – Terminating CAT5 with RJ45/T568B Jacks

<u>Lab Objective</u>

To properly terminate CAT5 UTP cable onto CAT5-compliant RJ45 termination jacks, using the T568B configuration

<u>Materials Needed</u>

- Lab 5 work sheet
- Cable, CAT5 UTP cable, 100-ohm, non-plenum (spool)
- Electricians' scissors
- Pliers, diagonal cutters
- Termination jacks, CAT5/RJ45, T568B (2)
- Tension scale/Tape measure

<u>Instructions</u>

Before beginning this lab, read through all of the following steps at least once, thoroughly.

1. Observing the proper safety procedures, use the diagonal cutters, or their equivalent, to cut approximately 5 feet of CAT5 cable from the spool.

2. Being careful not to nick or cut any of the individual conductors, strip approximately 2.5 inches of cable sheath (not to exceed 3 inches) from one end of the 5-foot section, and cut off the strain-relief threads as well, using the electricians' scissors.

3. Untwist each of the cable pairs for a length of approximately 2 inches.

TIP: Keep the individual pairs grouped together for ease of identification, because some tip wires may not have any visible trace of color and may appear to be solid white.

4. Using the TIA/EIA T568B configuration, insert the untwisted CAT5 cable conductors into a CAT5-compliant termination jack cap.

TIP: Look on the jack cap first for a manufacturer's color code. If you find one, use it, because some jacks are manufactured with internal cross connections. For stop-end jacks, insert the conductors so that the cable jacket resides inside of the jack cap, and the tips of the wires reach to the very end of the cap.

5. For straight-through jacks, insert the CAT5 conductors until they protrude past the front of the jack, and the cable jacket is inside of the jacket cap.

6. Use the electricians' scissors to trim the excess cable flush with the head of the cap, as shown in Figure 1-54.

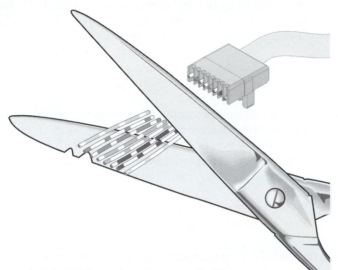

Figure 1-54: Trimming Excess Cable

7. For snap jacks, press the jacket cap onto the housing, until you hear a click or snap, as shown in Figure 1-55.

TIP: For stop-end jacks, insert the jack cap/housing combination into the crimping tool, apply pressure to the cable to keep the conductors inserted to the very ends of the jack, and apply firm, even pressure to the crimping tool to attach the jack, as shown in Figure 1-56.

8. Repeat steps 2 through 7 on the opposite end of the CAT5 cable with the remaining termination jack.

9. If necessary, repeat this lab exercise to gain the desired level of proficiency.

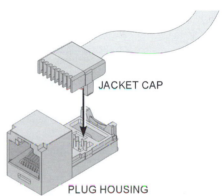

JACKET CAP

PLUG HOUSING

Figure 1-55: Crimping the Jack Housing

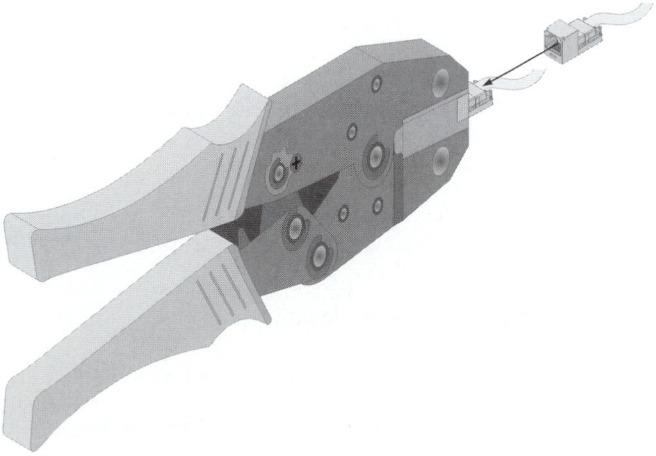

Figure 1-56: Using the Crimping Tool

TIP: Snap jacks can be pried apart, using a slotted-head screwdriver, and reused for practice. Stop-end jacks must be clipped off, and thrown away, and replaced in order to practice on the same piece of cable. If you must clip the cable ends to remove stop-end jacks for practice purposes, under no circumstances allow the cable to become shorter than four feet (48 inches) long. This cable will be used for later labs. If necessary, practice using pieces of scrap cable.

10. When you are satisfied, use the tension scale/tape measure to check the length of the cable. Make sure that it is at least 48 inches in length.

TIP: The tension scale/tape measure is exactly one meter long. Measure the cable to the 3-foot marker, place the start of the measure at that point, and measure at least 12 more inches beyond it. If the total length of the cable is less than four feet, you will have to discard it (or add it to the practice cable scrap pile) and try again using a new piece of cable. You may reuse the jacks if they are the snap type.

11. Save the T568B-configured cable (at least 48 inches long), on which you have just mounted CAT5-compliant termination jacks, for later use.

12. Check to be sure that all of the tools, and parts, have been returned to their storage area, or previous locations.

13. Before leaving your lab work area, check to be sure that you are leaving it in a clean and orderly condition.

Lab 5 Questions

1. List the advantages/disadvantages of straight-through Vs. stop-end termination jacks.

2. List the pin/pair configuration for the TIA/EIA T568B standard.

Lab 6 – Terminating CAT5 with RJ45/T568A Plugs

Lab Objective

To properly terminate a CAT5 UTP cable with two RJ45 modular plugs, using the T568A configuration

Materials Needed

- Lab 6 work sheet

- Cable, CAT5 UTP, 100-ohm, non-plenum (spool)

- Cable, CAT5 UTP T568A jack-terminated, > = 4 feet in length

- Cable, CAT5 UTP T568A plug-terminated, > = 4 feet in length

- Electricians' scissors

- Pliers, diagonal cutters

- Termination plugs, RJ45, unshielded (2)

- Crimping tool, RJ45

- Tension scale/Tape measure

- Cable tester, continuity-type

Instructions

Before beginning this lab, read through all of the following steps at least once, thoroughly.

1. Observing the proper safety procedures, use the diagonal cutters to cut 5 feet of CAT5 cable from the spool.

2. Being careful not to nick or cut any of the individual conductors, strip approximately 2.5 inches of cable sheath (not to exceed 3 inches) from one end of the 5-foot section, and cut off the strain-relief threads as well, using the electricians' scissors.

3. Untwist each of the cable pairs for a length of approximately 2 inches.

TIP: Keep the individual pairs grouped together for ease of identification, because some tip wires may not have any visible trace of color and may appear to be solid white.

4. Arrange the cable colors according to the TIA/EIA T568A pin configuration.

5. Using the electricians' scissors, trim the CAT5 conductors to approximately 5/8 of an inch, as shown in Figure 1-57. This includes the strain-relief threads as well.

Figure 1-57: Trimming the CAT5 Conductors

6. Insert the CAT5 conductors into one of the RJ45 modular plugs.

7. Make sure that the conductors are inserted as far as possible into the plug, as shown in Figure 1-58.

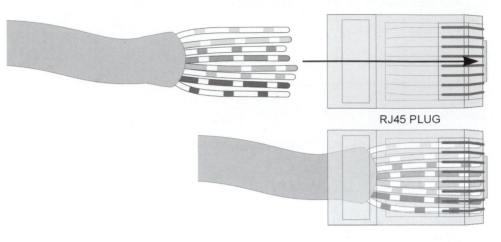

RJ45 PLUG

Figure 1-58: Inserting the CAT5 Conductors

8. Remove the conductors momentarily to examine their ends for evenness.

9. If the wires in the center of the bunch appear to be longer than the wires at the ends, use the cable cutters to snip enough to make all of the conductors even, as shown in Figure 1-59. Then, reinsert the trimmed wires into the RJ45 modular plug.

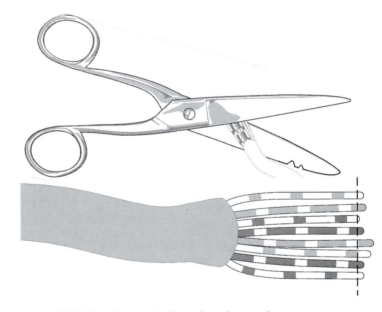

Figure 1-59: Trimming the CAT5 Conductors

10. Place the wired RJ45 modular plug into the crimp tool.

11. Before crimping, ensure that the CAT5 conductors are fully inserted into the plug, and that the cable sheath is resting above the crimp edge inside of the plug, as shown in Figure 1-60.

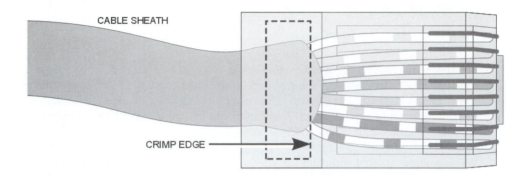

CABLE SHEATH

CRIMP EDGE

Figure 1-60: Resting the Cable Sheath

12. Apply pressure to the cable to make sure that the conductors remain inserted to the ends of the plug, and use firm, even pressure with the crimping tool to attach the RJ45 modular plug.

13. Remove the crimped end of the cable from the crimping tool, with the RJ45 modular plug attached.

14. Repeat steps 2 through 13 for the opposite end of the CAT5 cable, using the second RJ45 modular plug.

15. Using the CAT5 cable tester, check the completed cable for continuity, and proper sequencing of the wire pairs.

16. If and when the cable passes the test for CAT5 T568A configuration, disconnect the end attached to the remote section of the tester.

17. Connect the CAT5 T568A cable with termination jacks from a previous procedure to the free end you just disconnected.

18. Borrow a known good CAT5 T568A-configured cable terminated with plugs (from another student perhaps), and connect one of its plugs into the remaining jack on the second cable.

19. Finally, connect the remaining plug end of the borrowed cable to the remote section of the cable tester. The test setup should look like Figure 1-61.

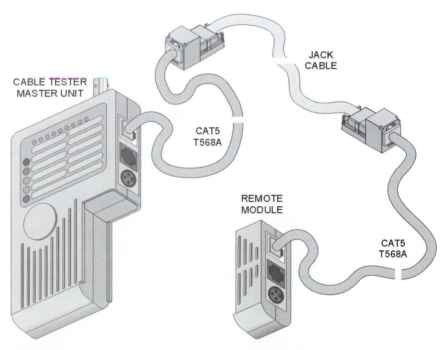

JACK CABLE

CABLE TESTER MASTER UNIT

CAT5 T568A

REMOTE MODULE

CAT5 T568A

Figure 1-61: Testing CAT5 Plug and Jack Cables for T568A

20. Using the CAT5 cable tester, check the three connected T568A cable sections (plugs-jacks-plugs) for continuity, and proper sequencing of the wire pairs.

21. If necessary, correct any problems that are revealed by the tester, for either plug or jack versions of the CAT5 T568A cables you have made.

22. Repeat this lab exercise until you have gained the desired level of proficiency.

TIP: In order to practice on the same piece of cable, a crimped RJ45 modular plug or jack must be clipped off, thrown away, and replaced. If you must clip the cable ends to remove RJ45 plugs or jacks for practice purposes, under no circumstances allow the cable to become shorter than four feet (48 inches) long. This cable will be used for later labs. If necessary, practice making terminations using pieces of scrap cable.

23. When you are satisfied with your work, use the tension scale/tape measure to check the lengths of the plugged or jacked cables. Again, make sure that they are at least 48 inches in length.

TIP: The tension scale/tape measure is exactly one meter long. Measure the cable to the 3-foot marker, place the start of the measure at that point, and measure at least 12 more inches beyond it. If the total length of any cable is less than four feet, you will have to discard it (or add it to the practice cable scrap pile) and try again using a new piece of cable. You must discard the used RJ45 modular jacks or plugs, if they have been crimped.

24. Save the T568A-configured cable (at least 48 inches long), on which you have just mounted CAT5-compliant RJ45 modular termination plugs, for later use.

25. Check to be sure that all of the tools, and parts, have been returned to their storage area, or previous locations.

26. Before leaving your lab work area, check to be sure that you are leaving it in a clean and orderly condition.

Lab 6 Questions

1. When arranging the cable colors for the TIA/EIA T568A configuration, what happens to pairs 1, 3, and 4?

2. When inserting CAT5 conductors into a RJ45 plug, is the tab facing up or down?

3. What differences exist in the cable color arrangements between a T568A plug configuration, and a T568A jack configuration?

4. What use can be made of a CAT5 cable terminated with a T568A plug at one end, and a T568A jack at the other?

Lab 7 – Terminating CAT5 with RJ45/T568B Plugs

<u>Lab Objective</u>

To properly terminate a CAT5 UTP cable with two RJ45 modular plugs, using the T568B configuration

<u>Materials Needed</u>

- Lab 7 work sheet
- Cable, CAT5 UTP, 100-ohm, non-plenum (spool)
- Cable, CAT5 UTP T568B jack-terminated, > = 4 feet in length
- Cable, CAT5 UTP T568B plug-terminated, > = 4 feet in length
- Electricians' scissors
- Pliers, diagonal cutters
- Termination plug, RJ45, unshielded (2)
- Crimping tool, RJ45
- Tension scale/Tape measure
- Cable tester, continuity-type

<u>Instructions</u>

Before beginning this lab, read through all of the following steps at least once, thoroughly.

1. Observing the proper safety procedures, use the diagonal cutters, to cut 5 feet of CAT5 cable from the spool.

2. Being careful not to nick or cut any of the individual conductors, strip approximately 2.5 inches of cable sheath (not to exceed 3 inches) from one end of the 5-foot section, and cut off the strain-relief threads as well, using the electricians' scissors.

3. Untwist each of the cable pairs for a length of approximately 2 inches.

TIP: Keep the individual pairs grouped together for identification, as shown in Figure 1-62. Some tip wires may not have any visible trace of color, and may appear to be solid white.

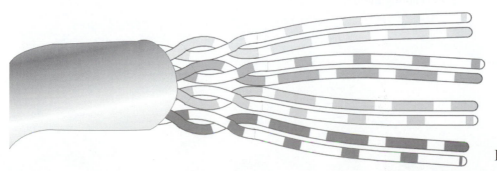

Figure 1-62:
Grouping
Individual Pairs

4. Arrange the cable colors according to the TIA/EIA T568B pin configuration.

5. Using the electricians' scissors, trim the CAT5 conductors to approximately 5/8 of an inch. This includes the strain-relief threads as well.

6. Insert the CAT5 conductors into one of the RJ45 modular plugs.

7. Make sure that the conductors are inserted as far as possible into the plug.

8. Remove the conductors momentarily to examine their ends for evenness.

9. If the wires in the center of the bunch appear to be longer than the wires at the ends, use the cable cutters to snip enough to make all of the conductors even. Then, reinsert the trimmed wires into the RJ45 modular plug.

10. Place the wired RJ45 modular plug into the crimp tool.

11. Before crimping, ensure that the conductors are fully inserted into the plug, with the cable sheath resting above the crimp edge inside of the plug, as in Figure 1-63.

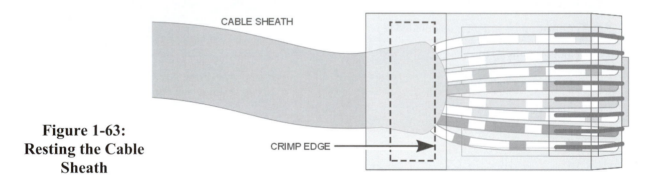

**Figure 1-63:
Resting the Cable
Sheath**

12. Apply pressure to the cable to make sure that the conductors remain inserted to the ends of the plug, and use firm, even pressure with the crimping tool to attach the RJ45 modular plug.

13. Remove the crimped RJ45 modular plug from the crimping tool.

14. Repeat steps 2 through 13 for the opposite end of the CAT5 cable, using the second RJ45 modular plug.

15. Using the CAT5 cable tester, check the completed cable for continuity, and proper sequencing of the wire pairs, as shown in Figure 1-64. Be sure the cable tester is set for T568B configuration.

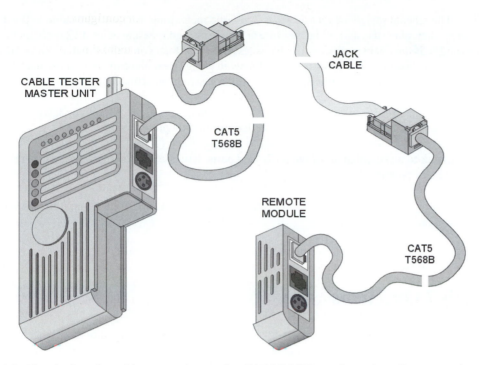

CABLE TESTER
MASTER UNIT

JACK
CABLE

CAT5
T568B

REMOTE
MODULE

CAT5
T568B

Figure 1-64: Testing CAT5 Plug and Jack Cables for T568B

16. If and when the cable passes the test for CAT5 T568B configuration, disconnect the end attached to the remote section of the tester.

17. Connect the CAT5 T568B cable with termination jacks from a previous procedure to the free end you just disconnected.

18. Appropriate a known good CAT5 T568B-configured cable terminated with plugs, and connect one of its plugs into the remaining jack on the second cable.

19. Finally, connect its remaining plug end to the remote section of the cable tester.

20. Using the CAT5 cable tester, check the three connected T568B cable sections (plugs-jacks-plugs) for continuity, and proper sequencing of the wire pairs.

21. If necessary, correct any problems that are revealed by the tester, for either plug or jack versions of the CAT5 T568B cables you have made.

22. Repeat this lab exercise until you have gained the desired level of proficiency.

TIP: In order to practice on the same piece of cable, a crimped RJ45 modular plug or jack must be clipped off, thrown away, and replaced. If you must clip the cable ends to remove RJ45 plugs or jacks for practice purposes, under no circumstances allow the cable to become shorter than four feet (48 inches) long. This cable will be used for later labs. If necessary, practice making terminations using pieces of scrap cable.

23. When you are satisfied with your work, use the tension scale/tape measure to check the lengths of the plugged or jacked cables. Again, make sure that they are at least 48 inches in length.

TIP: The tension scale/tape measure is exactly one meter long. Measure the cable to the 3-foot marker, place the start of the measure at that point, and measure at least 12 more inches beyond it. If the total length of any cable is less than four feet, you will have to discard it (or add it to the practice cable scrap pile) and try again using a new piece of cable. You must discard the used RJ45 modular jacks or plugs, if they have been crimped.

24. Save the T568B-configured cable (at least 48 inches long), on which you have just mounted CAT5-compliant RJ45 modular termination plugs, for later use.

25. Check to be sure that all of the tools, and parts, have been returned to their storage area, or previous locations.

26. Before leaving your lab work area, check to be sure that you are leaving it in a clean and orderly condition.

Lab 7 Questions

1. When arranging the cable colors for the TIA/EIA T568B configuration, are there any instances of white wires without any visible alternate color?

2. Why is it necessary to apply pressure to the cable before attaching the RJ45 modular plug with the crimping tool?

3. How can you visually identify which RJ45 wiring scheme (TIA/EIA T568A or T568B) a given cable is using?

4. What is the biggest difference between snap-type and crimp-type RJ45 jack or plug connectors?

5. Which cable would be less likely to have a problem, a single CAT5 cable that is long enough to reach from a computer to a wall jack, or a composite cable made up of several shorter sections connected through CAT5 plugs or jacks?

CHAPTER SUMMARY

This chapter has provided an accurate review of basic cabling standards and practices. Just as communication itself has evolved, the standards by which modern communications cabling are now judged continue to undergo radical changes within shorter periods of time.

You learned about the creation and evolution of the various organizations responsible for deciding the important issues involved in setting the standards by which cabling manufacturers must adhere. These organizations, both in the United States and the international theater, were identified.

Then, you reviewed the benefits brought about by the standardization process, and how important it is to have interface consistency between equipment, connectors, structures, and cabling in order to provide a high degree of competence when planning and executing a telecommunications installation.

This chapter also provided information about the basic structure of a telephone company, and how its operations are conducted. Terminology, familiar to those in the telephone industry, was explained in detail, as were the details about how a telephone call is placed, and how the equipment involved in completing the call operates. You learned the difference between voice and data, and why the successful transmission of data has become such an important aspect of modern telecommunications.

You were shown how the global trend towards computerization has led to the proliferation of more sophisticated equipment and strategies to deal with the ever-expanding world of telecommunications, and how that world is becoming digital in nature. Consequently, you learned about the binary numbering system, and how it compares to the decimal numbering system. You also converted several decimal numbers into binary notation, to see how the system works.

You now know why it is so important to keep a square wave square, whether it is transmitted along a communications cable, or through a satellite channel. Various methods of dealing with noise, distortion, crosstalk, or signal loss have been mentioned, as well as the benefits of keeping data in the digital realm, as opposed to analog.

This chapter examined the development of telephone wire, from the flat, non-twisted type used to carry strictly voice signals, to the CAT5-compliant twisted-pair cable used to send voice and data across modern telephone equipment.

Various parameters that serve to degrade digital data were examined, as well as the color codes used to distinguish between various wire pairs. You were also introduced to several types of connectors that have been standardized for telephone service wall jacks.

In addition to telephone cabling, you were presented with alternate cabling types, such as co-axial and fiber optic, which are used to provide high-speed communications services across computer networks not necessarily dependent upon telephone services. Lab 1 served to familiarize you with the various types of cabling used in network installations, and the separate parts of a Thicknet active tap kit.

The various networking topologies, including star, bus, and ring were discussed, as well as the advantages and disadvantages of each. A list of basic installation practices was presented, as well as a review of the basic cabling grades and termination methods used in telecommunications. Lab 2 provided the opportunity to examine the physical differences between the various cabling types used in this course. A review of basic tools was also included, along with some important tips for using them. Then followed Lab 3, which was designed to introduce you to the different tools associated with CAT5 cabling installations that are supplied with this course.

After learning about modular plugs, jacks, and telecommunications outlets, you reviewed some installation tips for modular cables and jacks, and some suggestions for making quality installations. Then, Lab 4 provided an opportunity for you to properly terminate CAT5 cable onto CAT5-compliant termination jacks, using the T568A configuration, while Lab 5 used the TIA/EIA T568B configuration in providing practice for properly terminating CAT5 cable onto CAT5-compliant termination jacks. In Lab 6, you learned how to properly terminate a CAT5 cable with two RJ45 modular plugs, using the TIA/EIA T568A configuration, and in Lab 7, you used the TIA/EIA T568B configuration to properly terminate a CAT5 cable with two RJ45 modular plugs. Finally, you used a cable tester to verify the continuity of all the cables constructed during these labs.

REVIEW QUESTIONS

The following questions test your knowledge of the material presented in this chapter:

1. State two benefits of standardization.

2. What are the two lines of a telephone system called?

3. Describe two methods of implementing color coding.

4. What is the maximum pulling force recommended for pulling a 4-wire cable through a conduit?

5. How far should a telecommunication cable be located away from a fluorescent light?

6. What type of topology is employed for residential telecommunication wiring?

7. Should cabling be run under carpeting? Explain.

8. How are type-66 IDC blocks for CAT4 and CAT5 different?

9. When using a binding post connection, in which direction should the wire be oriented?

10. What action should be taken after all of the communication components have been installed?

CHAPTER 2

CABLE RATINGS AND PERFORMANCE

OBJECTIVES

Upon completion of this chapter and its related lab procedures, you should be able to perform these tasks:

1. Describe various Level Ratings that apply to telecommunication cables and jacks and identify where each is implemented.

2. Describe the various levels of the cabling Category Rating Systems.

3. Define terms associated with Category performance.

4. State the proper type of wiring system for a given network application.

5. Given a certain network type, state the maximum transmission speed and distance the network can handle.

Cable Ratings and Performance

INTRODUCTION

In the early days of telecommunications, systems were installed with no regard as to how much, or how fast, information could be transferred. Today, however, those two factors are very important when deciding which type of cable to use as a transmission medium. With the development of faster, more powerful telecommunications equipment and techniques, enhancements have been taking place in the design and manufacturing of various cable types as well. To make sense out of the seemingly endless varieties of wire and cable available from today's manufacturers, several different cable rating systems have developed over time.

As the evolution of national and international standards organizations has progressed, the older cable rating systems have been examined, sifted, and reorganized into more universally recognized formats. The installation process for a given telecommunications system must now take into account a wide range of possibilities, not simply concerning the system's current function, but what future demands may be placed on it due to the rapid reorganization of business practices.

Among the many facets of telecommunications cabling systems, level and category ratings, performance criteria, and cost effectiveness must be taken into account when planning any installation. The quality of the materials used, the anticipated workload of the system, and the amount of money available to finance the work all play an important role in determining the success of an installation, and the satisfaction of a customer.

A thorough understanding of these variables will go a long way towards the achievement of successful cable installations, time after time. Along with this will come a business reputation for good workmanship, which is something that must be earned.

LEVEL RATING

Level rating was originally developed in an attempt to classify system cables, and jacks, according to their performance. For example, Anixter Inc. spearheaded a move toward creating a universally accepted UTP classification system, as far back as 1989. Then, Anixter began working with Underwriters Laboratories (UL), who had similar work in development. Together, they developed their "Level" program. There were 5 levels of cable quality in the UL Level classification system, with a higher number assigned to each level of increasing cable quality. Out of this original system, there are only two levels in general use today. They are Level 1 and Level 2, and are shown in Table 2-1.

As can be seen in Table 2-1, these ratings basically do not include the enhanced requirements for transmission cable that have recently become standardized for modern data communications purposes.

Table 2-1:
Rating Levels

LEVEL	MAXIMUM BANDWIDTH	WIRING TYPES	APPLICATIONS	TECHNICAL SPECIFICATIONS	SAFETY REQUIREMENTS
1	< 1 MHz	100 Ω UTP PREFERRED QUAD WIRE NOT RECOMMENDED FOR NEW INSTALLS	POTS RS232 & RS 422 ISDN BASIC RATE	ANSI/CEA S-80-576 ANSI/ICEA S-91-661 FCC PART 68 BELLCORE 48007	UL 1459 (TELEPHONE) UL 1863 (WIRE & JACKS) NEC, 1993 ARTICLE 800-4
2	4 MHz	100 Ω UTP	IBM TYPE 3 1.544 MBPS T1 4 MBPS TOKEN RING	IBM TYPE 3 ANSI/CEA S-80-576 ANSI/ICEA S-91-661 FCC PART 68 GA27-3773-1, IBM CABLING SYSTEM TECHNICAL INTERFACE	UL 1459 (TELEPHONE) UL 1863 (WIRE & JACKS) NEC, 1993 ARTICLE 800-4

Level 1

Level 1 is the minimum-quality cabling system recommended for analog voice service, or Plain Old Telephone Service (POTS). In the past, barbed wire running along a fence could be considered Level 1. Today, the definition of the type of systems that fall under Level 1 has been designated to protect the Telephone Company network. FCC part 68 defines the minimum requirements for jacks, plugs, and wire. Quad wire used to be acceptable for Level 1 wiring systems. However, today it is recommended that Unshielded Twisted Pair (UTP) be used for all new, and multi-line, installations. The termination method used for Level 1 is the screw terminal, or binding post block. **Insulation Displacement Connectors (IDCs)** exceed Level 1 termination requirements.

Level 2

Level 2 is actually defined as the IBM Type 3 cabling system. Obviously then, the requirements for Level 2 cabling systems are based strictly on IBM cabling methodology. Level 2 cabling systems run at low data speeds, and are not a big item in today's market. However, many such systems have been installed previously, and are still being operated. You will probably come into contact with such systems, and be required to repair or upgrade them accordingly.

Alternately, due to the rapid development in the data communications industry, modern cable installation companies must now take into account the transmission capabilities of telecommunications equipment not yet available, and plan for system upgrading.

Unfortunately, cable manufacturers no longer mention Level 2 cabling systems in their advertisements, news releases, or product reports. Instead, they provide technical information on their most recent cabling products, and indulge in the practice of prognostication. Predictions concerning newer, higher-quality cable types is forever forthcoming, as telecommunications hardware developments continue to raise the bar for faster data transfers, and wider information channels. The bottom line is this—virtually no new telecommunications installation is going to be geared towards Level 2 operation. If 4 MHz of bandwidth is sufficient for a given system (taking into account the hardware being used), so be it.

However, the cabling used in today's installations will far exceed the Level 2 classifications in terms of both transmission speeds and available bandwidth. Therefore, it is somewhat understandable if cable manufacturers are not falling all over themselves to provide Level 2 documentation. Of course, there could always be a spool of IBM Type 3 cable stashed away in a communications supply house somewhere that an eager sales rep is seeking to unload!

CATEGORY RATING SYSTEMS

Before you can begin to understand the different categories, and how they perform, the terms **Megahertz (MHz)** and **Megabits per second (Mbps)** need to be defined. Megahertz refers to the frequency, or speed, that the information can travel. Megabits per second refer to the actual amount of digital bits (data) that may be transmitted each second.

Analog communications, such as voice telephone, has traditionally been specified in terms of the overall bandwidth of a channel, in hertz (Hz). When modems became established as the normal method used to transmit digital data over the existing telephone system, they were generally classified by their **baud rate**. Baud rate was (and still is) used to describe the number of signal changes that occurred per second during the transfer of data. Using that definition, baud could be considered analogous to the transmission frequency of a given cable, which for a telephone line is limited to about 3 kHz. Because signal changes are so severely limited by the bandwidth of the telephone lines, the baud rate is the determining factor as to how much data can be transmitted.

Since the advent of digital communications however, the capacity of a communications channel is best described in terms of the number of data bits that can be carried each second. This is because the transmission of digital data is usually handled on a bit-by-bit basis, serially. In an effort to speed greater amounts of information along a given communications channel, various modulation schemes have been devised that allow the BPS rates to greatly exceed system transmission frequencies (or baud rates). Modem manufacturers have encoded data into different transmission formats (FSK, PSK, DPSK, and QAM) so that a number of data bits could be represented by a single signal change. In this way, the bit rates could be high, even though the maximum baud rate was relatively low.

Due to the development of extremely high-speed telecommunications equipment, telephone companies have opted to provide special communications channels and services that bypass the bottleneck conditions of the public telephone system. Consequently, the capabilities of internally networked telecommunications cables have been considerably upgraded as well.

The various ratings for telecommunication cable, although specified in bandwidths of 16 MHz, 20 MHz, 100 MHz, etc., are usually used to carry digital data. Depending on the hardware application making use of the cabling, the data transfer rates must always be taken into consideration when determining what constitutes an adequate cabling system.

For example, an older CAT3 cabling installation could never be called upon to implement an ATM application requiring transfer rates of 155 Mbps, because the maximum bandwidth rating for CAT3 cable is 16 MHz.

Megahertz (MHz)

Megabits per second (Mbps)

baud rate

The UL Level Classification system was combined with the TIA/EIA category system on January 3, 1994, and the nomenclature changed to use the term "category". The UL now classifies cabling as category (CAT) 1 through 5, while the use of the term "level" has been generally discontinued. The older Level 1 and Level 2 specifications correspond to CAT1 and CAT2. UL categories 3 through 5 now correspond exactly to the TIA/EIA T568A categories (CAT3, CAT4, and CAT5).

Some cable vendors are promoting their cables as rated to 350 Mhz. However, it should be noted that there is no ISO or TIA/EIA cable, connector, or test specification covering anything beyond 100 MHz. While there has been some initial work in Europe on CAT6, it does not yet exist. When the CAT6 concept gains support, it will take some additional time before cable, connector, and test specifications are approved.

TIA/EIA 568 defined the technical specifications and safety requirements for cabling applications up to bandwidths of 16 MHz, which was acceptable for many applications. However, newer cabling applications, requiring higher data rates, were not addressed. The TIA originally developed the category rating system in response to the demand for cabling applications using these higher data rates. The information concerning these higher data rates was released in two Technical Systems Bulletins (TSBs) called TSB 36, and TSB 40.

Basically, the TSBs are additions made to the TIA/EIA 568 standards. TSB 36 covers additional specifications for UTP cables, while TSB 40 covers additional specifications for connecting devices, such as jacks, crossconnect blocks, and patch panels. These two TSBs have been added into the main body of the TIA/EIA 568 standard document to form the TIA/EIA T568A standard, and the details of these standards and bulletins are included in the Appendix. There are three cabling categories that are currently recognized by TIA/EIA (CAT3, CAT4, and CAT5).

CAT3

Category 3 cabling, and cabling components, were designed and tested to cleanly transmit 10 Mbps of information at a maximum rate of 16 MHz. CAT3 cabling is the most common type of cabling used in today's voice applications.

CAT4

Category 4 cabling, and cabling components, were designed and tested to cleanly transmit 16 Mbps of information at a maximum rate of 20 MHz. Yet, because of the rapid development and deployment of more robust systems, CAT4 cabling systems have rarely been seen or used. It's considered the Edsel of the cabling industry.

In retrospect, CAT4 cabling systems appear to have been shortsighted and unnecessary. CAT4 was originally billed as a slight improvement in performance over CAT3 for premises network upgrades. However, CAT4 has had a rather bland, ill-fated existence, having never developed as a stepping stone to CAT5. Today, conversations about categories of cable for office buildings and residences start with CAT3, and jump to CAT5, with seldom a mention of CAT4.

CAT5

The last recognized category is CAT5. **Category 5** cabling/components are designed and tested to cleanly transmit 100 Mbps of information, at a maximum rate of 100 MHz. CAT5 is the most popular cabling system being installed in current telecommunications systems.

Category 5

In fact, CAT5 UTP cable has become so widespread, many people in the cabling industry find it difficult to imagine life without it. The industry is growing, with people and organizations constantly entering the field. For them, there never was life in cabling without CAT5 cable. Because it has captured the industry so quickly, many existing cabling manufacturers and premises wiring companies have had to refocus.

According to John Siemon, vice president of engineering at The Siemon Company, "Although the CAT5 cabling classification is more than five years old, it has certainly had the single largest technological impact on the cabling and telecommunications industry over the past five years by enabling high-speed data transport over a stable, inexpensive, and robust cabling infrastructure."

With its 100MHz specification, CAT5 cable has allowed users to migrate from network speeds of 10 Mbps to 100 Mbps. Users who installed new, 10Mbps networks selected CAT5 cable as the transmission medium, with the anticipation of making that migration. Perhaps as much as any other single factor, this kind of thinking has contributed to CAT5 becoming a fixture in premises network infrastructures.

While some of those early users have migrated to 100 Mbps networks, and not others, CAT5 cabling has successfully handled all of the applications that have so far been thrown at it. Because the group drafting the 1000baseT standard is committed to run that protocol over CAT5 cabling, you should expect CAT5 UTP cable to be around for quite some time.

PROPOSED STANDARDS

While CAT3, CAT4, and CAT5 are the three cabling categories currently recognized by TIA/EIA, there are three additional categories that have been proposed (CAT5E, CAT6, and CAT7), but not yet approved.

A proposed cabling standard is one that is under consideration by the TIA/EIA committee, and it is brought about by a series of steps. The first step is for an individual, or company, to present a technical document to the Engineering Committee. If there is enough support for this technical document, and there are enough people willing to work on it, then a **Project Initiation Notice (PIN)** application form is filled out, and submitted to the TIA for approval.

Project Initiation Notice (PIN)

After approval from the TIA, the Engineering Committee, and all those willing to work on the document, develop the technical parameters for the project. When the document is almost complete, the Engineering Committee circulates the draft of the document on a ballot called a "Committee Letter Ballot". The purpose of this document is to identify any unresolved issues, and to get agreement from the experts in the technical field under consideration. During this phase, the draft document is not released to the general public.

During the balloting period, any interested party may cast his/her vote. The ballots and the document are then sent to a review group in the TIA, called the **Telecommunications Standards Subcommittee** (**TSSC**). The TSSC reviews all of the balloting information, and with the approval of the **Board of Standards Review** (**BSR**), the document is approved for publication as a TIA Standard.

CAT5E

Category 5E

Category 5E is currently being proposed to TIA/EIA, and if CAT5E is indeed approved, it will have to maintain the existing requirements of CAT5. CAT5E will provide a higher performance level than CAT5, including an available bandwidth extending up to 160 MHz. It will also be capable of supporting a data rate of 1.2 Gbps, using the same technology as is under development for 100baseT.

Equal-Level FEXT
(ELFEXT)

Power-Sum NEXT

Formally known as TIA SP 4195A, the proposed Addendum 5 to TIA/EIA T568A specifies the Enhanced CAT5 (CAT5E) performance requirements. It is strongly recommended that new CAT5 cabling installations be specified to satisfy the minimum requirements of this document, and it is expected that TIA/EIA T568A-5 will emerge as the new de facto minimum standard for CAT5 cabling. It specifies the minimum **Equal-Level FEXT** (**ELFEXT**) loss and return loss requirements that are necessary to support newer developments in applications technology, and defines the minimum performance required to achieve a worst-case, four-connector channel to support applications that use full-duplex transmission schemes. To ensure additional crosstalk headroom for robust applications support, TIA/EIA T568A-5 also specifies **Power-Sum NEXT** and ELFEXT performance for CAT5E cables, links, and channels.

Proposed Addendum 5 to TIA/EIA T568A is a prescribing document and, unlike TSB95, it provides mandatory requirements, not recommendations.

Gigabit Ethernet (1000baseT in its copper implementation) theoretically should operate over today's widely-deployed cabling systems for high-speed LANS around the world (CAT5 UTP). Because Gigabit Ethernet is designed to operate over a frequency range of 100 MHz, enhanced Category 5 cabling (CAT5E) will be the preferred wiring solution for Gigabit Ethernet in new installations, because it will provide a higher performance margin.

Proposed Draft
Amendment (PDAM)

As an amendment to the ISO/IEC 11801 standard, the performance specifications in the **Proposed Draft Amendment** (**PDAM**) number 3 provide new requirements for return loss and ELFEXT loss to complement existing ISO Class D requirements. PDAM3 specifies return loss and ELFEXT requirements that agree with the values proposed in TIA/EIA T568A-5. However, the amendment does not specify any additional NEXT margin above the existing Class D requirements. PDAM3 also includes propagation-delay and delay-skew requirements for channels, and permanent links, that coincide with the requirements of TIA/EIA T568A.

The requirements of proposed Amendment 3 to ISO/IEC 11801 are prescribed, and this document is expected to become the de facto standard for new Class D cabling installations.

CAT6

Category 6 is also currently being proposed. Although this standard has also not been finalized, or released to date, the specifications for CAT6 are expected to include an available bandwidth of at least 200 MHz.

The proposed CAT6 and Class E standards under development by the TIA and ISO working groups describe a new performance range for UTP and **Screened Twisted-Pair (SCTP)** cabling. The goal of these working groups is to specify the best performance that UTP and SCTP cabling can be designed to deliver.

It is expected that these requirements will be specified in a frequency band that may reach as high as 250 MHz, and will be capable of supporting a positive **Attenuation-to-Crosstalk Ratio (ACR)** to 200 MHz.

For compatibility with existing category and class requirements, the standards groups have agreed upon the 8-position modular-jack interface as the prescribed interface for CAT6 and Class E work-area cabling topologies. This ensures that CAT6 and Class E specifications will be backward-compatible with applications running on lower categories or classes of cabling. In cases where components of a different category or class are combined with CAT6 and Class E components, the combination shall meet the transmission requirements of the lowest-performing component.

Collaborating closely on the development of CAT6 and Class E standards, the TIA/EIA, ISO/IEC, and **European Committee for Electrotechnical Standardization (CENELEC)** are submitting proposals and requirements that are very similar. It is therefore expected that CAT6 and Class E requirements will shortly be available for industry review. Barring any unexpected technical issues from the TIA or ISO, the industry could have access to published CAT6 and Class E standards soon after.

CAT7

Category 7 is the most ambitious category yet to be proposed with a suggested bandwidth requirement of 862 MHz. Proposed CAT7 and Class F standards are being developed for fully shielded (outer shield, plus individually shielded pairs) twisted-pair cabling. CAT7 and Class F will be supported by an entirely new interface design (socket and plug).

CAT7 and Class F will also be backward-compatible with lower-performing categories and classes, even though their implementation will be supported by a new connecting hardware interface. It is anticipated that CAT7 and Class F requirements will specify a bandwidth of no less than 600 MHz.

Although TIA is not actively developing a CAT7 standard, it will most likely mirror the ISO Class F requirements. Pending industry consensus on the selection of a CAT7 interface design, Class F requirements could be available simultaneously with the Category 6 and Class E specifications.

Category 6

Screened Twisted-Pair (SCTP)

Attenuation-to-Crosstalk Ratio (ACR)

European Committee for Electrotechnical Standardization (CENELEC)

Category 7

The currently existing categories, as well as those being proposed, are listed in Table 2-2. The table compares channel-performance data at 100 MHz and other frequency values of interest for the proposed TIA Category 5E, 6, and 7 and the corresponding European ISO Class D, E, and F performance standards.

Table 2-2: Cabling Category Ratings

Euro ISO Class	Category	Maximum Bandwidth	Wiring Types	Applications	Technical Specifications	Safety Requirements
A	3	16 MHz	100Ω UTP Rated Category 3	10Mbps Ethernet 4Mbps Token Ring	TIA/EIA T568A ANSI/ICEA S91-661 FCC Part 68 NEMA (Std Loss)	UL 1459 (Telephone) UL 1863 (Wire &Jacks) NEC, 1993 (Article 800-4)
B	4	20 MHz	100Ω UTP Rated Category 4	10Mbps Ethernet 16Mbps Token Ring	TIA/EIA T568A ANSI/ICEA S91-661 FCC Part 68 NEMA (ext dist)	UL 1459 (Telephone) UL 1863 (Wire &Jacks) NEC, 1993 (Article 800-4)
C	5	100 MHz	100Ω UTP Rated Category 5	100Mbps TPDDI 155Mbps ATM	TIA/EIA T568A ANSI/ICEA S91-661 FCC Part 68 NEMA (ext freq)	UL 1459 (Telephone) UL 1863 (Wire &Jacks) NEC, 1993 (Article 800-4)
D	5E	160 MHz	100Ω UTP Rated Category 5E	1.2Gbps 1000baseT High-Speed ATM	Addendum to TIA/EIA T568A FCC Part 68	UL 1459 (Telephone) UL 1863 (Wire &Jacks) NEC, 1993 (Article 800-4)
E	6	200-250 MHz	100Ω UTP Rated Category 6	1.2Gbps 1000baseT High-Speed ATM and beyond	Proposed Addendum to TIA/EIA T568A	UL 1459 (Telephone) UL 1863 (Wire &Jacks) NEC, 1993 (Article 800-4)
F	7	600-862 MHz	100Ω UTP Rated Category 7	1.2Gbps 1000baseT High-Speed ATM and beyond	Proposed Addendum to TIA/EIA T568A	UL 1459 (Telephone) UL 1863 (Wire &Jacks) NEC, 1993 (Article 800-4)

As the installer, you will need to select the strongest and most affordable cabling foundation to support the present and future networking needs of your customers when actually designing and installing their telecommunications systems.

It is most important to be as informed as possible to ensure that your installations will support the emerging technologies that use the latest advances in signaling schemes.

PERFORMANCE CRITERIA

In order for a telecommunications system to be determined as being compliant, the system must at least meet the TIA/EIA T568A electrical and mechanical specifications, which define end-to-end connectivity, and channel performance parameters, for structured cabling systems. A given system may perform at a specific level or category; however, for it to be compliant, it must also use the equipment specified by TIA/EIA T568A.

For example, imagine that you have installed a LAN in which all of the horizontal cabling is rated as CAT5, all of the modular plugs are configured to T568B, and all of the jacks are CAT5 rated. Now suppose that all of the tests that are performed on this LAN indicate a CAT5 level of performance. However, if your work-area patch cords are rated at CAT3, is the LAN truly CAT5-compliant? The amount of signal attenuation, near-end crosstalk (NEXT), far-end crosstalk (FEXT), reflections, and return loss determines the category of compliance for the system.

Attenuation

Attenuation is the decrease in the power of a signal as it travels through a medium. Various electrical factors can conspire to reduce signal power traveling through copper conductors, such as resistance and reactance (both capacitive and inductive).

Attenuation is measured in **decibels (dB)**, where the lower the number, the better the situation. The TIA/EIA T568A specification allows for maximum dB levels of attenuation for every 1,000 feet of cable at 10 different testing frequencies. A compliant installation will meet the expected criteria. The higher categories of cable are tested as higher frequencies.

Table 2-3 indicates the maximum attenuation permissible for CAT5, 100-ohm UTP cable, at various frequencies for any pair, through 1,000 feet of cable. Keep in mind that the TIA/EIA T568A specification does not permit any horizontal cable lengths to be longer than 100 meters within the installation. When considering attenuation, a lower number is better.

Attenuation

decibels (dB)

Table 2-3: CAT5 Maximum Attenuation Ratings

Frequency MHz	Attenuation dB
0.512	4.5
0.772	5.5
1.000	6.3
4.000	13
8.000	18
10.00	20
16.00	25
20.00	28
25.00	32
31.25	36
62.50	52
100.0	67

Near-End Crosstalk

Near-End Crosstalk (NEXT) is the interference of two signals traveling in opposite directions within the same cable sheath. This happens because most of today's telecommunications applications are designed to carry information over two adjacent pairs, in opposite directions. The stronger transmitted signal drowns out, or distorts, the weaker signal that is being received from the far end of the circuit.

NEXT ratings actually describe the emission and immunity properties of two adjacent pairs inside a channel. When measuring NEXT between pair 1 and pair 2, or vice versa, similar results should be obtained. A signal injector can be used to measure the signal leakage from the active pair (carrying the injected signal) into an inactive pair, over a set of different frequencies. The minimum allowable NEXT for the worst pair within a cable shall be greater than, or equal to, the values shown in Table 2-4. When considering NEXT, a higher number is better.

Table 2-4: Minimum NEXT Ratings

Frequency MHz	Attenuation dB
0.150	74
0.772	64
1.000	62
4.000	53
8.000	48
10.00	47
16.00	44
20.00	42
25.00	41
31.25	40
62.50	35
100.0	32

Inside a channel, NEXT becomes the dominant source of internally induced noise, because it's so much stronger at the beginning of the channel, whereas the incoming data signals are weaker due to attenuation. In addition, if all four UTP pairs are used simultaneously, additional noise is induced on each pair. Therefore, network interface cards that have been designed for 1000baseT installations are equipped with NEXT cancelers to suppress this type of noise. Obviously, as the overall noise level is decreased, higher data rates are achieved.

The most common causes of serious NEXT are crossed wire pairs and/or split wire pairs. While cable scanners can identify crossed pairs during a wire map test, they cannot help to identify split pairs. If twisted pairs are carelessly untwisted during the attachment phase to a crossconnect device, this situation could also lead to NEXT, as could the use of untwisted patch cables. If a cable is pulled too tightly around a sharp corner, some pairs could shift their positions inside of the jacket, and result in NEXT.

Attenuation-to-Crosstalk Ratio (ACR)

CAT5 cable NEXT ratings are slightly higher than those that take into account the crosstalk occurring through connectors (Class D link attenuation). These lower standards also take into account the NEXT contributions from outlets, distributors, and equipment connectors, which are aggregated due to the many interfaces through which the signals must pass. The **Attenuation-to-Crosstalk Ratio (ACR)** defines the difference between Class D link attenuation, and NEXT, for systems that include connectors and interfaces in their ratings.

Far-End Crosstalk

Far-End Crosstalk (FEXT) is the interference of two signals traveling in the same direction on two adjacent pairs within the same cable sheath. Again, the stronger signal distorts or completely drowns out the weaker one. Noise caused by FEXT can be canceled out, decreasing the noise level.

Although FEXT is not the dominant problem that NEXT is, there are high-speed systems now being developed that do such a good job of eliminating NEXT, that the cables designed for them use FEXT ratings as their test criteria for compliance.

Reflections

Emerging, high-speed telecommunication applications can carry information over all four pairs, in both directions, simultaneously. Network managers are concerned about how well their networks will perform using these higher frequencies/data rates, which will be required by the newer technologies. As a result, current test methods will have to change in order to accurately measure newly critical parameters such as impedance.

reflections

Impedance stability is more difficult to achieve at high frequencies, due to small variations in manufacturing, such as concentricity, diameter, quality of twisting, and wall thickness. This means that a cable pair's impedance varies very slightly along its length. These small impedance variations along each pair create signal losses, and increased noise, due to **reflections** of the energy transmitted along the pair. These losses are more significant at higher frequencies because as the signal wavelengths become shorter, small variations become more "visible" in the network. For example, at 300 MHz, imperfections on the order of 3 cm are visible to the signal, compared to 10 cm at 100 MHz, and 1 meter at 10 MHz.

Although reflections can appear when information is carried bi-directionally over a wire pair, there exist certain prescribed techniques that can cancel out these types of noise.

Another critical measurement is called **return loss**, which compares the impedance of a transmission line to the impedance at its termination point. Therefore, return loss measures all of the relative reflected power that is dissipated in a network. System return loss is so important for Gigabit Ethernet performance that it is one of the new test criteria now being defined for existing CAT5, enhanced CAT5 (also known as Category 5E), and CAT6.

return loss

Return loss is measured from the ideal 100-ohm impedance target for the network. Currently, the TIA/EIA-T568A standard requires that manufacturers measure structural return loss for cable, rather than for an entire network. This is a special case of return loss, where the measurement is based on an averaged, or smoothed, impedance line.

For successful Gigabit Ethernet implementation, a network's return loss becomes very important. Because the configuration calls for full-duplex (bi-directional) transmission on each wire pair, return loss is critical at higher frequencies, and any reflected energy will be interpreted as noise by the receiver. Without the use of echo cancellation to minimize this noise, a worst-case CAT5 installation would not support Gigabit Ethernet, or any other high-speed applications, using bi-directional pair transmission.

When considering a system, always be sure to compare the available test data to the requirements of the performance intended. For example, before attempting to run Gigabit Ethernet over existing CAT5 cabling installation, you should verify that its performance meets the minimum recommendations of Technical Systems Bulletin 95.

Existing channel configurations with three or four connectors, that satisfy the TSB 95 Equal-Level Far-End Crosstalk (ELFEXT) and return loss requirements, will also support Gigabit Ethernet. However, because the recommendations of TSB 95 are applicable for the qualification of existing installed cabling only, they should not be used as the minimum performance criteria for new CAT5 cabling.

More information on various TIA/EIA standard, and accompanying TSBs, can be obtained at http://www.eia.org.

NETWORKING

A network is a group of computers and other devices connected together. Networking is the concept of computers, connected together, sharing resources. Computers that are networked together can share data, messages, graphics, printers, fax machines, modems, and any other communication device that is tied to the network. The two major types of networks are the **Local Area Network (LAN)**, and the **Wide Area Network (WAN)**.

Local Area Network (LAN)

Wide Area Network (WAN)

A LAN is a short-distance communications network used to link computers, and other devices, under some form of standard control. Because LANs can operate independently of the telephone system, they are not constrained by the same speed and bandwidth limitations imposed by Telco operations. Network Interface Cards (NICs) can permit the transfer of files across the LAN as quickly as a user can click a mouse, with no preliminary dialing activities required.

A WAN connects computers and other devices from different cities or states. Wide area networks use common carrier lines to link LANs in remote buildings, or across the country. The tremendous growth in wide area networks has engendered a host of superior services by the common carriers to provide cleaner and wider channels for high-speed telecommunications. This growth has reduced the need for paper communications, and has provided all types of data for network users who need it.

Network Applications

Although there are many different networking applications that can be used to achieve the desired results, the choice as to which system will be installed at any job site will ultimately rest with the customer. However, your knowledge about how the customer expects the network to function, coupled with a good understanding as to how different networking applications perform their tasks, should give you the ability to make informed suggestions to the customer about any performance criteria related to a proposed telecommunications installation. In the long run, your guidance will be much appreciated by the customer.

Consider it your duty to inform a customer when you recognize a problem that may occur either immediately, or in the near future, due to the initial choices made with regards to equipment, cabling, software, wiring closet location, etc. A customer is not going to be upset to learn from you that an early idea was not the best one to implement, considering how important it is for you to provide an installation that will lend itself to some painless upgrading at some future date.

Ethernet

Ethernet

International Data Corporation (IDC)

The **Ethernet** topology was the first widely accepted, nonproprietary, standardized multiple-access networking application. It allowed many computers or workstations to share a high-speed communications line without regard to customary hierarchy. According to the **International Data Corporation (IDC)**, by the end of 1996, more than 83 percent of all installed network connections were Ethernet, representing over 120 million interconnected PCs, workstations, and servers. A combination of Token Ring, Fiber Distributed Data Interface (FDDI), Asynchronous Transfer Mode (ATM), and other protocols made up the remaining network connections. Add to that the fact that all popular operating systems and applications are Ethernet-compatible, as are upper-layer protocol stacks such as Transmission Control Protocol/Internet Protocol (TCP/IP), IPX, NetBEUI, and DECnet.

Originally, Ethernet was first developed in the early 1970's by the Xerox Corporation at its Palo Alto Research Center. Within several years (late 1970's), Xerox Corporation, Digital Equipment Corporation (DEC), and the Intel Corporation agreed to promote Ethernet as an open standard for computing networks. Their initials were used to identify the 15-pin, DIX connector used to interface various Ethernet network components. This connector is also known today as an Attachment Unit Interface (AUI).

This standard eventually became IEEE 802.3, administered by the Institute of Electrical and Electronic Engineers. IEEE 802.3 has been revised and refined several times to where it now includes coax, twisted pair, fiber optics, and wireless connection methods.

When discussing the original Ethernet, the term Ethernet II is commonly used. This is because Ethernet II, and the IEEE 802.3 protocols used to describe modern Ethernet operations, use incompatible formats for data packets. In terms of cables, connectors, and electronics however, a relative compatibility does exist. Today, standard Ethernet refers primarily to a local area network, consisting of light to medium traffic, used to connect computers, servers, and printers within the same building or campus. It performs best using short bursts of data, and has become the most popular network application in many universities and government installations.

Ethernet uses a method of signaling called **Carrier Sense Multiple Access/Collision Detection (CSMA/CD)**. Each device on an Ethernet network listens for the transmissions from other devices. If a transmission is received that matches the devices address, it is processed. Transmissions sent to other addresses however, are ignored. When a device wishes to transmit, it first listens for another existing transmission (carrier sense). If there is no other transmissions present, it transmits the message, and then checks the voltage level on the network to determine if another device happened to transmit at the same time. When two different signals are sent at the same time, this results in a collision, which increases the voltage level (collision detection).

Standard Ethernet operates over coaxial cable, or twisted wire, at speeds up to 10 Mbps. The topology used for Ethernet is star or bus, and there are many different types of Ethernet. But before we cover the different types of Ethernet, we must first review what the terms **baseband** and **broadband** mean. Baseband is defined by Newton's Telecom Dictionary as, "a form of modulation in which signals are pulsed directly on the transmission medium, without frequency division."

To simplify this definition, baseband is used to transmit a single, digital signal, using the entire capacity of the LAN. A broadband network divides the cable's capacity into many channels, which can simultaneously transmit many signals.

The various classes of Ethernet are referred to using a prefix number, a middle word, and a suffix, such as 10baseT. The 10baseT nomenclature means that an Ethernet network is operating at 10 Mbps, using baseband communication, and transmitting over twisted-pair cable. Notice that the prefix number defines the operating speed, the middle word defines the signaling techniques, and the suffix defines the transmission medium.

Another example of Ethernet is 10base5. This term refers to an Ethernet network operating at 10 Mbps, using baseband communication, and transmitting over thick coaxial cable. A final example of Ethernet is 10base2, which defines a network operating at 10 Mbps, using baseband communication, and transmitting over thin coaxial cable.

Token Ring

The Token Ring network was introduced by IBM in the mid 1980's as a token-passing Local Area Network (LAN). Now known as the IEEE 802.5 standard, it continues to be IBM's primary LAN technology supporting PC-based networks. Although the Token Ring topology originally used Shielded Twisted Pair (STP) cable, and a unique hermaphroditic connector (called the IBM Data Connector), more recent Token Ring networks have been installed using Unshielded Twisted Pair (UTP) cable. Experts in the telecommunications industry estimate that more than 12 million network nodes now exist, and that Token Ring networks are growing at a 20-percent annual rate. Today, Token Ring is second only to Ethernet in overall LAN popularity.

This existing infrastructure is proof of significant investment in Token Ring hardware and software. While expanding their networks to support the ever-increasing numbers of bandwidth-hungry users and applications, network managers are always seeking ways to leverage these rather large investments. This goal becomes even more important if their companies have selected Token Ring technology to support mission-critical applications.

IEEE 802.5 networks are completely compatible with IBM Token Ring networks. This should not be surprising, because the IEEE 802.5 standard was modeled after the original IBM Token Ring, and it continues to follow IBM's Token Ring development.

Token Ring Layout

Typically, Token Ring networks are bridged networks that carry protocols such as Systems Network Architecture (SNA), NetBIOS, Transmission Control Protocol/Internet Protocol (TCP/IP), and IPX. While employing a hierarchical design using single- or dual-backbone rings, bridged to end-user floor rings, as shown in Figure 2-1, these networks often consolidate their bridges and servers at a single location (a large equipment room or telecommunications closet) to simplify troubleshooting and maintenance.

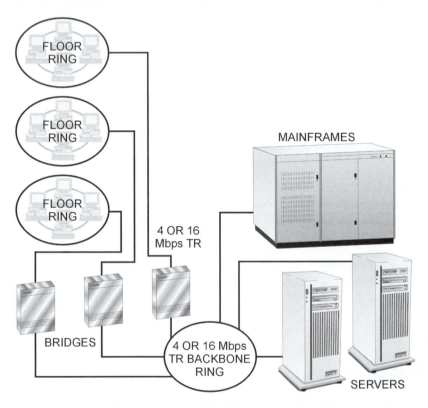

Figure 2-1: Hierarchical Token Ring

Token Ring Specifications

In short, Token Ring is a 4- or 16-Mbps LAN access mechanism, running over a star topology, in which all stations actively attached to the bus listen for a broadcast token, or supervisory frame. Stations wishing to transmit must receive the **token** first. After a station finishes transmission, it passes the token to the next node in the ring.

Token Ring implements an electrically continuous ring, using a cabling system that is installed in a hubbed-star configuration. Each horizontal run from the star is called a **lobe**. Each lobe is wired to a hub, also called a **Multi-Station Attachment Unit (MSAU)**, which allows signals from workstations, servers, or bridges to be looped to each workstation, until the signal is returned back to the original point of transmission. This process turns the star wiring into a loop, or ring, network. When a workstation is ready to send information onto the network it must wait for the so-called token. The token is passed from station to station, along with data, commands, and acknowledgments, allowing network signals to proceed around the ring in an orderly fashion. It operates like this:

token

lobe

Multi-Station Attachment Unit (MSAU)

- A station can only transmit when it has "the token."

- All transmitted frames are passed from station to station around the ring.

- All stations test each passing frame for messages addressed to them, process the information if it is theirs, and pass a marked token back around the ring.

- The original transmitting station releases the token when it returns around the loop.

Using a signaling structure called Differential Manchester Code, a token can circle a ring with a diameter of 200 meters, at a rate of 10,000 times a second. Table 2-5 summarizes the most important specifications common to Token Ring networks.

Table 2-5: Token Ring Specifications

Parameter	Specification
Maximum Number Stations	260 stations on one ring using STP 72 stations on one ring using UTP
Maximum Distance from Station to MAU	100m (328ft)—Type 1 cable in one contiguous segment 45m (150ft)—for segments joined by patch cables
Data Rate	4 Mbps or 16 Mbps
Maximum Total Ring Length	Variable from vendor to vendor
Transmission Media	Shielded Twisted Pair

Token Ring Design and Topology

When discussing the philosophy behind the Token Ring topology, its design characteristics must be taken into account, as well as its possible variations.

deterministic

The overall design of a Token Ring is inherently **deterministic**, unlike Carrier Sense Multiple Access/Collision Detect (CSMA/CD) networks, such as Ethernet. This characteristic allows for the calculating of the maximum times between transmissions, helping to promote more accurate traffic planning. The ability to incorporate deterministic design allows Token Ring networks to run those applications that require high-reliability environments. Although the original topology for IBM's Token Ring network specification was defined as a star, the IEEE 802.5 spec differs from the IBM Token Ring in that it does not specify a particular topology.

However, most IEEE 802.5 implementations are star-based, where all end stations are attached to a hub (MSAU). These hub-based Token Ring LANs should allow the choice between inexpensive, passive hubs for moderate-sized networks, and active technology solutions for high-performance, large network applications.

Growth Pangs

Token Ring network growth is currently being driven by LAN-based applications, while network administrators add rings to maintain their fault domains at comfortable levels (typically 50 to 80 users). As more users populate the network, more bridges are required to interconnect more rings. Bridges are simple to install and require little maintenance, but they lack fault tolerance and SNMP management, and consume expensive rack and floor space.

As the network expands, traffic on the backbone (the busiest ring on the network) increases, congesting bridges and affecting network performance. Traffic growth may result in the addition of superservers, requiring more bandwidth than Token Ring can offer. When there are excess traffic Token Ring throughput problems, servers respond sluggishly to application requests. While waiting for acknowledgments from the far end of a connection, a session can die, degrading user productivity. To keep pace with continued network growth, the Token Ring technology must continually improve LAN-to-LAN performance, at higher speeds.

Lab 8 – Installing UDC Connectors on 150-Ohm STP

Lab Objective

To understand how to install UDC connectors on 150-ohm STP cable

Materials Needed

- Lab 8 work sheet

- Cable, Type-1A STP, 150-ohm, non-plenum (spool)

- Universal data connector (UDC), kit (2)

- Adapter, UDC/RJ45 (2)

- Patch cable, RJ45 (2)

- Utility knife

- Wire stripper, CAT5-type

- Pliers, diagonal cutters

- Nippers, end-cutting

- Pliers, groove-joint

- Screwdriver, slotted-head, small

- Cable tester, continuity-type

Instructions

Before beginning this lab, read through all of the following steps at least once, thoroughly.

The earliest token ring cables were designed to IBM specifications, and provided highly dependable connections, even in very noisy electrical environments. Although modern token ring systems operate very well using UTP cable and RJ45 connectors among others, as a cable installer, you should know how to install the **Universal Data Connector (UDC)** on 4-wire STP cable (Type-1A). This lab should give you some practice doing just that.

When you hear someone talking about a Token Ring connector, or an IBM connector, be aware that the UDC connector is what is being referred to.

You'll need to consult the Appendix for the applicable instruction sheets. Once the STP cable has been outfitted with the UDC connectors, you'll use the UDC/RJ45 adapters to conduct various tests for shorts, opens, and reliable grounding.

1. Consult Appendix A - Manufacturer Instruction Sheets for the following documents, and read them thoroughly.

   • Instruction Sheet 408-3195, page 331

   • Instruction Sheet 408-3321, page 348

   • Instruction Sheet 408-3344, page 353

TIP: You should take special note of the various connector and cable types specified, as well as their configurations.

2. Use the diagonal cable cutters to cut a 5-foot piece of Type-1A STP, 150-ohm cable from the spool.

3. Keep the AMP 408-3195 instruction sheet nearby.

4. Locate a UDC dust cover and strain relief from one of the kits and thread it over a free end of the 150-ohm STP cable, as shown in Figure 2-2.

5. Using the remaining parts from the UDC kit, follow the instructions provided in the AMP 408-3195 instruction sheet to install a universal data connector on the end of the cable containing the strain relief.

TIP: Remember to trim the four wires to 5/8 of an inch, as indicated in the instruction sheet. While following the instructions, keep in mind that you are using Type-1A cable.

6. Once you have completed installing the UDC connector on one end of the 5-foot section of 150-ohm, STP cable, repeat the procedure on the remaining free end, using the second UDC kit.

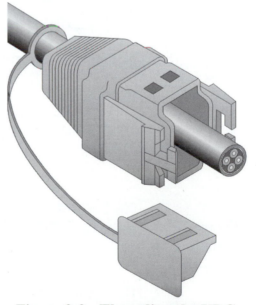

**Figure 2-2: Threading the UDC
Strain Relief and Dust Cover**

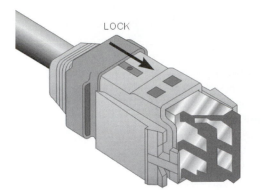

LOCK

Figure 2-3: Removing the UDC Lock

7. With the 5-foot section of 150-ohm, STP cable now outfitted with UDC connectors on either end, locate the two UDC/RJ45 adapters.

8. After removing its dust cover, prepare a UDC/RJ45 adapter for use by first flipping the UDC lock up towards the UDC part of the adapter, and continuing to rotate it up until it comes completely out of its mated slots, as shown in Figure 2-3.

TIP: Notice that once the lock is removed, the UDC release tabs on either side of the adapter are free to flex.

9. Plug the UDC/RJ45 adapter onto one of the UDC connectors on the 5-foot STP cable you just created, as shown in Figure 2-4.

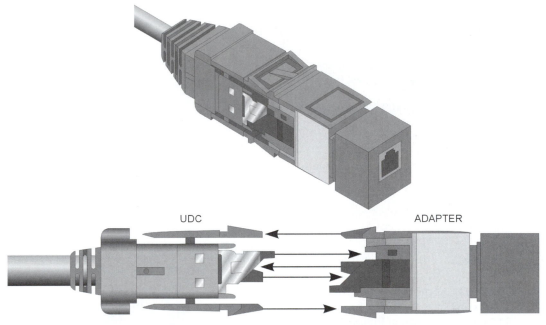

UDC   ADAPTER

Figure 2-4: Connecting the UDC and Adapter Together

10. With the adapter and the UDC connector firmly mated, reinstall the adapter's UDC lock by placing its tabs into the mated slots on either side of the adapter, and firmly pressing it down towards the RJ45 end, as shown in Figure 2-5.

LOCK   UDC

Figure 2-5: Reinstalling the UDC Lock

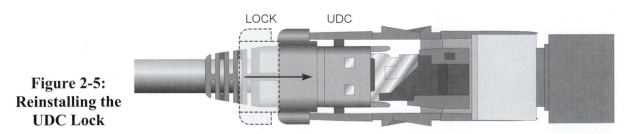

11. Repeat steps 8 through 10 for the other UDC connector on the cable.

12. Locate two RJ45 patch cables that were constructed in previous labs.

TIP: For the following steps, make sure that both RJ45 patch cables conform to either T568A or T568B, but not one of each. If necessary, borrow a cable from another student.

13. Connect one of the RJ45 patch cables to each end of the newly constructed 150-ohm, STP token ring cable.

14. Using the cable tester, connect the entire cable assembly as shown in Figure 2-6, terminating the cable ends in the RJ45 jacks of the tester and its remote unit.

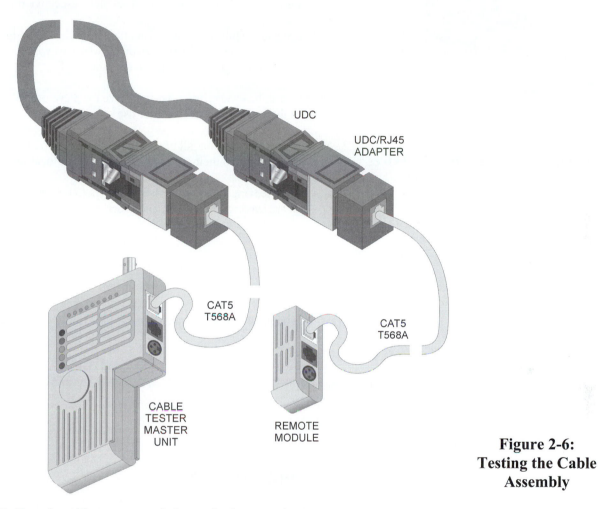

Figure 2-6:
Testing the Cable
Assembly

15. Turn the cable tester on and observe its front panel.

TIP: If everything is working properly, the cable tester should indicate that lines 1 through 6 are connected, while lines 7 and 8 are not. If your tester indicates a short between lines 3 and 6, this is normal because both lines are grounded.

16. If the cable tester indicates a problem, locate and correct the problem. All cables should check good on the cable tester before moving on.

17. Disconnect the UDC/RJ45 adapters, and store the 150-ohm, Type-1A STP cable for possible use later.

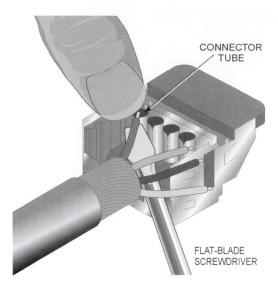

CONNECTOR TUBE

FLAT-BLADE SCREWDRIVER

Figure 2-7: Removing the Wires

TIP: Check with your instructor in case he/she wants you to disassemble this cable, and save the parts for reuse. If your instructor indicates that there is no need to reuse the UDC connectors, skip to step 21.

18. Cut the UDC connectors off from the ends of the cable.

19. Remove the crimp rings and disassemble the connectors by carefully disengaging the wires from their respective tubes.

TIP: If you hold a tube down with your thumb, as shown in Figure 2-7, you can use a slotted-head screwdriver to pry its wire up and out without inflicting any damage on the tube itself.

20. Discard the used crimp rings, and save the piece of cable, and the connector parts for later use.

21. Check to be sure that all of the tools, cables, and parts have been returned to their storage area, or previous locations.

22. Before leaving your lab work area, check to be sure that you are leaving it in a clean and orderly condition.

Lab 8 Questions

1. How many signal wires are supplied with 150-ohm, Type-1A STP cable?

2. Which pins are grounded in a UDC/RJ45 adapter?

3. Is it possible to disconnect the UDC/RJ45 adapter from a UDP connector without first removing the UDC lock?

100-Ohm STP Wiring Systems

Shielded Twisted Pair (STP) cable is capable of transmitting data at very high speeds. For example, STP is used extensively by the telephone company for moving large groups of digitized telephone conversations over distances of 6,000 feet between repeaters. Signals are retransmitted down the next span of cable covering a distance of several miles between telephone company switching stations. This so-called T2 connection involves digital data transmission at speeds of 6.312 Mbps. In fact, high-quality STP has been applied by the telephone companies for transmission rates as high as 8.448 Mbps in Europe.

Originally, STP cable was developed for 1 Mbps Token Ring networks. Later, IBM extended the frequency to 20 MHz for 4 Mbps and 16 Mbps Token Ring networks.

For small Token Ring networks (supporting up to 96 workstations/file servers) that can be relocated when necessary, IBM Type 6 STP cable is used. This cable uses stranded conductors and, as you would expect, is fairly flexible. This makes it suitable for use as wiring closet patch cords. Its weakness however, is that it is useful for only limited distances (150 feet maximum). IBM Type 1 or Type 2 cable is suitable for larger, non-movable networks, supporting up to 260 workstations/file servers. These STP cables use a solid-wire conductor and can carry signals much further than Type 6 varieties.

STP cables used in Token Ring applications are typically 2-pair layouts. The "Halloween Pair" (Orange and Black) is used for the transmit (Tx) function, while the "Christmas Pair" (Red and Green) is used for the receive (Rx) function.

Due to the growing preference of UTP cable for premises wiring installations, the use of STP cable is expected to drop in the short term, in both inter-building and intra-building applications. UTP cable is becoming more technically advanced, and fiber optic installations are gaining in popularity, as well.

However, not everyone needs one million bits per second to his/her desktop. It is possible that the industry, and end-users, will begin to embrace STP cable in its next incarnation. If and when that happens, the pundits will be exclaiming that the IBM cabling system wasn't so bad after all.

100VG-AnyLAN

100VG-AnyLAN is a 100Mbps local area network, utilizing the Demand Priority Protocol, as described in the IEEE 802.12 standard. Previously known simply as AnyLAN, before 1996, the term VG stands for **Voice Grade**, and means that voice grade Unshielded Twisted Pair (UTP) cable is generally used as the transmission medium.

AnyLAN means that any LAN network can be accommodated. For example, 100VG-AnyLAN can utilize cabling media installations of CAT3, CAT 4, CAT5 (using 4-pair UTP or 2-pair STP), and fiber optics. This allows it to support frame formats for IEEE 802.3 (Ethernet) and 802.5 (Token Ring). As a result, 100VG-AnyLAN permits an easy transition /upgrade path from less robust network topologies. In short, 100VG-AnyLAN seeks to provide a high-speed, shared-media LAN to replace slower-speed network protocols, while still utilizing the existing media.

A Demand Priority network consists of end nodes, repeaters, switches, bridges, routers, and network links. The fundamental components of Demand Priority are end nodes and repeaters. An end node is typically a client or server workstation, or a computer, and it is connected to the network through an interface card that plugs into the system's bus. The repeater is the device that interconnects end nodes and the other components.

A basic Demand Priority configuration consists of a small number of local nodes, or end nodes (usually between 6 and 32), that are connected to a single repeater. A larger network can be constructed by cascading several repeaters.

The Demand Priority Protocol allows for a variety of cabling types between networks that are implemented by different vendors. However, within the Demand Priority network, IEEE 802.12 specifies which cable formats can be used. CAT3 and CAT4 (4-pair UTP or 2-pair STP) can be used in lengths of 100 meters, or less. CAT5 4-pair UTP can be used in lengths of 100 meters. CAT5 fiber optic cable can be used in lengths of up to 2,000 meters.

100VG-AnyLAN

Voice Grade

100baseT

100baseT

Fast Ethernet

100baseT is a 100Mbps LAN, also known as **Fast Ethernet**. Fast Ethernet is a supplement to the existing 802.3 protocols, and represents one of the best choices available for those customers who are interested in high-speed networking. This is because there are already many millions of 10Mbps Ethernet users in the world today, backed and supported by an existing multibillion-dollar industry.

From this 10 Mbps world has evolved the 100baseT technology. With the essential characteristics of Ethernet technology (CSMA/CD) remaining unchanged in the 100Mbps world, the body of Ethernet expertise developed over the years can continue to benefit cabling customers, planners, manufacturers, and installers.

Already known about Ethernet are the good characteristics, as well as the difficulties of using it. Knowing about both sets of behavior is important to network planners and installers in order to know how to exploit the benefits, while avoiding the negative aspects. This knowledge gives cablers an inside track in creating successful and reliable network installations, and can only be obtained after a specific technology has been widely deployed, for many years. This is why newer technologies are considered for a time to be a gamble, because there is no practical way to anticipate their behavior during the early stages of development.

Among the high-speed LAN technologies available today, Fast Ethernet has become the leading choice. Building on the near-universal acceptance of 10baseT Ethernet, Fast Ethernet technology provides a smooth, non-disruptive evolution to 100Mbps performance. Offering ten times the performance of 10baseT for twice the price, 100baseT uses 0.45 micron chips, representing an almost eight-fold reduction in chip size, due to advances in integrated circuit chip technology. As chips get smaller, they run faster, use less energy, and are cheaper to produce. Early Ethernet controllers were made in 1.2-micron chips.

There are three distinct cabling systems used for 100baseT Fast Ethernet: 100baseT4, 100baseTX, and 100baseFX (UTP, STP, and fiber optic cables). Each one works in the same manner, except for the transmission medium (Physical layer) schemes. The standard uses two transmission schemes, referred to as 100baseX and 100baseT4. Because of the evolution of the Ethernet standard, 100baseX supports two types of media: twisted-pair and fiber optic cables. These are referred to as 100baseTX and 100baseFX respectively. Within any organization, these systems can be combined to address virtually any cabling needs.

The underlying technology is Carrier Sense Multiple Access/Collision Detection (CSMA/CD), exactly the same access method used in 10Mbps Ethernet networks. As a workgroup network technology, the collision domain is limited to a geography of 205 meters in diameter, matching the two main building wiring standards ISO/IEC 11801 and TIA/EIA T568A. All of the connections are point-to-point link segments.

100baseT4

Using 4-pair UTP cables, a 100baseT4 transceiver is designed for CAT3, CAT4, and CAT5. In each case, the length of a twisted-pair segment (running between the computer and the wiring closet) may be up to 100 meters (328 feet), which is the same distance used by 10baseT links. Cable bundles such as 25-pair cables cannot be used.

The 100baseT4 standard was originally designed to migrate existing CAT3 installations to beyond the 10Mbps limit. However, older CAT3 installations had poor noise performance above 25 MHz, and did not meet FCC or European emission standards. Therefore, in order to work on CAT3 wire, all four pairs of wire had to be utilized. The data signal was split among the eight wires, and encoded, using a block code known as 8B6T. This got the resulting link up to the 100m standard, but, once again, 25-pair bundle cables could not be used.

Splitting the signal among the four pairs of standard-quality twisted pair wiring resulted in one pair being dedicated to transmission, and one pair being dedicated to reception. The remaining two pairs were operated bi-directionally. The 100Mbps data signal was divided equally over three cable pairs (in each direction). Although the effective data rate remained at 100 Mbps, the cable frequency was much lower (33.3 Mbps), allowing the use of a less-sophisticated voice-grade cable (CAT3 UTP). It is very important to remember, however, that 100baseT4 is not limited to voice-grade, CAT3 cable. It can be used equally well on CAT4 UTP, CAT5 UTP, and Type 1 STP.

The advantages of using 100baseT4 include:

- It operates on virtually any preexisting twisted pair cabling.

- Its adapters are about 10% less expensive than those for other systems.

- Less expensive CAT3 cabling can be used.

The disadvantages of using 100baseT4 include:

- It cannot support Full Duplex mode (unneeded in workstations).

- It requires four pairs of wiring to operate.

One conclusion that can be drawn is that 100baseT4 is best suited for use in workstations, where cost is an important issue.

100baseTX

The 100baseTX method requires two pairs of high-quality twisted pair wiring (one for transmission and one for reception). This cable type can be either Unshielded Twisted Pair (UTP) or 150-ohm Shielded Twisted Pair (STP), providing it meets the required performance specifications. The most common examples of cable that meet these requirements are CAT5 UTP and IBM Type 1 STP. For UTP installations, 100baseTX is designed for CAT5 cabling exclusively. The CAT5 UTP cables are usually fitted with RJ45 plugs, while the 150-ohm STP cables normally attach to DB9 connectors.

The advantages of using 100baseTX include:

- It uses only two pairs of twisted pair wiring.

- It supports Full Duplex mode for up to 200 Mbps in network servers.

The disadvantages of using 100baseTX include:

- All of the patch panels and jumper blocks must be CAT5-compatible.

- Adjacent cable pairs cannot be used for other applications.

- Distance limitations per link is 100 meters, or 200 meters with a single repeater.

- Full Duplex mode requires Full Duplex switching apparatus.

100baseTX is more desirable in network servers that can take advantage of Full Duplex Fast Ethernet (FDFE) mode, and where the quality of the cabling can be easily controlled.

100baseFX

100baseFX is designed for fiber optics, and so far, we have not considered the role of fiber optic cabling in a Fast Ethernet network. Fiber has the same advantages in 100baseFX as it does in FDDI networks. It extends the working distances of the network, it displays electromagnetic immunity, and it operates with increased data security. It also uses the same type of fiber optic cable as FDDI. Similar to 10baseT, 100baseT allows a maximum distance between a repeater (hub) and a node of 100 meters. Using fiber optic cable, 100baseFX increases this distance to a maximum of 185 meters. Between a server and workstation (with no repeater), the maximum distance is increased to 400 meters, or as much as 2 kilometers when using Full Duplex mode.

The standard cable type for 100baseFX is multi-mode fiber, with a 62.5-micron core and a 125-micron cladding. Only one pair of fibers is required—one fiber for transmission and one fiber for reception. In the interconnection of repeaters, to form a fiber optic backbone, the 100baseFX standard will best apply. A typical installation will contain Fast Ethernet repeaters on each floor, or in each department, that supports 100baseTX or 100baseT4 workstations. The 100baseFX links are then interconnected using repeaters. When connecting repeaters on different floors, the fiber optic cabling will provide protection from electromagnetic noise that is often associated with elevators, and will also enable longer cable runs between buildings.

One reason why 100baseT is such a popular and cost-effective high-speed network technology is because it is designed to combine with other networks, without disturbing them.

Gigabit Ethernet

As applications such as scientific modeling and engineering, medical data transfer, publication, Internet/Intranet transfers, data warehousing, network backup, and video conferencing have proliferated, ever greater shares of bandwidth at the desktop are in demand. As the total number of network users continues to grow, organizations must transfer critical portions of their networks to higher-bandwidth technologies.

A clear need for higher-speed network technology, at the backbone and server level, has been created with the growing use of 100baseT connections to servers and desktops. Thankfully, such technology can be implemented without the need for extensive user retraining, costly equipment upgrades, or bumpy network transitions. Truly, we have been experiencing a bandwidth crunch!

The most appropriate solution, recently developed, is **Gigabit Ethernet**. Gigabit Ethernet is an emerging LAN transmission standard that will provide 1Gbps bandwidths (that's a data rate of 1 billion bits per second) for preexisting networks, with Ethernet simplicity, and at a much lower cost than other technologies of comparable speed. It offers a natural upgrade path for current Ethernet installations, and takes advantage of existing end stations, management tools, and training.

Gigabit Ethernet

Several key considerations for choosing a Gigabit Ethernet high-speed network include:

- Straightforward migration to higher performance levels without disruption.

- Low cost of ownership—including both purchase and support.

- Capability to support new applications and data types.

- Network design flexibility.

The Fast Ethernet standard, approved in 1995, established Ethernet as a scaleable technology. Now, more recent development of Gigabit Ethernet extends that scaleability even further. Independent market research indicated a strong interest among network users in adopting Gigabit Ethernet technology. It's designed to run primarily on fiber optic cable, with very short distances possible on copper media. It specifies a multi-mode fiber to operate over 550 meters of cable, and single-mode to operate over 3,000 meters.

Standard Gigabit Ethernet is defined in the IEEE 802.3z standard, and the first product versions of it are now available. Gigabit Ethernet is used as an enterprise backbone, where existing Ethernet LANs, using 10Mbps and 100Mbps cards, can feed into it.

Because of these attributes, as well as support for Full-Duplex operation, Gigabit Ethernet is an ideal backbone interconnect technology for use between 10/100baseT switches as a connection to high-performance servers, and as an upgrade path for future high-end desktop computers requiring more bandwidth than 100baseT can offer.

The IEEE 802.3z task force has worked feverishly to formulate a definition for the Gigabit Ethernet Physical layer standards, based on the **International Organization for Standardization (ISO)** model. This ensures that Gigabit Ethernet implements a functionality that adheres to a Physical layer standard. In general, the Physical layer is responsible for defining the mechanical, electrical, and procedural characteristics for establishing, maintaining, and deactivating the physical link between network devices. For Gigabit Ethernet communications, several Physical layer standards (PHYs) have emerged from the IEEE 802.3z effort.

International Organization for Standardization (ISO)

Two of them provide Gigabit transmission over multi-mode, fiber-optic cabling. For example, the 1000baseSX standard is targeted at low-cost, multi-mode fiber horizontal runs and in short backbone applications. The 1000baseLX standard is targeted at longer, multi-mode building fiber backbones and single-mode campus backbones. For multi-mode fiber, these standards define gigabit transmission over distances of 300 and 550 meters, respectively.

Single-mode fiber, which is covered by the long-wavelength standard, is defined to cover distances of 3 kilometers.

There are also two standards that have been developed for Gigabit Ethernet transmission over copper cabling. The first copper link standard defined by the 802.3z task force is referred to as 1000baseCX, which supports interconnection of equipment clusters in which the physical interface is short-haul copper. It can be implemented using a switching closet, or a computer room, as a short jumper interconnection for 25-meter distances. This standard uses the Fiber Channel-based 8B/10B coding, at the serial line rate of 1.25 Gbps, and runs over 150-ohm, balanced and shielded specialty cabling assemblies, known as **twinax**. This copper Physical layer standard has the advantage that it can be generated quickly and is inexpensive to implement.

twinax

1000baseT

The second copper link standard is intended for use in horizontal copper cabling applications. In March, 1997 a **Project Authorization Request (PAR)** was approved by the IEEE Standards Board, enabling the creation of a separate but related committee referred to as the 802.3ab task force.

Project Authorization Request (PAR)

1000baseT

This group was chartered with the development of a proposed **1000baseT** Physical layer network standard, providing 1Gbps Ethernet signal transmission over 4-pair, CAT5 UTP cable, covering distances of up to 100 meters, or networks with a diameter of 200 meters.

This standard describes communications using horizontal copper runs on a floor within a building. The idea here is to use structured generic cabling, allowing network managers to take advantage of the existing UTP cabling already deployed. In order to complete this standard, new technology and new coding schemes were required in order to achieve compatibility with the previous Ethernet, and Fast Ethernet, standards.

1000baseT extends the Gigabit Ethernet envelope by enabling operation over the large base of legacy CAT5 cabling systems, as well as the CAT5 cabling systems currently being installed. The distribution of 4-pair, CAT5 cabling extends from the work area to the equipment room, as well as between various elements in the equipment room. This enables connectivity to switched and shared gigabit services for both high-bandwidth work area computers, and the associated server farms.

Transmitting a 1000Mbps data stream over four pairs of CAT5 UTP wires presents several design challenges to the transmission system designer (and standards developer). The challenges are due to the electrical parameters of wire, discussed in earlier sections of chapter 1: Twisted Pair Cable, and Shielded Twisted Pair Vs. Unshielded Twisted Pair. Among these parameters are signal attenuation, echo, return loss, and crosstalk, as well as electromagnetic emissions and susceptibility.

The characteristics of CAT5 cabling, and the regulatory constraints that govern the maximum amounts permitted for the wire parameters identified above, create a significant challenge in defining the 1000Mbps Physical layer. Fortunately, digital communications techniques developed in recent years can be used to design robust transceivers that are capable of achieving reliable operations in a CAT5 cabling environment. The 1000baseT application takes advantage of several of these techniques to transform the desired bit rate into an acceptable baud rate (signaling events per second) for operation over 4-pair, CAT5 cabling.

These techniques include:

- Using existing 4-pair, CAT5 cable that conforms to TIA/EIA T568A.

- Using all four pairs in the cable to keep symbol rate at, or below, 125 Mbaud.

- Using PAM5 coding to increase the amount of information sent per symbol.

- Using 4D 8-state Trellis Forward Error Correction to limit noise and crosstalk.

- Using pulse shaping to condition the transmitted spectrum.

- Using DSP equalization for noise, echo, crosstalk, and BER compliance.

When used with **Full-Duplex Repeaters** (**FDRs**), 1000baseT provides highly cost-effective, shared-gigabit service, because FDRs (also called buffered distributors) offer low-cost, full-duplex, shared-media operation. Coupled with 1000baseT, FDRs provide an easy-to-manage, high-burst-rate, shared-media solution capable of supporting both end users and server farms—a cost-effective combination of gigabit server support.

Full-Duplex Repeaters (FDRs)

Because the startup cost for copper networking solutions has traditionally been lower than for fiber-based solutions, many Gigabit Ethernet technology enthusiasts are already calling for a gigabit UTP copper solution. This is in addition to the fact that it will allow them to use their existing copper cable plant at gigabit speeds. Cable plants conforming to current TIA/EIA T568A specification should be able to support 1000baseT operation. The individual links should be tested against TIA/EIA TSB 67, along with the additional test parameters for FEXT (ELFEXT) loss, and return loss, included as an addendum to TIA/EIA T568A.

Cabling test tool manufacturers have had these capabilities added to their products prior to the release of the 1000baseT standard. Any CAT5 cabling that does not pass TSB 67 testing (including addendum parameters) should be suspected, as it may not have been correctly installed. For example, the amount of untwisting in a pair, as a result of termination to connecting hardware, cannot exceed 13 mm (0.5 in), or the connecting hardware may not meet current TIA/EIA T568A requirements, and may need upgrading (or at least the wiring aspect of the installation). All CAT5 links should be tested before using either 1000baseT or 100baseT. Obviously, these efforts have taken somewhat longer to achieve than the 802.3z Gigabit Ethernet solution, and the approval by the TIA/EIA has taken even longer, because of the new networking tests that have needed to be developed. However, considering the huge base of CAT5 cable already installed, the wait has been well worth the effort.

ATM

An alternative technology that competes with Gigabit Ethernet is **Asynchronous Transfer Mode** (**ATM**). ATM is a very high-speed backbone transmission network that uses an advanced implementation of packet switching (fast packet oriented transfer mode), and is based on asynchronous time division multiplexing. It has been accepted universally as the transfer mode of choice for the **Broadband Integrated Services Digital Network** (**BISDN**). These high-speed data transmission rates are achieved by sending fixed-size packets over broadband and baseband systems.

Asynchronous Transfer Mode (ATM)

Broadband Integrated Services Digital Network (BISDN)

**ATM Adaptation
Layer (AAL)**

As a broadband cell-relay method, ATM transmits data in 53-byte cells, rather than in variable length frames, consisting of 48 bytes of information, coupled with 5 additional bytes for each frame header. The header identifies cells belonging to the same virtual channel, for appropriate routing. Cell sequence integrity is preserved for each virtual channel. An **ATM Adaptation Layer** (**AAL**) supports various services, providing service-specific functions. AAL-specific information is contained in the ATM's information field. This results in a more consistent and uniform data packet. Error control is not done on the information field of ATM cells inside the network, and the cells are carried transparently in the network.

ATM can accommodate voice, data, fax, real-time video, CD-quality audio, imaging, and multi-megabit data transmission. Clearly, its reason for existence is the promise of seamless integration of legacy and advanced technologies used to deliver high-bandwidth networks. In fact, the term **broadband**, when used to describe the transmission platform for digital services, is practically synonymous with ATM. ATM provides a good bandwidth flexibility and can be used efficiently from desktop computers to local area and wide area networks.

broadband

To allow existing applications to operate unchanged over newly installed ATM backbones, **LAN Emulation** (**LANE**) software is used. LANE allows the bridging of multiple operating protocols without making major changes to the existing network infrastructure, or to the applications that already may be running over it. With the large investment in networking equipment, protocols, and software, the ATM developers use LANE to ease the transition, so existing network applications /protocols run smoothly during the migration. This results in a flexible mechanism that allows the evolution of legacy networks into ATM networks of the future, without having to modify protocols and applications that are currently well-understood and performing properly.

**LAN Emulation
(LANE)**

ATM's flexibility provides a quality of service that can be scaled to fit the customer's needs. The LANE specification, while supporting backbone implementations, directly attached ATM servers, and high-performance workgroups, is also well-suited to futuristic ATM-specific applications leading to complete worldwide connectivity. Data rates for ATM, depending on which medium/transmission system is used, range between 155-622 Mbps.

There are several practical applications using ATM Technology. It appears that ATM is destined to become the backbone network for many broadband applications, including the information superhighway. Some of its key applications include:

- Video Conferencing.

- Desktop Conferencing.

- Multimedia Communications.

- ATM Over Satellite Communications.

- Mobile Computing over ATM for Wireless Networks.

Coaxial Cable

At one time, coaxial cable was the most commonly used medium for interconnection of network devices because it was inexpensive, light, flexible, and easy to work with. Coax consists of a solid core of copper, surrounded by insulation, within a braided metal shielding, then, bound in a plastic sheath.

Connecting hardware used with coaxial cable for data transmission is attached through what are known as **British National Connector**, or **Bayonet-Neil-Concelman** (**BNC**) connectors. A BNC is a bayonet locking connector, and there are several different types.

A BNC barrel connector can be used to join two lengths of thin coaxial together, while a BNC T connector can be used to connect two lengths of coaxial together, as well as connect a device to the network, as shown in Figure 2-8.

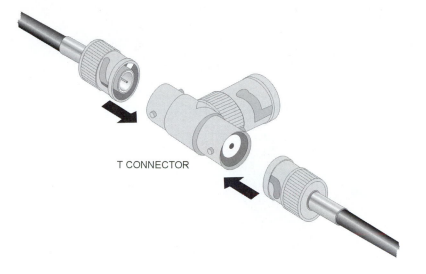

T CONNECTOR

**Figure 2-8: BNC
T Connector**

Lab 9 – Placing a BNC Connector on a Coaxial Cable

Lab Objective

To understand how to place a BNC connector on a coaxial cable

Materials Needed

- Lab 9 work sheet

- Cable, RG58, 50-ohm, coax (spool)

- Tension scale/Tape measure

- Multimeter, digital

- Pliers, needle-nose

- Pliers, diagonal cutters

- Crimping tool, RG58/RG6

- Cable stripper, RG58/RG6

- Hex wrench

- Utility knife

- Connector, crimp, BNC (2)

Before beginning this lab, read through all of the following steps at least once, thoroughly.

1. Use the diagonal cable cutters to cut a 5-meter piece of RG58 cable, and a separate shorter piece (approximately 6 inches) from the spool.

2. Examine Figure 2-9 carefully.

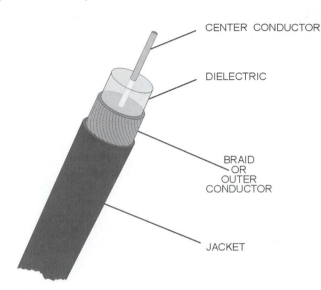

CENTER CONDUCTOR

DIELECTRIC

BRAID
OR
OUTER
CONDUCTOR

JACKET

Figure 2-9: Coaxial Components

3. Make sure that both the 5-meter cable section and the 6-inch cable section have an even cut on both ends, and then temporarily set the 5-meter section aside.

TIP: You will first use the 6-inch piece of RG58 cable, or a piece from the scrap heap, for practice purposes before working on the 5-meter section.

4. If necessary, adjust the cable stripper for working with RG58 cable (setting "8"), as shown in Figure 2-10.

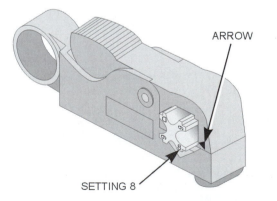

ARROW

SETTING 8

**Figure 2-10:
Setting the Cable
Stripper for RG58**

TIP: The V-block insert will slide out if you push it from the back side, using the Allen wrench stored in the bottom of the stripper, so that you can rotate it to the correct setting before reinserting. Table 2-6 lists the V-block settings for various cable types.

Table 2-6: V-Block Settings

Cable	Setting
RG6	6
RG58	8
RG59/RG62	9
Special	D

5. After checking and setting the V-block as required, remove the center pin from the cable stripper, so that its top can be removed and temporarily set aside.

TIP: Again, use the Allen wrench to push the large center pin out just as you did with the V-block insert. Cup the top in your hand as the pin is removed to prevent it from jumping off the assembly. Set the center pin safely aside.

6. Carefully lift the top of the cable stripper off the assembly, along with the spring.

7. Then study Figure 2-11, which depicts the interior view of the stripper with the various parts identified.

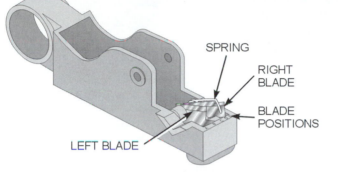

SPRING
RIGHT BLADE
BLADE POSITIONS
LEFT BLADE

Figure 2-11: Interior Parts and Locations of Stripper

8. Identify the five blade positions A, B, C, D, and E in the cable stripper.

TIP: The blades may need to be adjusted somewhat for stripping RG58 cable. The blade positions A and B are intended to adjust the position of the blade that strips to the inner conductor of the cable. Blade positions C, D, and E are intended to adjust the position of the blade that strips to the braid (or alternately, to the center insulation through the braid).

9. Observe the possible blade settings, and their corresponding dimensions (in millimeters), as shown in Table 2-7.

Table 2-7: Cable Stripper Settings

BC	BD	BE	AE
4 mm	6 mm	8 mm	12 mm

10. If the blades are already set to "BC", don't change anything.

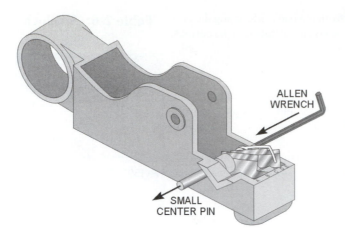

Figure 2-12: Removing Small Center Pin

11. If either, or both, of the blades require a change of position, use the Allen wrench in the same way you did to remove the large center pin earlier, only this time, use it to push the small center pin out, as shown in Figure 2-12.

12. Set the small center pin aside, temporarily, and locate the needle-nose pliers.

13. Using the needle-nose pliers as required, place the left blade at position B in the stripper, and the right blade at position C, just to the left of the spring (for a dimension of 4 mm), as shown in Figure 2-13.

TIP: The dimensions refer to the distance between a stripped center conductor and an intact cable jacket. Take care to keep the spring oriented properly once the blades have been positioned.

14. Once you have the blades positioned properly, carefully reinsert the small center pin into the stripper. Use the Allen wrench to push the pin as needed.

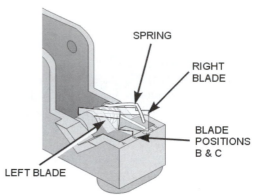

Figure 2-13: Positioning the Cutting Blades

TIP: This all sounds easy enough, but the blades need to be aligned perfectly before the small center pin will reinsert correctly. If you need help, see your instructor.

15. Carefully replace the top of the cable stripper, along with the spring, on the assembly, and press it down until the holes for the large center pin line up.

16. When you can see through the large center pin holes, insert the large center pin back through the stripper assembly until it lines up evenly on both sides.

17. With the stripper reassembled and calibrated for RG58 cable, locate the 6-inch piece of RG58 cable you cut earlier, and position the stripper over the cable, as shown in Figure 2-14.

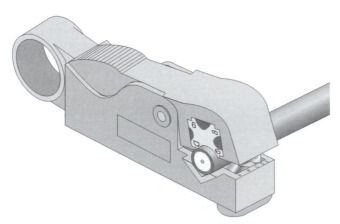

Figure 2-14: Positioning the Stripper

18. If the cable feels too loose, adjust the depth of the blades until you can feel that the cutting action will occur. Use the Allen wrench as shown in Figure 2-15.

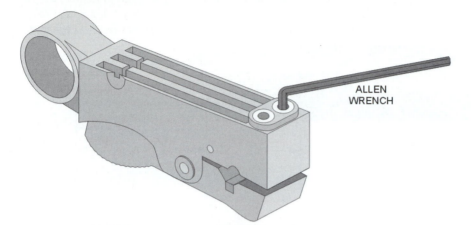

ALLEN
WRENCH

Figure 2-15:
Adjusting the Depth
of the Blades

TIP: Keep in mind that the blade on the left strips to the center conductor, while the blade on the right strips to the braid, or alternately, through the braid to the center insulation. In this case, you'll want to strip through the braid to the center insulator. When trying to perfect the depth of cuts provided by the stripper, you should use the 6-inch piece you cut from the spool (or a piece of RG58 from the scrap heap), and practice adjusting the blade heights so that neither the center conductor nor the insulation is damaged by the cut. Obviously, a two-bladed stripper will require you to perform additional work to trim the braid properly.

19. Release the grip on the stripper so that the blades grip the cable on their own.

20. Rotate the stripper clockwise (in the direction of the arrow on the handle) for two or three turns.

TIP: Do not rotate the stripper too many times with thin RG58 cable, because the friction between the center conductor and insulation may serve to twist the center conductor off, or the continuous spring tension on the blade may cause it to cut too deeply with extra rotations.

21. Then, using the utility knife, carefully strip 5/16 of an inch more of the jacket to reveal the braid below. Table 2-8 reveals the approximate lengths of each section.

Table 2-8: Cutting Measurements

Jacket	Braid	Dielectric	Conductor
3/4" (1.905cm)	5/16" (0.79375cm)	5/32" (0.397cm)	3/16" (.476cm)
± 1/8" (.3175cm)	± 1/16" (.15875cm)	± 1/32" (.07938cm)	± 1/64" (.03969cm)

22. Lastly, trim the length of the center conductor to exactly 3/16 of an inch.

TIP: All of the cuts are to be sharp and square. Do not nick the braid, the dielectric, or the center conductor. Use the tension scale/tape measure to check the measurements.

23. Compare your work with Figure 2-16, where the strip dimensions are shown in both inches and centimeters (in parentheses).

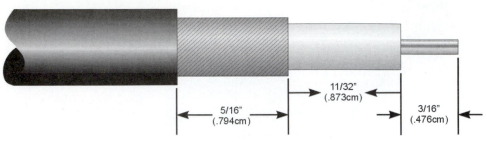

Figure 2-16: Strip Dimensions

5/16"
(.794cm)

11/32"
(.873cm)

3/16"
(.476cm)

24. If necessary, continue to practice on the shorter (scrap) piece of RG58 cable until you are satisfied with your stripping technique.

25. When you are ready, strip both ends of the 5-meter RG58 cable you cut previously, to the dimensions shown in Table 2-8 and Figure 2-16.

26. Examine the BNC components displayed in Figure 2-17, and be sure you can recognize them.

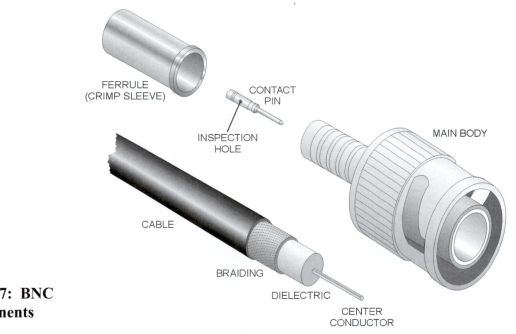

FERRULE
(CRIMP SLEEVE)

CONTACT
PIN

INSPECTION
HOLE

MAIN BODY

CABLE

BRAIDING

DIELECTRIC

CENTER
CONDUCTOR

Figure 2-17: BNC Components

27. Refer to Figure 2-18, and slide a ferrule (or crimp sleeve) over one end of the 5-meter cable you have previously stripped.

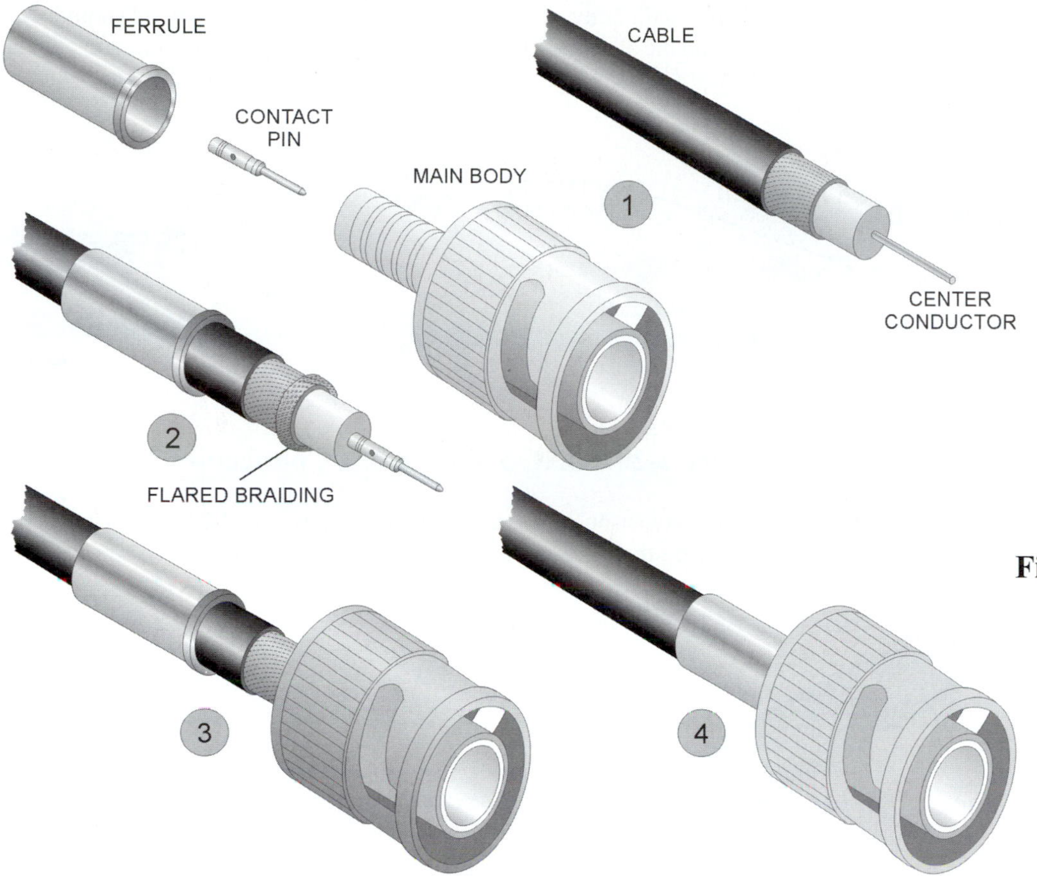

FERRULE

CONTACT PIN

MAIN BODY

CABLE

CENTER CONDUCTOR

1

2

FLARED BRAIDING

Figure 2-18: BNC Installation

3

4

TIP: Be sure to orient the ferrule so that the flared end is facing towards the cable end that you are currently preparing.

28. Place a contact pin on the end of the center conductor.

TIP: As you slide the contact pin on, you should check its inspection hole to verify that the center conductor is visible.

29. With the contact pin pushed back on the center conductor as far as the dielectric material (center insulator), use the crimping tool's .068 die position to crimp the contact, as shown in Figure 2-19.

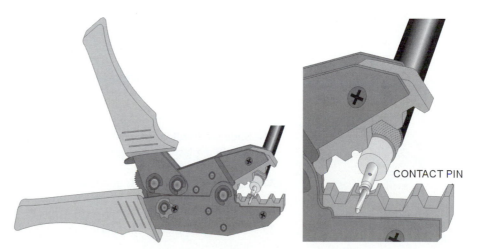

CONTACT PIN

Figure 2-19: Crimping the Contact Pin

30. Slightly flare the braiding, and then push the main body onto the contact pin. Be sure that it seats well, with no part of the braiding touching the center conductor.

TIP: As you slide the main body onto the cable, the heel of the main body should slide between the braiding and the dielectric. If you are using a snap-type connector, you will feel the snap as you push the main body and contact pin together. Check Figure 2-20 and Figure 2-21, if necessary.

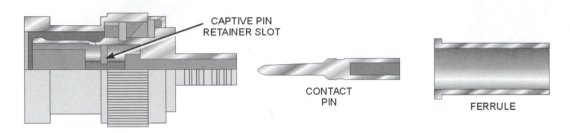

CAPTIVE PIN
RETAINER SLOT

CONTACT
PIN

FERRULE

Figure 2-20: BNC 3-Piece Coax Connector

TIP: If you are using a twist-on connector, as depicted in Figure 2-21, the main thing to remember is to make sure that no braid is touching the center conductor.

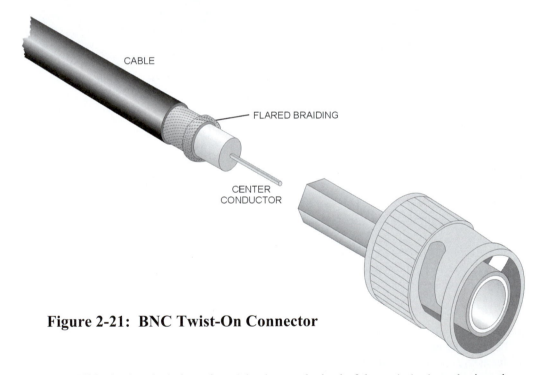

CABLE

FLARED BRAIDING

CENTER
CONDUCTOR

Figure 2-21: BNC Twist-On Connector

31. Slide the ferrule (crimp sleeve) back over the heel of the main body and crimp the ferrule in place, using the crimping tool's .213 die position, as shown in Figure 2-22.

32. Locate the second BNC connector, and install it on the remaining end of the 5-meter section of RG58 cable by repeating steps 27 through 31.

33. Utilizing the multimeter, perform a continuity check on the BNC cable you just created.

34. In addition to the continuity checks between the center contacts at each end of the cable, and between the grounded main bodies at either end, be sure to check for an expected open circuit between the center conductor and the grounded main bodies.

TIP: The multimeter should show an infinite resistance between the center conductor and the grounded main bodies of the RG58 connectors. If it does not, the cable is not correctly prepared, and you may have to repeat this procedure. If the cable checks out good, continue with the remaining steps.

35. Once the 5-meter RG58 coaxial cable passes the continuity checks, set it aside for later use.

Figure 2-22: Crimping the Sleeve

TIP: Recall that taps can be used on backbone cables in order to add nodes to an existing network. These taps provide two-way communications just as if the tap line were an original part of the network.

Tapping into an RG58-type cable is called a **Thinnet tap** due to the rather flexible characteristic of the cable and its comparatively thin diameter. The much fatter RG8 cable can also be tapped into, and such a tap is referred to as a **Thicknet tap**. Because you will perform a procedure for tapping into a RG8 cable later in the course, you will only be required to read about a Thinnet tap at this point.

Thinnet tap

Thicknet tap

36. Consult Appendix A for the following instruction sheet, and read it thoroughly.

- Instruction Sheet 408-9365

37. Check to be sure that all of the tools, cables, and parts have been returned to their storage area, or previous locations. Before leaving your lab work area, check to be sure that you are leaving it in a clean and orderly condition.

Lab 9 Questions

1. What will happen if the braid is touching the center conductor after the installation?

2. Why must all of the cuts to the braid, dielectric, and center conductor be accomplished without nicks?

3. What should the resistance be between the center conductor and the braid shield in a RG58 coaxial cable?

4. What happens when the cable stripper is rotated too many times?

5. Where is the allen wrench stored?

Lab 10 – Installing F Connectors on a Coaxial Cable

<u>Lab Objective</u>

To understand how to install an F connector on an RG6, 75-ohm coaxial cable

<u>Materials Needed</u>

- Lab 10 work sheet
- Cable, RG6, 75-ohm, coax (spool)
- Tension scale/Tape measure
- Multimeter, digital
- Pliers, needle-nose
- Pliers, diagonal cutters
- Crimping tool, RG58/RG6
- Cable stripper, RG58/RG6
- Hex wrench
- Connector, crimp, F (2)

<u>Instructions</u>

Before beginning this lab, read through all of the following steps at least once, thoroughly.

1. Using the diagonal cable cutters, cut a 5-meter piece of RG6 coaxial cable and a separate shorter piece (approximately 6 inches) from the spool.

2. Make sure that both the 5-meter cable section and the 6-inch cable section has an even cut on both ends, and then temporarily set the 5-meter section aside.

TIP: You will first use the 6-inch piece of RG6 cable, or a piece from the scrap heap, for practice purposes before working on the 5-meter section.

3. If necessary, adjust the cable stripper's V-block for working with RG6 cable (setting "6"), as shown in Figure 2-23.

TIP: Slide the V-block insert out by pushing it from the backside with the hex wrench stored in the bottom of the stripper, rotate it to the correct setting, and then reinsert. RG6 cable is the same type that is used for cable TV transmission in the home, and the F connector is the same type that connects to the back of your television set.

4. Determine if the blades in the cable stripper are still set to the "BC" position (from the previous procedure). If they are, jump ahead to step 14. If they are not, continue on to the next step.

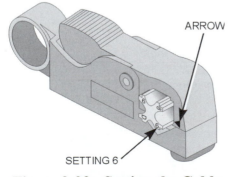

ARROW

SETTING 6

Figure 2-23: Setting the Cable Stripper for RG6

5. After checking and setting the V-block as required, remove the large pin in the center of the cable stripper. Push it out using the hex wrench, and cup your hands over the top of the assembly so it doesn't jump up when the pin slides out. Temporarily remove its top and spring, and set them aside.

6. If necessary, review Table 2-7 and Figure 2-11 for the settings and locations of the stripping blades.

TIP: The locations of the blades for RG6 cable will be the same as for RG58. However, the depth of the blades will have to be altered, because RG6 is a larger diameter cable than RG58.

7. If the blades are already set to "BC", don't change anything.

8. If either blade requires a change of position, use the hex wrench in the same way you did to remove the large center pin earlier, only this time, use it to push the small center pin out, as in the previous procedure (Figure 2-12).

9. Set the small center pin aside, temporarily, and locate the needle-nose pliers.

10. Using the needle-nose pliers as required, place the left blade at position B in the stripper, and the right blade at position C, just to the left of the spring (for a dimension of 4 mm). See Figure 2-13 as before.

TIP: Remember that the dimensions refer to the distance between a stripped center conductor, and an intact cable jacket. The spring needs to be oriented properly once the blades have been positioned.

11. Once you have the blades positioned properly, carefully reinsert the small center pin into the stripper. Use the hex wrench to push the pin as needed.

TIP: The small center pin will reinsert correctly only if the blades are perfectly aligned. This is not always as easy to accomplish as it sounds. If you need help, see your instructor.

12. Carefully replace the top of the cable stripper, along with the spring, on the assembly, and press it down until the holes for the large center pin line up.

13. When you can see through the large center pin holes, insert the large center pin back through the stripper assembly until it lines up evenly on both sides.

14. With the stripper reassembled and calibrated for RG6 cable, locate the 6-inch piece of RG6 cable you cut earlier, and position the stripper over the cable, in the same way as in the previous procedure for RG58 (Figure 2-14).

15. If the cable feels too tight, adjust the depth of the blades until you can feel that the proper cutting action will occur. Use the hex wrench as shown in Figure 2-15.

TIP: Again, the blade on the left strips to the center conductor, while the blade on the right strips to the braid, or alternately, through the braid to the center insulation. In this case, you'll want to strip down through the braid, to the foil shield just below it. Use the 6-inch piece you cut from the spool (or a piece of RG6 from the scrap heap), and practice adjusting the blade heights so that neither the center conductor nor the foil shield is damaged by the cut.

16. Release the grip on the stripper so that the blades grip the cable on their own.

17. Rotate the stripper clockwise (in the direction of the arrow on the handle) for two or three turns (or, until you feel the resistance reduced).

TIP: Even though RG6 cable is thicker than RG58, do not rotate the stripper too many times or damage will occur to either the center conductor or the foil shield. All of the cuts are to be sharp and square. Do not nick the foil shield, the dielectric, or the center conductor. You do not want any of the braid showing along the foil shield.

18. Compare your work with the stripped RG6 cable shown in Figure 2-24.

Figure 2-24: Properly Stripped RG6

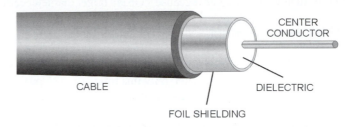

19. If necessary, continue to practice on the shorter (scrap) piece of RG6 cable until you are satisfied with your stripping technique.

20. When you are ready, strip both ends of the 5-meter RG6 cable you cut previously, to the dimensions provided by the BC settings of the stripper.

21. Locate one of the F connectors, and mount it on one end of the 5-meter cable you previously cut and stripped. Refer to Figure 2-25.

Figure 2-25: Mounting an F Connector

22. Push the F connector onto the end of the RG6 cable until the center insulator's front surface is even with the F connector's inside surface (Figure 2-25).

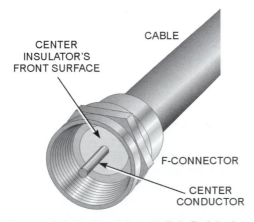

Figure 2-26: Pushing RG6 Cable into an F Connector

TIP: This sounds much easier than it happens to be. The main problem occurs when trying to fit the inside collar of the F connector between the foil shield and the outer jacket of the cable. You are having to overcome the resistance of the braid shield just below the jacket.

If you are having trouble doing this, try using the crimper tool (in its closed position) as a platform to push against. If you place the F connector face down over one of the die holes, you will gain the leverage necessary to force the cable down and into the collar, as shown in Figure 2-26. Be sure to orient the F connector so that the center conductor is allowed to protrude through the front end of the connector and into one of the die holes of the crimper. This keeps the center conductor from being bent as the cable is forced into the collar.

23. When you have reached the required depth (center insulator's front surface is even with the F connector's inside surface), check to be sure that the center conductor extends at least 1/8 of an inch, or more, from the front of the connector (as viewed from the side).

TIP: If it does not, you will have to pull the connector, cut the end from the cable, and repeat the stripping procedure so as to meet this requirement. Before doing so, check the settings on the stripper to be sure that the dimension setting is BC, and allow a short length of cable to protrude beyond the front edge of the stripper, rather than placing it flush. This should give you more than 1/8 of an inch of protrusion for the center conductor from the F connector.

24. When the RG6 cable and the F connector have been properly mated, locate the crimping tool and place it in its open, or relaxed, position.

25. Position the F connector/RG6 cable combination (with center insulator and the connector's inside surface flush) under the .319 crimper die, as shown in Figure 2-27.

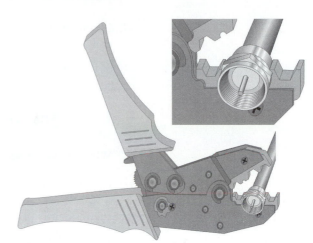

TIP: Once you begin to make the crimp in the F connector, something may cause you to want to stop before the crimping has completed (wire not positioned properly, etc.). If this happens, you can cause the crimping tool to release the connector by pushing up on the release lever between its handles.

26. Look at Figure 2-27 again, and make sure that there is at least a 1/8 inch-buffer space between the crimper tool and the edge of the connector's barrel.

Figure 2-27: Crimping an F Connector

TIP: The 1/8-inch buffer will ensure that the nut on the F connector will turn freely once the crimp has been completed. This is important because F connectors are attached by screwing the male nut onto a threaded female barrel.

27. With the F connector/RG6 cable combination properly positioned in the crimping tool, slowly squeeze the handles together to complete the crimp. A stronger squeeze will cause the crimper to release and open.

28. Check the crimp by spinning the nut and ensuring that it is free.

29. Repeat steps 21 through 28 for the second F connector on the remaining end of the 5-meter section of RG6 cable.

30. Use the multimeter to perform a continuity check on the BNC cable you just created.

31. In addition to the continuity checks between the center contacts at each end of the cable, and between the grounded main bodies at either end, be sure to check for an expected open circuit between the center conductor and the grounded main bodies, as shown in Figure 2-28.

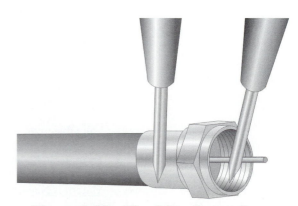

Figure 2-28: Checking for an Open Circuit

TIP: The multimeter should show an infinite resistance between the center conductor and the grounded main bodies of the RG6 connectors. If it does not, the cable is not correctly prepared, and you may have to repeat this procedure. If the cable checks out good, continue with the remaining steps.

32. Once the 5-meter RG6 coaxial cable passes the continuity checks, set it aside for later use.

33. Check to be sure that all of the tools, cables, and parts have been returned to their proper storage areas.

34. Before leaving your lab work area, check to be sure that you are leaving it in a clean and orderly condition.

Lab 10 Questions

1. Why should the nut on a crimped F connector be free to spin?

2. How can you interrupt the crimping tool before a crimp has been completed?

3. What should the resistance be between the center conductors on each end of the RG6 coaxial cable?

4. When crimping an F connector, how much uncrimped distance should be allowed between the edge of the crimper and the edge of the connector's crimp barrel?

5. Where does the collar on the rear of the F connector fit when a properly stripped RG6 cable is pushed into it?

Lab 11 – Installing N Connectors on a Coaxial Cable

Besides the fact that RG8 cable (called Thicknet) is much thicker than RG58 (called Thinnet), it also is marked on the cable jacket (yellow or orange) every 2.5 meters to indicate minimum distances allowed between taps. The normal practice is to install all taps and all end connectors directly on these markers in order to avoid violating these requirements.

The markers indicate the quarter-wave points so that taps and connections can be set up so as to avoid the loss of transmission packets due to impedance mismatches.

Lab Objective

To understand how to install an N connector on a RG8, 50-ohm coaxial cable

Materials Needed

- Lab 11 work sheet

- Cable, RG8, 50-ohm, coax (spool)

- Tension scale/Tape measure

- Multimeter, digital

- Pliers, groove-joint

- Pliers, diagonal cutters

- Soldering iron and stand*

- Crimping tool, pin

- Cable stripper, RG8

- Utility knife

- Connector kit, N, compression (2)

* The use of soldering equipment is recommended, because the act of crimping the center pin renders the connector pin non-reusable. Due to its cost, reusing an N connector is advised.

Instructions

Before beginning this lab, read through all of the following steps at least once, thoroughly.

1. Using the diagonal cable cutters, cut a 5-meter piece of RG8 coaxial cable, and a separate shorter piece (approximately 12 inches) from the spool.

TIP: The 5-meter piece should be cut using the 2.5 meter markings on the cable. This will result in a marker at either end, and a marker in the middle. If there is an odd length left over, use this for the shorter piece.

2. Make sure that the 5-meter and 12-inch (or odd length) cable sections have even cuts on both ends, and temporarily set the 5-meter section aside.

TIP: As with previous procedures, you will first use the 12-inch piece of RG8 cable, or a piece from the scrap heap, for practice purposes before working on the 5-meter section.

3. Examine the specifically designed RG8 cable stripper, as shown in Figure 2-29.

TIP: Notice that the stripper slides apart slightly, retracting the blade far enough to allow the insertion of a RG8 cable for stripping.

4. Retract the blade by pulling on the ring of the RG8 cable stripper with your middle finger, while pushing up on the finger tabs with your forefinger and thumb.

5. Locate the 12-inch section of RG8 cable (or selected scrap piece), insert one end into the retracted stripper, and flush with the first notch of the measuring tab, as shown in Figure 2-30.

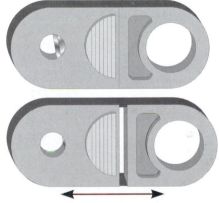

Figure 2-29: Retracting the RG8 Stripping Blade

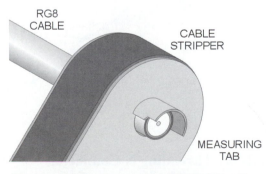

Figure 2-30: Inserting RG8 into the Cable Stripper

6. Release the blade and rotate the stripper in a clockwise direction for at least 6 rotations, or until the stripper begins to spin freely.

7. Now rotate the stripper in a counterclockwise direction for 2 rotations.

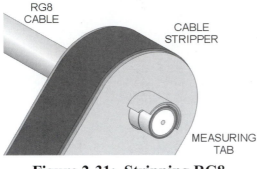

Figure 2-31: Stripping RG8

8. Retract the blade once again, and push the RG8 cable further into the stripper until the cut you just made appears at the first notch on the measuring tab, as shown in Figure 2-31.

9. Release the blade and turn the stripper counterclockwise for 10 rotations before removing the cable from the stripper.

TIP: Even though the required cuts have been made, you may find that the insulation is difficult to remove from the center conductor. If this is the case, use the pliers to get a good grip on the cut portion of the insulation before rotating it loose.

10. From the RG8 cable's first cut area (3/16 of an inch), pull off the jacket along with the braided and foil shields and the center insulator.

11. From the second cut area, remove only the outer jacket. If necessary, use the utility knife to remove it.

TIP: The friction that exists between the jacket and the braided shielding is larger than that of the other cable types you have been using, even though the stripper may have made a good cut around the jacket. If the jacket still refuses to slide, the utility knife may be needed to cut a slit in the outer jacket lengthwise. Then the jacket can simply be peeled off from the cable.

12. Practice stripping on the 12-inch section of RG8 cable (or selected scrap piece) until you are satisfied with your level of performance.

13. When you are ready, locate the 5-meter section of RG8 cable you cut previously and strip both ends in the same manner as you practiced on the 12-inch section of RG8 cable (or selected scrap piece).

14. Examine Figure 2-32 closely, and then thread the nut, flat washer, gland, and ferrule onto the stripped end of the cable exactly as shown.

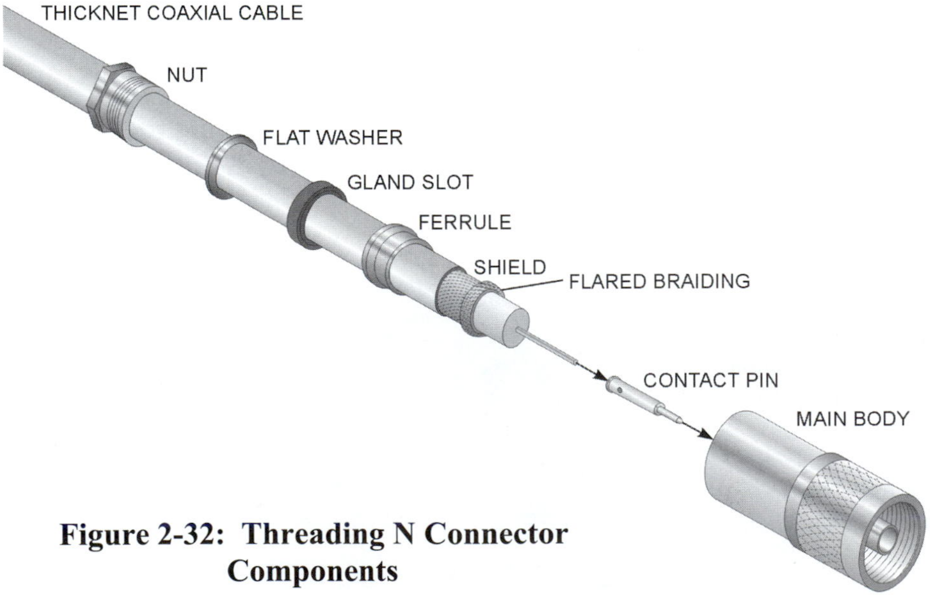

Figure 2-32: Threading N Connector Components

15. Spread the braided shield back to cover the ferrule.

16. Check the length of the center conductor to be sure it is no longer than 3/16 of an inch. If necessary, trim it accordingly.

TIP: Professional installations require that the contact be soldered to the 12 AWG center conductor. If soldering equipment is not available, the pin crimper tool (10-14 slot) can alternately be used to make an emergency connection.

17. Locate the contact pin and slide it over the center conductor, previously trimmed to 3/16 of an inch.

18. Depending on the equipment available to you, solder or crimp the contact in place on the center conductor.

19. Once the contact has been attached to the center conductor, push it into the N connector's body until it snaps into position.

20. Thread the nut into the connector's barrel and tighten it with the pliers.

TIP: You do not need to tighten the nut with all your strength! Firm is tight enough! Once you have an N connector installed on one end of the cable (or any type of cable where soldering is used), it's a good idea to test it before working on the other end of the cable. It will make it easier to troubleshoot the cable knowing that one end has already checked out good.

21. Make an inspection of the opposite end of the cable to ensure that the shielding (braid or foil) is not making contact with the center conductor.

22. Use the multimeter to perform a continuity check on the RG8 cable as it now stands, with one N connector installed.

TIP: Do this for both the center conductor at both ends and the connector bodies or braided/foil shielding at both ends. In both cases, the meter should indicate a short.

23. In addition to the continuity checks, be sure to check for an expected open circuit between the center conductor and the grounded main body, as shown in Figure 2-33.

24. Repeat steps 14 through 23 on the free end of the 5-meter section of RG8 cable.

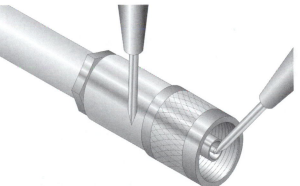

Figure 2-33: Checking for an Open Circuit

25. Once the 5-meter RG8 coaxial cable passes the final continuity checks, set it aside for later use.

26. Check to be sure that all of the tools, cables, and parts have been returned to their storage area, or previous locations.

27. Before leaving your lab work area, check to be sure that you are leaving it in a clean and orderly condition.

Lab 11 Questions

1. Why should the center pin on an N connector be soldered, rather than crimped?

2. What wire gauge is the center conductor of a RG8 cable?

3. Why should the connectors be mounted at the marked locations on a RG8 cable?

4. If the jacket of a RG8 cable won't slide after being properly cut with the stripper, what should you do?

5. How does the cutting blade retract in an RG8 wire stripper?

Lab 12 – Installing a Thicknet Tap

RG8 cable (called Thicknet) is most often used as a network backbone. It's used as an Ethernet bus-type network with a maximum operating distance of 500 meters and a maximum data transfer rate of 10 Mbps.

Connections to the network are accomplished using Thicknet taps. These taps must be attached at one of the marks on the cable jacket (yellow or orange) that occur every 2.5 meters. Recall that these markings indicate minimum distances allowed between taps (quarter-wave points). In this procedure, you will mount a Thicknet tap at the marker on the 5-meter RG8 cable you created previously, and test its continuity.

Lab Objective

To understand how to install a Thicknet tap on a RG8, 50-ohm coaxial cable

Materials Needed

- Lab 12 work sheet

- Cable, RG8 coax, 5-meter (with N connectors installed)

- Multimeter, digital

- Thicknet tap kit

- Thicknet application tool

- Hex wrench (from tap kit)

- Electrical tape

Instructions

Before beginning this lab, read through all of the following steps at least once, thoroughly.

1. Consult Appendix A (page 362) for the following instruction sheet, and read it thoroughly.

   • Instruction Sheet 408-6814

2. On the instruction sheet, identify the braid terminator and the probe assembly. Then, examine these parts from the Thicknet tap kit for excessive wear or deterioration.

3. If you feel that any of these parts should be replaced, notify your instructor.

4. Locate the marker in the middle of the RG8 cable on which you installed the N connectors earlier.

5. On the instruction sheet, perform the Assembly Procedure (beginning at heading 3).

6. When you get to step 6 in the Assembly Procedure, thread the button-head socket screw down until the first edge of the pressure block touches the clamp assembly frame.

TIP: There are four edges to the pressure block, two at each end. As you use the hex wrench to tighten the button-head socket screw, keep an eye on all of them to see which one touches the frame first.

7. When you observe any edge of the pressure block touch the clamp assembly, as shown in Figure 2-34, stop tightening the button-head socket screw.

8. Only complete as far as step 9 in the instruction sheet, because you are not installing this Thicknet tap onto a circuit board.

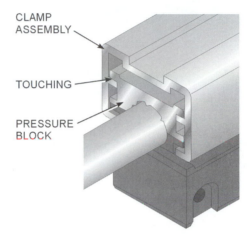

Figure 2-34: Touching Frame with Pressure Block

9. Protect the application tool by placing the cover on its exposed end.

10. Use the multimeter to test the cable to be sure that there is an open circuit between the center conductor and the connector body, as in a previous procedure.

TIP: Although the cable has already passed this test before the tap was attached, it's important to make sure that this measurement has not been compromised by installing the tap.

11. Refer to Figure 2-35 and use the multimeter to test for a short circuit between the center conductor on the RG8 cable, and the center pin of the probe assembly.

12. Next, use the multimeter to test for an open circuit between the main body of an N connector and the braid terminators in the Thicknet tap assembly.

13. Make sure that the multimeter has been turned off and returned to its storage area.

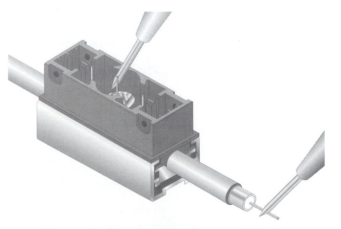

Figure 2-35: Testing Tap and Cable Continuity

14. Slip the dust cover over the tap assembly to protect the contacts inside.

15. Notify your instructor that you have completed this procedure and check to see when you should disassemble the tap from the 5-meter section of RG8.

16. When instructed to do so, disassemble the Thicknet tap kit and return all of its parts to the proper container.

17. Use the electrical tape to wrap the tap area on the RG8 cable, as shown in Figure 2-36, so as not to expose its braid or center conductor to the outside air.

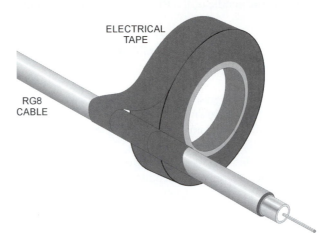

Figure 2-36: Patching the Tap Drill Hole

18. Set the 5-meter RG8 coaxial cable, with installed N connectors, aside for later use.

19. Check that all of the tools, cables, and parts have been returned to their storage area, or previous locations. Before leaving your lab work area, check to be sure that you are leaving it in a clean and orderly condition.

Lab 12 Questions

1. Where should the Thicknet tap be physically located on a RG8 backbone cable?

2. How many braid terminators are there in the tap body?

3. Why must the drill hole be carefully inspected before threading the probe assembly into the tap body?

4. How do you avoid drilling the hole too deeply when using the coring tool?

TP-PMD

Twisted Pair–Physical Medium Dependent (TP-PMD) is a wiring application that provides a way to obtain CAT5 performance by using only the two outside pairs of an 8-position jack, as shown in Figure 2-37. The performance on the two outside pairs will be more than sufficient to meet the requirements of 100Mbps applications (with lower performance ratings on the two unused inside pairs). The physical separation of the two outer pairs reduces near-end crosstalk to a level that meets the TIA/EIA T568A transmission specifications.

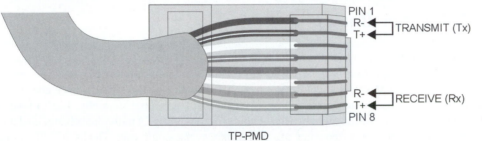

Figure 2-37: TP-PMD Wiring

Though TP-PMD provides CAT5 performance, it is not considered to be CAT5 compliant. It is used for jacks that do not comply with CAT5, to support 100Mbps applications. TP-PMD is a copper version of the **Fiber Distributed Data Interface (FDDI)** Network, using 100-ohm UTP cable and a counter-rotating Token Ring topology. TP-PMD is currently under review of the **American National Standards Institute (ANSI)** to qualify as a standard.

T-Carrier Telecommunication Services

WANs use common carriers (telephone companies) in order to link LANs located in remote buildings, or across the country. For customers who require this linkage, T-Carrier telecommunications have become a standard implementation. Although relatively expensive, **leased** and **dedicated** T-Carrier lines deliver the clean, wide, and continuous telecommunication channels that are needed for today's high-speed networking environments.

History

The earliest voice telecommunications ran a pair of wires from one location to another. Because a call was routed by the use of switchboards, an individual wanting to talk to someone in another city would wait for the local switchboard operator, working at the local Central Office (CO), to connect to a remote switchboard operator (working in the remote CO), who would then complete the connection. This type of operation caused several major problems, one being the maintaining of a usable signal to send over the wires. The development of analog signal amplifiers took care of this problem, but it also created a few of its own. The voice signal was not the only thing that was amplified—the interfering noise was also.

Another concern was the need for large amounts of wire running between the widely separated central offices. With the development of **multiplexing**, many signals (individual calls) could be transmitted over a single pair of wires, greatly reducing the amount of wire used for telecommunications. Subsequent technological advances within the telecommunications industry have brought about the use of digital signal transmission, as opposed to the older and noisier analog systems. Digital telecommunications has provided the age of clean and clear signal transmission, because the telephone receiver needs to distinguish only between on and off signal conditions. This is where the T-Carrier multiplexing scheme comes in.

T-Carrier Basics

The long distance carriers have reconfigured their networks to carry both digital voice and data. The multiplexing scheme chosen to handle the thousands of long-distance calls is **Time-Division Multiplexing (TDM)**, named **T-Carrier** by the long-distance companies. T-Carrier is a hierarchical system, with T1 as the first level of the hierarchy. The T1 level, in its simplest terms, can be thought of as a cable box containing 24 individual channels (data or voice) transmitting at a frequency of 1.544 Mbps, over a single wire. The basic T1 line then, consists of a 1.544 Mbps composite digital signal, divided into 24 equal parts. These 24 equal parts are referred to as **Digital Signal Level 0 (DS0)** channels, which make up one T1 signal. T1 TDM assigns different time intervals for each of the 24 signals to be sent.

Imagine two Ferris wheels, side by side, rotating in opposite directions. Then imagine one person on each Ferris wheel. The two people would only be able to communicate when they were directly across from each other. Now imagine that there are many people on both Ferris wheels, continually talking to each other, with each conversation composed of bits and pieces. These bits and pieces can be compared to **Pulse-Coded Modulation (PCM)**. PCM is produced by converting an analog waveform into a digital bit-stream, and is the most widely used digital modulation technique for multiplexed hierarchies of the common carriers.

If the Ferris wheel begins to speeds up, eventually the two people might feel as if there was no gap between the bits and pieces of their conversation. That is, they would be able to communicate normally, in spite of the fact that their conversation is actually composed of bits and pieces. The spinning of the two Ferris wheels can be compared to the multiplexing operation of TDM, while the creation of the bits and pieces can be compared to the PCM process of converting an analog signal into a digital signal. This digital signal, in the form of binary numbers, is what is transmitted over the T1 carrier.

The T1 digital communications system, introduced in the 1960's, was originally designed to carry 24 voice channels between two central offices. The central offices would first use PCM to convert the analog voice signals into digital. To do this, each analog signal would be sampled at 8,000 times per second, with the amplitude of the signal being converted into an 8-bit, binary numbering pattern, and then sent across the T1 carrier.

If there are 24 sampled channel slots per frame, with each channel containing 8 bits (plus one additional bit per frame for synchronization), this makes a total of 193 bits per frame (24 sample channels X 8 bits, plus 1 bit per frame for synchronization). In order for a receiver to reproduce an analog voice signal, the information in each channel must be updated 8,000 times per second. This means that 8,000 frames must be sent for each 1-second time span, resulting in a T1 data rate of 1.544 Mbps (8,000 frames X 193 bits). That is the basic theory of T1 operation. The T-Carrier multiplexing scheme doesn't end with T1, however.

T-Carrier Multiplexing

As shown in Figure 2-38, the T1 level contains 24 channels (either data or voice) transmitting at a frequency of 1.544 Mbps.

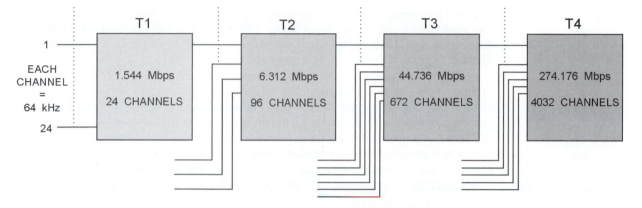

Each channel is composed of digitized information sampled at a rate of 8 kbps. Four T1 carriers are multiplexed together to form the 96-channel T2 level, transmitting at 6.312 Mbps. The T3 level is composed of seven T2 carriers, and handles 672 channels transmitting at 44.736 Mbps. The highest level of the hierarchy is T4, operating at 274.176 Mbps. It contains six T3 carriers multiplexed together for a total count of 4,032 channels.

Figure 2-38: T-Carrier Multiplexing Scheme

CHAPTER SUMMARY

This chapter dealt with the various methods used to classify the different types of telecommunications cable, and the performance capabilities of each type. You learned that the quality of the materials used in a cabling installation, the amount of work expected from the system, and the total cost to the consumer must all be taken into account.

You read about how level ratings were originally conceived, as a method of classifying cables and jacks according to their performance characteristics. You learned some details about Level 1 and Level 2 ratings, even though modern cabling installations use cables with much higher capabilities.

While introducing the newer category rating systems, this chapter explained to you the difference between Megahertz (the speed that data travels), and Megabits per second (the actual amount of data transmitted each second). You read how the original UL level system became the TIA/EIA category system, and how telecommunications cabling is now classified as category (CAT) 1 through 5, with UL categories 3 through 5 corresponding exactly to the TIA/EIA T568A categories (CAT3, CAT4, and CAT5).

You read about how the various standards organizations go about examining and adopting the proposed standards for telecommunications, and the way in which subsequent technical systems bulletins modify the existing standards. You not only completed a review of CAT3, CAT4, and CAT5 cabling specifications, but you also had an opportunity to learn about the expected performance capabilities of several proposed standards, such as CAT5E, CAT6, and CAT7.

You learned that unless the selected cabling was used with the prescribed connectors and associated equipment, the installation itself could not be considered as being compliant with its rated category. You were shown how the amounts of signal attenuation, near-end crosstalk (NEXT), far-end crosstalk (FEXT), reflection, and return loss determine the category of compliance for the system.

You were introduced to the two major types of networks (LANs and WANs), and were acquainted with details concerning the major differences between them. The various networking applications (Ethernet, Token Ring, 100-ohm STP, 100VG-AnyLAN, 100baseT, Gigabit Ethernet, 1000baseT, ATM, Coaxial Cable, and TP-PMD) were described and compared.

Following the discussion on token ring networks, you performed Lab 8, which gave you some experience installing IBM-type UDC connectors on 150-ohm STP cable. The discussion on coaxial cable immediately led to Lab 9, where you learned how to install and test a BNC connector on some RG58 coax. After practicing on the BNC connectors, you went to Lab 10 to learn how to install F connectors on RG6, 75-ohm coax and test the created cable for continuity. Continuing with Lab 11, you learned how to install N connectors on RG8, Thicknet cable and to test the cable you made.

For Lab 12, you installed a Thicknet tap on the RG8 cable you made during the previous lab. You also tested the continuity between the cable and the tap, before disassembling the tap and returning it to its storage area.

After examining the TP-PMD wiring application, you read about the T-Carrier multiplexing scheme used by the common carriers.

REVIEW QUESTIONS

The following questions test your knowledge of the material presented in this chapter:

1. Which Telco Level Rating applies to the standard telephone system?

2. Which wiring category cleanly transmits 100 Mbps at 100 MHz?

3. What is the maximum bandwidth of a Level 1 telecommunications system?

4. What type of wire is typically used for Level 1 wiring?

5. Where would Level 2-rated components be used?

6. What type of termination method is typically used for Level 1 wiring?

7. Are a 100MHz line and a 100Mbps transmission rate the same thing?

8. What is attenuation, and what is its unit of measurement?

9. What is ATM and where is it used?

10. What type of cable is specified for use in a 100baseT network?

CABLE INSTALLATION AND MANAGEMENT

Upon completion of this chapter and its related lab procedures, you should be able to perform these tasks:

1. Define Horizontal cabling.

2. Describe Backbone cabling.

3. Define the term Ground and describe different types of grounds.

4. List different types of backbone cable systems, and describe their attributes.

5. Describe a crossconnect panel.

6. List the characteristics of a proper telecommunications closet.

7. Describe procedures associated with proper cable management.

Cable Installation and Management

INTRODUCTION

Telecommunications installations today would be exactly as they were before the 1970's, if there weren't any standards. Standards allow for any modern, specification-aware company to arrive at a job site, to perform an installation, or to successfully troubleshoot an existing cabling system. This situation allows the customer to be confident that the job is going to be done correctly, and because the standards are practiced throughout the industry, just about any company selected will be able to come in and maintain an installation that was performed by another. This customer confidence is an important consideration in the telecommunications industry today, especially with the greater degree of competition in the market than ever before.

Standards are carefully studied recommendations that if followed will extend the life of the system without the need for extensive amounts of maintenance. The life of a properly installed system will extend way beyond that of one conceived in the absence of any standards.

A system that was simply "slapped up on the wall" has relatively little chance in gaining that all-important customer satisfaction. In order to gain the reputation for performing quality installations, it is very important for you to understand exactly what the standards are, and to follow them as closely as possible. Be sure that you know which standard deals with what situation, so that you, as an installer, may refer to the applicable standard needed for the particular installation that you are currently working on.

TIA T568A COMMERCIAL BUILDING COMPLIANCE

Large differences exist between the telecommunications cabling requirements intended for residential installations and those intended for use within commercial buildings. Perhaps, as the need, or desire, for more robust data services grows within the residential sector, the cabling requirements for private homes may someday rival those already being standardized in the business community. But for now, the governing bodies within the telecommunications industry have concentrated on organizing the more advanced cabling standards for the commercial environment. This is probably because the data communications industry itself is a commercial entity, and the bottom line is vitally dependent on the quick and reliable transfer of all forms of information.

When you are involved with the cabling installation for commercial buildings, whether they are single units or many interconnected structures, your knowledge of the applicable standards now generally applied by the cabling industry is a crucial aspect of your job performance. For work in the United States, the ANSI/TIA/EIA T568A standard will dictate what you can or cannot do, and for European work sites, the ISO/IEC 11801 directives will apply.

Structured Cabling

Structured cabling

In the past, buildings could contain different cabling systems for various forms of communications. For example, there might have been block wiring for voice, coaxial for Ethernet, and multi-pair for RS232 interfaces. **Structured cabling** is a means of replacing all of the different cabling scenarios with a single cabling system, which covers an entire building, or a related group of buildings, for all voice and data (including cable TV and video) requirements.

A structured cabling system consists of modular wall outlets, providing the user with RJ45 interfaces. These user outlets are outfitted with either one or two RJ45 connectors mounted in a standard, single-gang face plate. Alternately, they can exist as single snap-in modules fitted into floor boxes, using single-gang face plates (up to two modules) or dual-gang face plates (up to four modules). If only one RJ45 interface is supplied, a second interface type is usually prescribed (such as an F video connector) in order to provide a minimum of two types of service to each outlet. Figure 3-1 depicts a generalized structured cabling system.

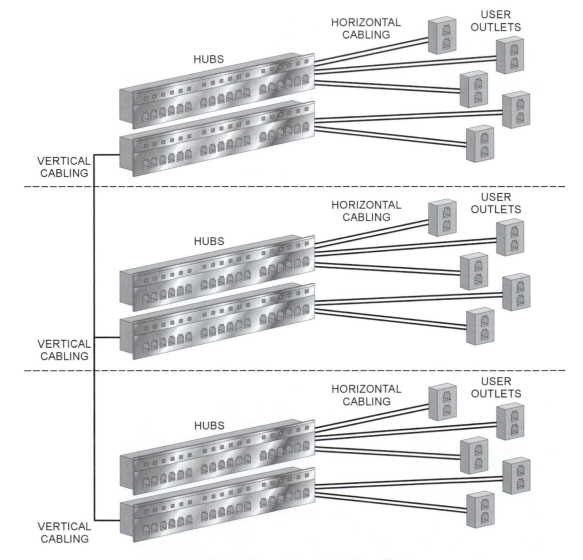

Figure 3-1: Structured Cabling System

Notice that each user outlet is wired back to a hub using an individual cable. This cable, known as the "horizontal cabling," contains four twisted pairs, and in most cases meets the CAT5 specifications. Horizontal cabling can be either Unshielded Twisted Pair (UTP), or Shielded Twisted Pair (STP). The cable itself is connected to the back of the user outlet by means of an **Insulation Displacement Connector (IDC)**. If necessary, an external adapter can be attached to the outlet to provide any required format conversion for the equipment being attached to it.

In a true structured cabling system, the horizontal cabling and user outlets are identical regardless of the services being utilized, so that any outlet can be configured for voice, Ethernet, RS232, video, or other service. When a user's requirements change, the service provided at the outlets can be changed as well by simply repatching the configuration in the **Equipment Room (ER)** or the **Telecommunications Closet (TC)**, where the hubs are located. The individual hubs are connected using "vertical or riser" cable (also called backbone).

Vertical cabling consists of either UTP, STP, or fiber optic cable (multi-mode or single-mode). In most modern systems, optical cables are used for the data backbone cables, and multi-pair copper cables are used for the voice backbone cables.

When a structured cabling system is installed, the floors are usually "flood wired," so that the outlets are positioned using a grid layout. The layout will specify the number of outlets and the distances between them, rather than map them to individual user positions. This provides the flexibility to quickly rewire a workstation when future changes are made to the layout of the building, without having to recable the original installation.

Insulation
Displacement
Connector (IDC)

Equipment Room (ER)

Telecommunications
Closet (TC)

TIA/EIA T568A

First released in 1991, TIA/EIA T568A specifies a generic telecommunications cabling system for commercial buildings that will support a multi-product, multi-vendor environment. Since that time, TSB 36, TSB 40A, and TSB 53 (addendums, including cable and connecting hardware performance, and additional STP specifications) were released. This information, as well as additional information on optical fiber and link performance were added to TIA/EIA 568, circulated for industry review (as SP 2840 then SP 2840A drafts), and approved for release as TIA/EIA T568A.

The establishment of this standard is recognized as a substantial accomplishment, because it addresses cabling topologies, distances, channel media, and connectors. Its global acceptance includes our own federal government, as well as Canadian and other international standards organizations. The latest edition of TIA/EIA T568A was released in 1995.

The overall scope of the TIA/EIA T568A specifications are intended for "office oriented" telecommunications installations. One of the main requirements for a structured cabling system calls for a usable life in excess of 10 years. Other criteria covered include the quality of the media used (cable and connecting hardware), the performance of the system, the network topology, the maximum cabling distances allowed, the accepted installation practices, the user interface, and the performance of each communication channel.

The purpose of the TIA/EIA T568A standard is to enable the planning and installation of a structured cabling system for commercial buildings.

It establishes the performance and technical criteria for various generic voice and data telecommunications cabling system configurations for interfacing their respective elements. And because the standard provides direction for the design of telecommunications equipment and cabling products intended to specifically serve commercial enterprises, these configurations must support multi-product, multi-vendor environments. The products must be capable of supporting the diverse telecommunications needs of a building's occupants.

Although the performance and technical criteria for various types of cable, the connecting hardware, and the design/installation of a cabling system are specified, keep in mind that the TIA/EIA T568A specification concerns itself solely with the cabling system located in a building. Other issues must be directed towards those standards that pertain directly to them.

HORIZONTAL CABLING

Horizontal cabling is defined by the TIA/EIA as being the cabling between, and including, the Telecommunications Closet (TC) or Equipment Room (ER), the Horizontal Crossconnect (HC) located in the TC, and any Telecommunications Outlet or connector (TO) located in any Work Area (WA). Basically, it extends from the telecommunications outlet/connector to the horizontal crossconnect. Horizontal cabling is designed as a generic cabling system that will accommodate most applications. For the European ISO/IEC 11801 specifications (roughly equivalent to TIA/EIA T568A), the equivalent cabling element to the Horizontal Crossconnect (HC) is called the **Floor Distributor** (**FD**). It's a good idea for cabling technicians to learn the abbreviations (acronyms) to these various terms, because many cabling diagrams label the elements or components using them. Figure 3-2 depicts the structural components of a horizontal cabling system, and the configuration of a multi-floor network.

Beginning where the user plugs in his/her terminal and ending at a centrally located point called a **distribution frame** in the TC, the horizontal cabling includes the horizontal cables themselves, the premises equipment cords, the telecommunications outlet/connector in the work area, the mechanical termination, the patch cords or jumpers (located in the telecommunications closet) that comprise the horizontal crossconnect, the transition point equipment, and the work area equipment cords.

Telecommunications Closets and Equipment Rooms

A **Telecommunications Closet** (**TC**) is a small room, or area, which houses the telecommunications wiring and its accompanying equipment. The TC contains the backbone-to-horizontal crossconnect, and it may also contain the network demarcation, or the Main Crossconnect (MC). It is usually an enclosed space, within the confines of a building, used to locate all of the telecommunications cable terminations and any of its associated hardware. Normally, there is a minimum of one telecommunications closet per floor, and that closet is set up to serve that floor only.

By contrast, an **Equipment Room** (**ER**) is a centralized space for telecommunications equipment serving the occupants of the entire building. Its design aspects are specified in TIA/EIA 569, and it usually houses equipment of higher complexity than that of a TC. Any or all of the functions of a telecommunications closet may be provided by an equipment room.

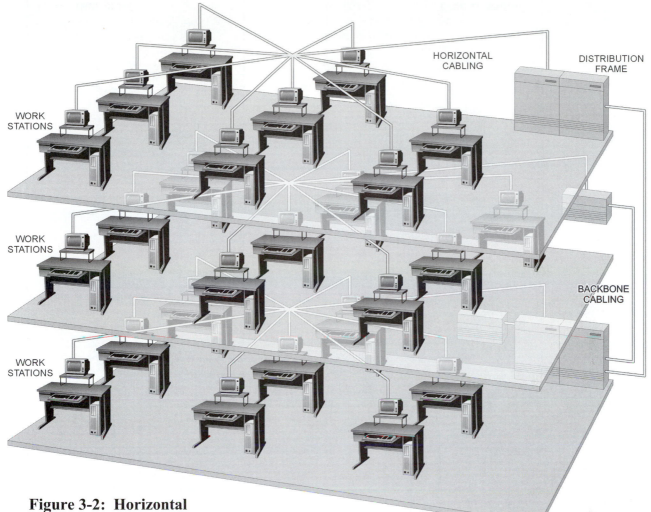

HORIZONTAL CABLING

DISTRIBUTION FRAME

WORK STATIONS

WORK STATIONS

BACKBONE CABLING

WORK STATIONS

Figure 3-2: Horizontal Cabling System Components

A typical telecommunications closet/equipment room layout is shown in Figure 3-3, on the following page.

In addition to being centrally located, a telecommunications closet must have 3/4-inch thick plywood backboard, of eight feet in height, installed on at least two of its walls, for mounting the equipment. The closet itself must be built to accommodate the current and future telecommunications needs of the building. There should be no false ceilings, and the lighting must be sufficient for a technician to be able to read the labels on cables and distinguish between the different colors that code the wires. The **TIA/EIA 569** standard contains all of the design specifications for a telecommunications closet, which should be dedicated strictly to the telecommunications functions and the accompanying support facilities. Although the minimum of one telecommunications closet per floor is standard, additional closets should be provided for each service area up to 10,000 square feet (930 square meters), or when the required horizontal cabling distance exceeds 300 feet (90 meters).

In cases where there is the need for multiple TCs on a single floor, the recommendation is to interconnect the closets with at least one trade size 3 conduit (3/4-inch), or equivalent. The closet doors must be at least 36 inches wide to allow for the entry and exiting of various pieces of telecommunications equipment.

TIA/EIA 569

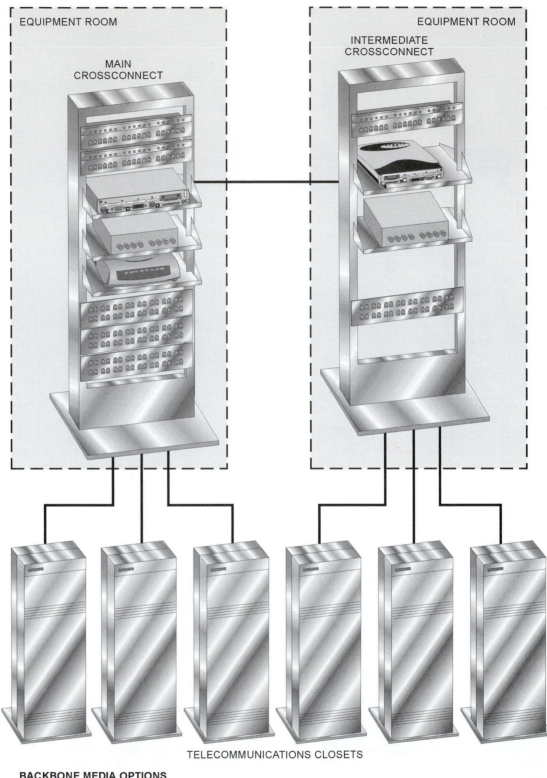

These doors must open outwards, to comply with standard electric codes, and they should possess locks to allow entry only to those personnel authorized to be there. All TCs must be kept neat to allow unfettered access to each of the different components located there.

A certified electrician must install all of the electrical outlets, and there must be enough dedicated outlets to support the equipment that is housed in each closet (a minimum of two dedicated, duplex electrical outlets on separate circuits, placed at least 6 feet/1.6 meters apart). Additional outlets should maintain the 6-foot interval around the perimeter walls.

Telecommunications Closet Sizing

In Table 3-1, the typical TC floor space requirements are displayed for a telecommunications closet serving various office areas, based on one workstation per 100 square feet (9.3 square meters) of office floor space.

Serving Area		Telecommunications Closet	
Square Meters	Square Feet	Square Meters	Square Feet
930	10,000	10.2	110
744	8,000	8.4	90
465	5,000	6.5	70

**Table 3-1:
TC Vs. Serving Area
Floor Space**

Horizontal Crossconnects

In general, a **crossconnect** is a facility that enables the termination of cable elements and their interconnection, and/or crossconnection, primarily by means of a patch cord or jumper. The general idea behind a crossconnection describes the wiring scheme that exists between cabling runs, subsystems, and equipment cabinets. Crossconnect panels utilize various patch cords/jumpers to attach the connecting hardware on each end of a termination.

Typical crossconnect facilities consist not only of the crossconnect jumpers or patch cords, but also include the terminal blocks and patch panels that are wired directly to the horizontal or backbone cabling. Actually, there are three types of crossconnects: the **Main Crossconnect (MC)**, which is usually physically located in the ER, the **Intermediate Crossconnect (IC)**, and the **Horizontal Crossconnect (HC)**. The IC and HC can both be physically located either in the same telecommunications closet, or in separate ones.

These crossconnect types use equipment cables that consolidate several ports on a single connector, such as a 25-pair hub, and are terminated on dedicated connecting hardware (system specific). At the hub, the individual 4-pair cables from the user outlets are terminated on patch panels with rear IDCs for terminating the 4-pair cables and providing an RJ45 presentation on the front for patching.

crossconnect

Main Crossconnect (MC)

Intermediate Crossconnect (IC)

Horizontal Crossconnect (HC)

The patch panels are usually mounted in standard 19-inch racks, wall-mounted or freestanding. Although RJ45 patch panels usually come in multiples of 16 connectors, panels containing 32 and 48 RJ45 connectors are also common.

Crossconnect panels are usually mounted on a plywood backboard in the TC, and are organized by cable type: backbone, horizontal, and equipment. TIA/EIA 606 recommends that the TC crossconnect should use the color scheme as shown in Table 3-2.

Table 3-2: Wire Colors for TC Crossconnect

Color	Crossconnect Field
Orange	Central Office Cable (demarcation)
Green	Customer Side of Demarcation
Purple (USA) White/Silver (CANADA)	Common Equipment (PSBs, LANs)
White (USA) Purple (CANADA)	Main Crossconnect Backbone Cable
Gray	Intermediate Crossconnect Backbone Cable
Blue	Horizontal Cable
Brown	Inter-Building Backbone
Yellow	Auxiliary Circuits, Alarms, Security
Red	Key Telephone Systems

The horizontal crossconnect is a linkage point of horizontal cabling to other cabling types (such as horizontal or backbone), or equipment, as shown in Figure 3-4, and normally serves one floor of a building.

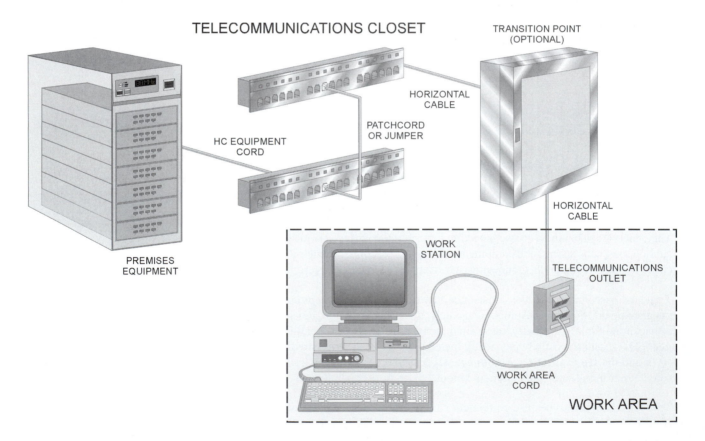

Figure 3-4: Horizontal Crossconnect

Horizontal Crossconnect Distances

basic link

The **basic link** begins at the work area wall plate and ends at the horizontal crossconnect in the TC. The maximum distance permitted for this basic link horizontal cabling is 90 meters (295 feet). These are the cables that run from the termination points, at the horizontal crossconnect in the telecommunications closet, to the outlets and/or connectors mounted in the walls at the various work areas on a single floor of a building, as shown in Figure 3-5.

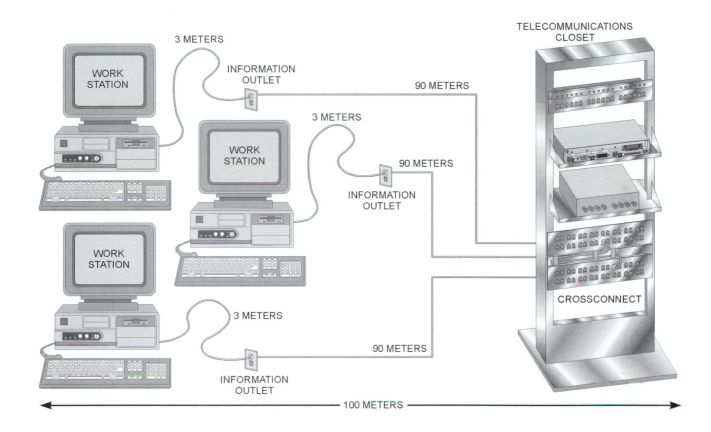

Figure 3-5: Horizontal Cable Run Distances

Horizontal Jumpers and Patch Cords

Jumpers

Jumpers can be single, or an assembly of, twisted pairs, without connectors, that are used to join telecommunication circuits/links at the crossconnect panels. Since they have no connectors of their own, they are punched down in order to make a connection. They can also be in the form of conductive clips, for making connections at the termination blocks.

Patch cords

Patch cords are lengths of cable, with connectors at one or both ends, also used to join telecommunication circuits/links at the crossconnect panels. The allowance made for work area equipment cords is 3 meters (9.8 feet), while that for the combined length of patch cords, jumpers, equipment cables and cords in the horizontal crossconnect, including the Work Area (WA) equipment cords, is 10m (33 feet).

Horizontal crossconnect jumpers/patch cords in the telecommunications closet should be less than 7 meters (23 feet) long. However, in order to maximize the reliability of the installation, it is recommended that no one cord should exceed 6 meters (20 feet) in length. A maximum of two patch cords is allowed per horizontal cable run. Keep in mind that although patch cords are a vital link in any cabling installation, inferior devices have been known to wreak havoc on data transmission. They have also been found to cause crosstalk and attenuation problems.

Horizontal Cabling Types

TIA/EIA T568A requires that there are at least two cable runs from the telecommunications closet to each individual wall outlet and/or connector. Most outlets will accommodate both voice and data telecommunications within the lifetime of the system.

The types of cables that are recognized by TIA for horizontal cabling are as follows:

- 4-pair, 100-ohm unshielded twisted-pair (UTP) cables.

- 2-pair, 150-ohm shielded twisted-pair (STP) cables.

- 2-fiber, 62.5/125-micron optical fiber cable.

Telecommunications Outlets and Connectors

As mentioned above, the usual configuration for a **Telecommunications Outlet/Connector (TO)** is for each work area to be provided with a minimum of two information outlet ports, one for voice and one for data. The outlet cabling choices are shown in Figure 3-6.

Recall that the horizontal wiring from the telecommunications closet to each user outlet must be capable of providing voice and data capabilities. If possible, enhanced CAT5 cable (currently proposed as a standard to TIA/EIA) should be used for each individual jack. If video services are included, its wiring should be with RG59.

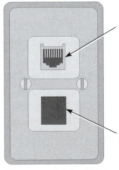

100 OHM UTP 4-PAIR FOR VOICE T568A OR T568B WIRING CAT3 OR HIGHER

100 OHM UTP 4-PAIR, 150 OHM STP 2-PAIR OR 62.5/125mm FIBER FOR DATA 2 STRAND

Figure 3-6: Telecommunications Outlet Cabling Choices

Whenever possible, the wiring from the outlet to the ceiling space should be concealed within the wall. If this objective is not attainable, the wiring should be run behind surface-mounted plastic raceway. Wiring extended above accessible ceilings may be run in conduit, cable tray, or "J" hooks. However, any wiring that is extended above plaster or drywall ceilings must be in placed in conduit, with accessible junction boxes as specified by the applicable electric codes.

As shown in Figure 3-7, the 8-position modular jack pair assignments for UTP are described by the TIA/EIA T568A and TIA/EIA T568B standards. The difference between the two is the locations for pairs 2 and 3, with TIA/EIA T568B being the preferred arrangement in situations that require high data throughput.

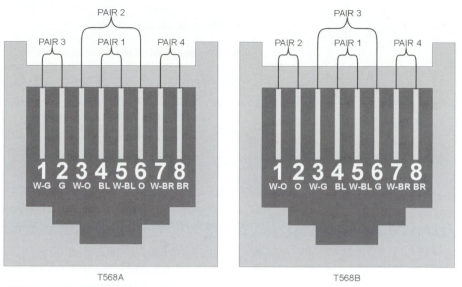

Figure 3-7: UTP Pair Assignments for 8-Position Modular Jacks

The outlets for telecommunications services should be provided throughout a facility, and they should all sport a common face plate style, regardless of the combination of voice, data, and video jacks required. Given the current and near-future telecommunications requirements, it's difficult to imagine any justification for voice and data jacks being less than CAT5-compatible, with the video jacks being "F" connectors. This will be doubly true when, shortly, new installations will require, as a minimum-quality standard, enhanced CAT5 cabling throughout.

The exact location for each telecommunications outlet is best determined during the design process, although most modern designs call for many more outlets than will actually be used at any one time.

Lab 13 – Installing a Punch-down Modular Outlet

Lab Objective

To understand how to install an RJ45 punch-down modular outlet on a section of 100-ohm UTP cable

Materials Needed

- Lab 13 work sheet
- Cable, CAT5 UTP, 100-ohm, non-plenum (spool)
- Punch-down outlet kit, RJ45/T568A (1)
- Termination plug, RJ45, unshielded (1)
- Cable, CAT5 UTP, 100-ohm, non-plenum (with RJ45 plugs installed)
- Wall plate, dual, RJ45
- Stuffer caps (2)
- Punch-down tool, plastic

- Punch-down tool, steel

- Crimping tool, RJ45

- Cable tester, continuity-type

- Pliers, diagonal cutters

- Wire stripper, CAT5

Instructions

Before beginning this lab, read through all of the following steps at least once, thoroughly. As you perform the steps in this procedure, you should vary the way in which you make the connections so that you learn how to use both the plastic and steel-blade punch-down tools.

Because the stuffer caps allow you to install the wires without a punch-down tool, it is recommended that you use only one of the stuffer caps for punch-down purposes, in order to get the experience using the two types of punch-down tools. Afterwards, you can put the remaining stuffer cap in place.

The RJ45 modular outlet that you are wiring in this procedure will be configured for the T568A specification. The internal wiring of the outlet does not necessarily match the order of wires that are punched down. Instead, the internal configuration is marked on the underside of the modular jack itself and on the tiny printed circuit board inside its body.

1. Use the diagonal cable cutters to cut a 5-foot piece of CAT5 UTP cable from the spool. Then, set the cable section aside.

2. Find the following instruction sheet in Appendix A (page 357), and read the pages thoroughly.

    - Instruction Sheet 408-3354

3. Once you have read the instruction sheet thoroughly, go ahead and install one end of the 5-foot section of 100-ohm, CAT5 UTP cable to the RJ45 modular punch-down outlet jack, as described.

TIP: The end of the cable needs to be prepared first. Once you have done that, simply disregard the instructions that pertain to connector or cable types that are not included in your parts inventory. Keep in mind that the unshielded twisted pair cable does not require the grounding clip to be installed. Use the stuffer cap for one side of wires, and the punch-down tools for the other side. Once the wires have been punched down, go ahead and install the second stuffer cap so that both halves of the jack look identical.

4. Prepare the free end of the 5-foot, 100-ohm UTP cable, and install an RJ45 un-shielded modular plug according to the T568A specification.

5. Locate the 5-foot cable with RJ45 plugs installed on both ends, which you made in Lab procedure 6.

6. Plug one end of the RJ45 cable from Lab procedure 6 into the modular jack that you just installed.

7. Plug its remaining end into the cable tester's main body.

8. Plug the remaining end of the modular jack cable (with the RJ45 plug installed) into the cable tester's detachable terminator, and check Figure 3-8 for the setup wiring.

9. Turn the cable tester on and check the modular cable that you just created.

TIP: Recall that the RJ45 cable from Lab procedure 6 was already checked. Any problems noted on this test should point to the newly created cable.

10. Make any repairs that are necessary until all readings are good on the cable tester before going on.

11. Locate the wall plate, and orient it so that the "UP" arrow points in the direction shown in Figure 3-9.

12. Position the modular jack block as shown in Figure 3-9, and gently push it into the top opening of the wall plate until it locks or snaps into place.

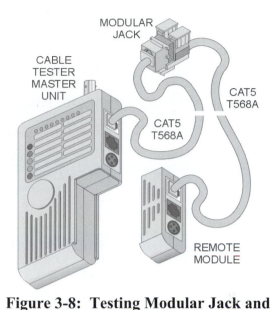

Figure 3-8: Testing Modular Jack and Cable Continuity

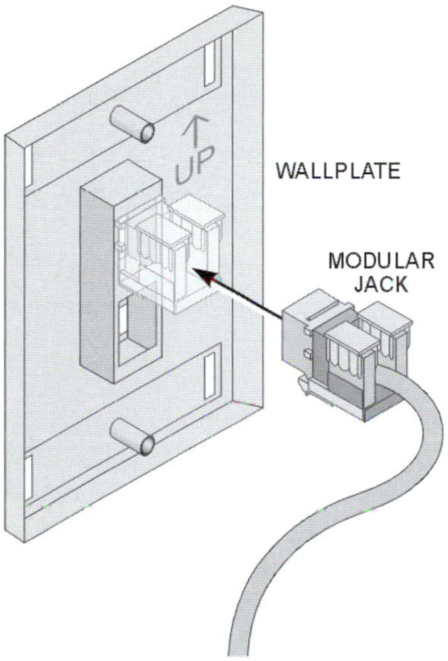

TIP: Don't worry about installing the modular jack upside-down, because the design of the wall plate will prevent this.

13. Set the RJ45 modular jack cable and wall plate combination in a safe place, for later use.

14. Check to be sure that all of the tools, cables, and parts, have been returned to their storage area, or previous locations.

15. Before leaving your lab work area, check to be sure that you are leaving it in a clean and orderly condition.

Figure 3-9: Fitting the Outlet into the Wallplate

Lab 13 Questions

1. What does the instruction sheet 408-3354 indicate about keeping the wires twisted to within specific distances from the termination?

2. When using the steel-tip punch-down tool, what impact setting is recommended?

3. Are the wires cut on the inside or the outside of the wiring block?

4. How are dimensions indicated on the instruction sheet?

Lab 14 – Assembling Edge-Connector Modular Outlets

Lab Objective

To understand how to assemble an RJ45 edge-connector modular outlet on a section of 100-ohm UTP cable

Materials Needed

- Lab 14 work sheet

- Cable, CAT5 UTP, 100-ohm, non-plenum (spool)

- Edge-connector outlet kit, RJ45/T568A (1)

- Edge-connector insert, RJ45/T568A (1)

- Edge-connector modular housing, RJ45/T568A (1)

- Termination plug, RJ45, unshielded (1)

- Cable, CAT5 UTP (with RJ45 plugs installed)

- Stuffer caps (2)

- Punch-down tool, steel

- Crimping tool, RJ45

- Cable tester, continuity-type

- Pliers, diagonal cutters

- Wire stripper, CAT5

- Pliers, groove-joint

Instructions

Before beginning this lab, read through all of the following steps at least once, thoroughly.

Again, the RJ45 modular outlet that you are wiring in this procedure will be configured for the T568A specification. As was the case with the punch-down modular outlet, the internal wiring of the edge-connector outlet does not necessarily match the order of wires that are punched down. Instead, the internal configuration is indicated on the separate label included with the bag containing the jack. The underside of the modular jack itself, or the tiny printed circuit board inside its body, may or may not contain configuration coding.

1. Use the diagonal cable cutters to cut a 5-foot piece of CAT5 UTP cable from the spool. Then, set the cable section aside.

2. Find the following instruction sheet in the Appendix A (page 343), and read the pages thoroughly.

    - Instruction Sheet 408-3232

3. Once you have read the instruction sheet thoroughly, locate the RJ45 modular edge-connector outlet and its matching housing from the kit.

TIP: Although you will be using 100-ohm UTP cable, this type of RJ45 modular outlet can accommodate STP cable of both the 100-ohm, 8-wire and 150-ohm (Token Ring), 4-wire systems. When installing the connector within the matching housing, your parts may either snap or screw together. For screw-type connections, follow the alignment of parts shown in Figure 3-10 when removing or installing the connector.

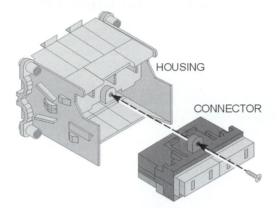

Figure 3-10: Aligning the Connector and Housing

4. Prepare one end of the 5-foot, 100-ohm CAT5 UTP cable you set aside earlier, as shown in Figure 3-11, by removing 2 inches of jacket.

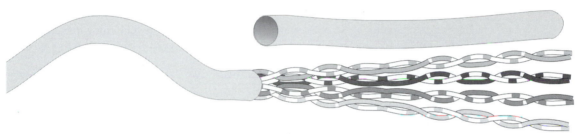

Figure 3-11: Removing 2 Inches of UTP Cable Jacket

5. Locate the edge connector and its housing to identify which type of edge connector you have been supplied.

TIP: A 110Connect arrangement comes with the edge connector already installed in the port housing and remains this way during the wiring. The AMP-BARREL arrangement requires that the edge connector be terminated prior to its being mounted into the housing. Both types of installations are covered in the instruction sheet.

6. Determine which type of connector you are using, and follow the instructions from sheet 408-3232 in Appendix A (page 343) for that type.

TIP: Instructions for the 110Connect arrangement are shown first on the instruction sheet.

7. Regardless of which type of edge connector you are using, be sure to follow the color coding marked on the edge connector or stuffer cap.

TIP: Don't forget to install the two stuffer caps on the edge connector, regardless of how the wires were punched down. They help to protect the connections against exposure.

8. When you have completed the wiring, and secured the edge connector to the housing (already done for you with 110Connect types), install the jack insert by pushing it into the port, as shown in Figure 3-12, aligning its latches with tabs on the housing.

TIP: As a rule, any housing that will not currently carry an active service would be outfitted with a blank insert, until a future need for expansion occurs. You will not be installing a face plate as part of this lab procedure.

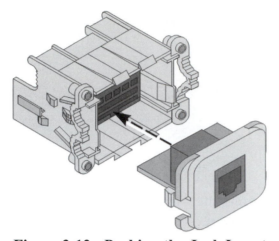

Figure 3-12: Pushing the Jack Insert

9. Prepare the free end of the 5-foot, 100-ohm UTP cable, and install a RJ45 unshielded modular plug according to the T568A specification.

10. Locate the 5-foot cable with RJ45 plugs installed on both ends, which you made in Lab procedure 6.

11. Plug one end of the RJ45 cable from Lab procedure 6 into the modular jack outlet that you just installed.

12. Plug its remaining end into the cable tester's main body.

13. Plug the remaining end of the modular jack cable (with the RJ45 plug installed) into the cable tester's detachable terminator, and check Figure 3-13 for the setup wiring.

14. Turn the cable tester on, and check the modular cable that you just created.

TIP: Recall that the RJ45 cable from Lab procedure 6 was already checked. Any problems noted on this test should point to the newly created cable.

15. Make any repairs that are necessary until all readings are good on the cable tester before going on.

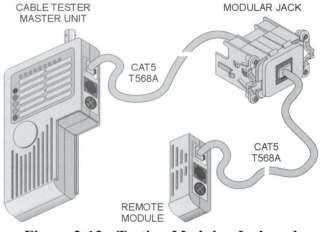

Figure 3-13: Testing Modular Jack and Cable Continuity

16. Set the RJ45 modular jack cable/edge connector in a safe place, for later use.

17. Check to be sure that all of the tools and parts have been returned to their storage area, or previous locations.

18. Before leaving your lab work area, check to be sure that you are leaving it in a clean and orderly condition.

Lab 14 Questions

1. After reading the instruction sheet 408-3232, which edge connector (110Connect or AMP-BARREL) would appear to be the easiest to install?

2. Which type of edge connector (110Connect or AMP-BARREL) requires termination prior to being installed in the housing?

3. Why is it a good idea to lace the middle wires into the edge connector first?

4. Which type of edge connector comes already installed in the port housing?

Work Area Cabling

The Work Area (WA) components, shown in Figure 3-14, extend from the telecommunications (information) outlet to the station equipment. The wiring interconnections are designed to be relatively simple, so that moves, adds, and changes within the work area are easily managed. This allows the telecommunications user outlet to serve as the main work area interface to the cabling system. A maximum distance of 3 meters (9.8 feet) is recommended from the user outlet to the station device.

The work area equipment, and the cables used to connect to the telecommunications user outlet are currently outside the scope of TIA/EIA T568A and ISO/IEC 11801, but are expected to be specified in the next edition of these standards. In the work area, adapters and application-specific devices, such as the balun, through which the telephone and the fax machine are shown connected in Figure 3-14, should remain external to the telecommunications outlet.

The important specifications related to work area cabling include:

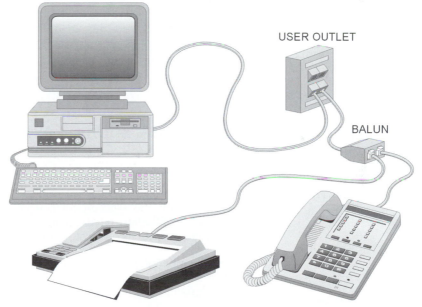

Figure 3-14: Work Area Components

- Equipment cords should have the same performance capabilities as patch cords of the same type and category.

- When adapters are used, they should be compatible with the transmission capabilities of the equipment to which they connect.

- Maximum horizontal cable lengths are specified using a maximum cable length of 3 meters (9.8 feet) for work area equipment cords.

For establishing the maximum horizontal link distances, a combined maximum length of 10 meters (33 feet) is allowed for patch cables/jumpers and equipment/equipment cables in the work area, and in the telecommunications closet.

Horizontal Topology

The Physical topology used by a horizontal cabling system is exclusively a star, with each telecommunications outlet/connector having its own mechanical termination position at the horizontal crossconnect in the telecommunications closet. For each floor, the star topology utilizes this central point of control, usually a hub in the TC. In other words, the telecommunication closet will be located on the same floor as its associated outlets and/or connectors.

Each station or device in the telecommunications system communicates via point-to-point wiring to its own central link, depending on which floor it resides. Normally, the address recognition chore is the responsibility of the central control point, which then directs the information to the cabling path of the device associated with that address. Floor-to-floor addressing is accomplished at the main crossconnect facilities, located in the ER, which can also be arranged in a star configuration.

Because each station's cabling is run directly out from the TC to the appropriate work area, a star topology is considered the easiest to design and install. Its adaptability with other topologies is another reason for this topology's wide acceptance. In combination with logical "Ring" and "Bus" topologies, the star provides a basis for additional configurations that will best meet user requirements.

Horizontal Cabling Key Points

The important points to remember about horizontal cabling subsystems include:

- Any application-specific components (such as splitters or baluns) cannot be installed as part of the horizontal cabling system. If used, they are located external to the telecommunications outlet or horizontal crossconnect.

- The proximity of horizontal cabling to sources of EMI must be taken into account.

- Horizontal cable types that have been recognized by TIA/EIA T568A include:
  - ♦ 4-pair, 100-ohm UTP.
  - ♦ 2-pair, 150-ohm STP-A.
  - ♦ 2-fiber (duplex), 62.5/125µm, or a multi-mode optical fiber such as 50/125µm, which is permitted under TIA/EIA T568B.

- ISO/IEC 11801 permits 120-ohm UTP, and 50/125µm multi-mode optical fiber.

- Multi-pair and multi-unit cables are permitted, as long as they meet the hybrid/bundled cable requirements of TIA/EIA T568A-3.

- Grounding must conform to all applicable local/federal building codes, as well as TIA/EIA 607.

- A minimum of two telecommunications outlets are required for each individual work area, such that:

    ◆ The first outlet is wired to 100-ohm, twisted-pair.

    ◆ The second outlet is wired to 100-ohm, twisted-pair, or to 150-ohm STP-A, or to 62.5/125µm multi-mode fiber.

- One **Transition Point (TP)** is allowed between different forms of the same cable type, such as in situations where under-carpet cable connects to round cable. This term is broader by definition in ISO/IEC 11801 than in TIA/EIA T568A. In addition to the transitions for under-carpet cabling, it also includes **Consolidation Point (CP)** connections.

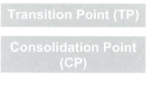

- Although 50-ohm coax is recognized by TIA/EIA T568A, it is not recommended for new cabling installations.

- Although additional outlets may be provided, they are considered to be in addition to, rather than replacements for, the minimum requirements of the standard.

- For copper-based horizontal cabling, bridged taps and splices are not permitted. Splices are allowed however, for fiber.

BACKBONE CABLING

Hubs are connected together to the crossconnect facilities or to the ER using "riser" or **backbone cabling**, which can either be copper or optical. A backbone cabling system is dependent on the type of building and communication being used, and other issues that are detailed in the TIA/EIA 569 specification, which covers the backbone recommendations.

The backbone cabling provides interconnection between Telecommunication Closets, Equipment Rooms, and Entrance Facilities. It consists of the backbone "vertical" cables, Intermediate and Main Crossconnects, mechanical terminations, and patch cords or jumpers used for backbone-to-backbone crossconnection. This definition includes the vertical connection between floors (risers), as well as the cables between an Equipment Room and the building's cable Entrance Facilities. In a campus environment, the backbone can also extend between various buildings. Figure 3-15 highlights the structural components of a backbone cabling system, in addition to depicting other components of the overall system.

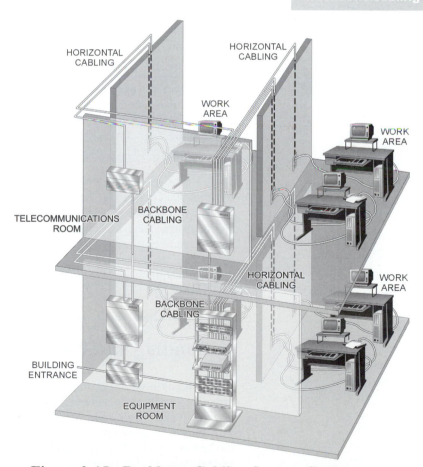

Figure 3-15: Backbone Cabling System Components

In any information transport system, the backbone cabling system normally carries the heaviest traffic. In most systems, optical cables are used for the data backbone cables, while multi-pair copper cables are used for the voice backbone cables. The dedicated hardware is then interconnected to horizontal or backbone terminations. This backbone cabling, also called "vertical cabling," carries the network throughout the building. It is the main data cabling trunk that commonly distributes communications vertically, from floor to floor.

Equipment Rooms and Entrance Facilities

As mentioned in an earlier section, the centralized location for telecommunications equipment serving the users on every floor of an entire building is the Equipment Room (ER). Because it usually houses equipment of higher complexity than that of a telecommunication closet, and because it is designed to serve every floor (as per TIA/EIA 569 specifications), the amount of floor space dedicated to the equipment room should be expected to be somewhat greater than that reserved for a TC. Considered distinct from TCs due to the nature or complexity of equipment they contain, ERs can provide any, or all, of the functions of a TC. They provide a controlled environment and a centralized space to house the telecommunications equipment, its connecting hardware, various splice closures, the grounding/bonding facilities, and the applicable protection apparatus.

The equipment room may contain the MC, several ICs, an HC for portions of the building, and it's not uncommon to locate the network trunk/auxiliary terminations there as well. The general practice is to house only the equipment that directly relates to the telecommunications system, and its environmental support facilities.

Entrance Facilities (EF)

Because the ER is where the telecommunications interface to the outside world occurs, you would expect it to be situated rather close to the **Entrance Facilities (EF)**. Normally, this proximity would dictate that the ER be located at ground (or basement) level, but this is not always the case. The same conditions that affect the overall architectural design of a structure can affect the choice as to where the telecommunications conduit enters the building. Wherever the entrance point is, the ER should be positioned nearby.

The EF consists of the telecommunications service entrance to the building, including the entrance point through the wall, and continuing to the entrance room or space (which could be the ER). The entrance facilities may also contain the backbone pathways that link to other buildings in a campus environment. At a minimum, the EF can be considered to consist of cables, connecting hardware, protection devices, and other equipment that may be needed to connect the outside service facility to the premises cabling. The design and installation of the entrance facilities should follow the TIA/EIA 569 specification.

Grounding and Bonding

grounding

For a cabling installer, **grounding** is an important issue that must be addressed. Proper grounding protects the equipment as well as personnel working with the equipment. Each building is equipped with a grounding busbar to which all electrical equipment, outlets, and associated wiring must be connected. Failure to do this can result in the presence of dangerous and potentially fatal voltages being bridged by electrical outlets, system components, and telecommunications cables (not to mention a pair of hands).

A grounding electrode, usually a rod, pipe, or plate, must make a direct contact with, and provides a low impedance connection to, the earth. This assures that all electrical connections and equipment within the building will be wired to a common ground potential, and prevents the existence of large voltage differences between components lying in close proximity.

Bonding is known as the permanent joining of metallic parts to form an electrically conductive path that will insure electrical continuity, and the capacity to conduct safely any current likely to be imposed on it.

Bonding

Grounding and bonding issues should be directed to TIA/EIA 607, and the National Electric Code (NEC).

TIA/EIA 607

Ratified in August, 1994, the TIA/EIA 607 specification is officially titled as the "Commercial Building Grounding and Bonding Requirements for Telecommunications." In order to achieve a uniform telecommunications grounding and bonding infrastructure within commercial buildings where telecommunications equipment is intended to be installed, TIA/EIA 607 specifies the requirements that system installers should follow.

The purpose of TIA/EIA 607 is to enable the planning, design, and installation of telecommunications grounding systems within a building with, or without, prior knowledge of the telecommunications systems that will subsequently be installed. This telecommunications grounding and bonding infrastructure is expected to support a multi-vendor, multi-product environment, as well as the grounding practices for various systems that may later be installed on the customer premises.

Telecommunications Main Grounding Busbar (TMGB)

This includes the specifications for a **Telecommunications Main Grounding Busbar (TMGB)**, **Telecommunications Grounding Busbar (TGB)**, bonding conductor for telecommunications, and **Telecommunications Bonding Backbone (TBB)** sizing and bonding.

Telecommunications Grounding Busbar (TGB)

TIA/EIA 607 specifies the requirements for:

Telecommunications Bonding Backbone (TBB)

- A ground reference for telecommunications systems within the telecommunications entrance facility, the telecommunications closet, and equipment room.

- Bonding and connecting pathways, cable shields, conductors, and hardware at telecommunications closets, equipment rooms, and entrance facilities.

As a cabling installer, you are responsible for following the standards related to the installation at hand. Grounding and bonding must conform to TIA/EIA 607. Many of the standards documents referenced in this text can be acquired via Global Engineering Documents.

TIA, EIA, and Global Engineering Documents have joined forces in an agreement appointing Global as the primary distributor of TIA/EIA standards and related publications. While TIA, EIA, and their associated groups continue their essential technical and editorial activities, Global is responsible for publication, sale, and distribution for the standards themselves. See your instructor about any of these documents that may be on hand.

National Electric Code

National Electric Code
(NEC)

The **National Electric Code** (**NEC**) is a document that describes the recommended safe practice for the installation of all types of electrical equipment. Revised every three years, the NEC is not a "legal document" unless it is so designated by a municipality as its own statute for safe electrical installations.

The code is "national" only in the fact that it is the only document of which all, or part, is accepted by all the states as an electrical guide, and further, because it is the only document of its kind written with national input supplied by over twenty "panels" of advisors containing several hundred electrical experts from all parts of the country.

National Fire
Protection
Association (NFPA)

National Electrical
Manufacturers
Association (NEMA)

International
Association of
Electrical Inspectors
(IAEI)

The sponsoring agency of the NEC is the **National Fire Protection Association** (**NFPA**). The NEC also uses input from the ANSI, UL, and the **National Electrical Manufacturers Association** (**NEMA**), plus other associations, such as the **International Association of Electrical Inspectors** (**IAEI**), individual "panel members," who may be from industry or the inspection community, and other sources. As of this printing, separation of telecommunications circuits from power circuits is being studied by the TIA/EIA.

Equipment Room and Entrance Facility Sizing

The amount of floor space needed to meet the requirements of the entrance facilities and the equipment room can easily be formulated when the specific equipment being housed there is known. If the equipment to be located here is unknown, you should plan for 0.75 square feet (0.07 square meters) of equipment room space for every 100 square feet (10 square meters) of workstation space. At a minimum, the ER/EF floor space must be at least 150 square feet (14 square meters). For special circumstances, the ER/EF floor space requirements must be based on the number of workstations being served.

The minimum amount of wall length reserved for the ER/EF equipment can be estimated , as shown in Table 3-3, depending on how much gross floor space is allocated for work areas.

Table 3-3: ER/EF Wall Length Vs. Serving Floor Space

Gross Floor Space		ER/EF Wall Length	
Square Meters	Square Feet	Meters	Inches
500	5,000	.99	39
1,000	10,000	.99	39
2,000	20,000	1.06	42
4,000	40,000	1.725	68
5,000	50,000	2.295	90
6,000	60,000	2.4	96
8,000	80,000	3.015	120
10,000	100,000	3.63	144

For larger amounts of gross floor space, the minimum dimensions reserved for the ER/EF can be estimated, as shown in Table 3-4.

Table 3-4: ER/EF Dimensions Vs. Serving Floor Space

Gross Floor Space		ER/EF Dimensions	
Square Meters	Square Feet	Meters	Feet
7,000	70,000	3.66 x 1.93	12 x 6.3
10,000	100,000	3.66 x 1.93	12 x 6.3
20,000	200,000	3.66 x 2.75	12 x 9
40,000	400,000	3.66 x 3.97	12 x 13
50,000	500,000	3.66 x 4.775	12 x 15.6
60,000	600,000	3.66 x 5.588	12 x 18.3
80,000	800,000	3.66 x 6.81	12 x 22.3
100,000	1,000,000	3.66 x 8.44	12 x 27.7

Vertical Crossconnects

Backbone cables are terminated on patch panels at the hub. If the telecommunications equipment is fitted with Telco connectors, as shown in Figure 3-16, then these must be connected to "equipment side" patch panels using mass termination cables (25-pair cables fitted with a "Telco"connector at one end, and no connectors at the other, so that they can be punched down onto IDCs) to provide an RJ45 connector for each data channel.

Copper cables are also terminated on the RJ45-type patch panels, one of which is shown in Figure 3-17. This is the identical type of panel that is used for the horizontal cabling crossconnects in the TC. These RJ45 panels provide an easy facility for crossconnecting selected channels over branches of the backbone.

To enable the patch leads to be routed neatly, a cable tidy (a 1U high, 19-inch rack panel, fitted with jumper rings) can be used in the hub cabinet between pairs of patch panels, as shown in Figure 3-18.

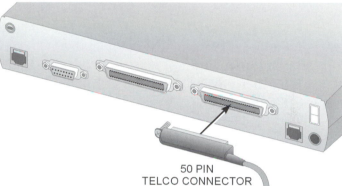

Figure 3-16: Telco Connector

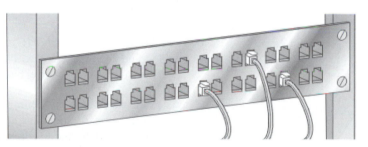

Figure 3-17: Patch Panel with RJ45 Connectors

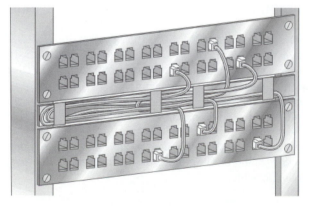

Figure 3-18: Cable Tidy Between RJ45 Panels

Alternately, individual cable ties may be used to bundle various wire and cable groups. When reference is made to rack measurements, it is understood that a 1U piece of equipment is 1.75 inches high.

Optical cables are terminated in patch panels that usually provide the user with an ST presentation, as shown in Figure 3-19. Fiber optic cables can be terminated by either fusion-splicing pigtails with factory-fitted connectors onto each fiber in the cable, or by directly fitting "field-mountable" connectors to each fiber.

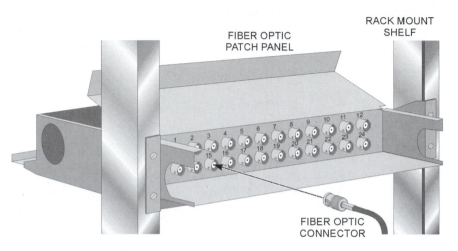

Figure 3-19: Patch Panel with ST Fiber Optic Connectors

Alternately, if an optical fiber backbone is used, the hubs must also contain equipment to enable the optical cable to interface with the copper cables.

Vertical Crossconnect Distances

Main Crossconnect (MC)

Intermediate Cross-connect (IC)

The **Main Crossconnect** (**MC**) is a crossconnect for first-level backbone cables, entrance cables, and equipment cables. The **Intermediate Crossconnect** (**IC**) is a crossconnect between first- and second-level backbone cabling.

As a rule, each backbone cabling run should be limited in length to 90 meters (295 feet) for any category-rated application and 800 meters (2,625 feet) for any voice application. However, backbone distances are application dependent. The maximum distances specified above are based on voice transmission for UTP and data transmission for STP and fiber. The 90-meter distance for STP applies to applications with a spectral bandwidth of 20 MHz to 300 MHz. A 90-meter distance also applies to UTP at spectral bandwidths of 5 MHz to 16 MHz for CAT3, 10 MHz to 20 MHz for CAT4, and 20 MHz to 100 MHz for CAT5.

Lower speed data systems such as IBM 3270, IBM System 36, 38, AS 400, and asynchronous (RS232, 422, 423, etc.) can operate over UTP (or STP) for considerably longer distances (typically from several hundred feet to over 1,000 feet). The actual distances depend on the type of system, data speed, the manufacturer's specifications for the system electronics, and the associated components used (baluns, adapters, line drivers, etc.). State-of-the-art distribution facilities include a combination of copper and fiber optic cables in the backbone.

For crossconnections between MC equipment in the ER, to HC equipment in the TC, distances of 2,000 meters (6,560 feet) for multi-mode fiber, and 3,000 meters (9,840 feet) for single-mode fiber are possible.

Vertical Jumpers and Patch Cords

Patch cords that are used to crossconnect between the MC and any ICs can be somewhat longer than those used in the HCs housed in the TCs. However, Main and Intermediate Crossconnect jumper/patch cord lengths should not exceed 20 meters (66 feet).

Avoid installing copper-based patch cords in areas where sources of high-level EMI/RFI may exist. Also be sure that the grounding for the associated equipment being crossconnected meets the electrical requirements as defined in TIA/EIA 607.

Vertical Cabling Types

As shown in Figure 3-20, the types of cables that are recognized by TIA for vertical backbone cabling are as follows:

- 4-pair, 100-ohm unshielded twisted-pair (UTP) cables.

- 2-pair, 150-ohm shielded twisted-pair (STP-A) cables.

- 2-fiber, 62.5/125-micron multi-mode optical fiber cable.

- 1-fiber, 9/125-micron single-mode optical fiber cable.

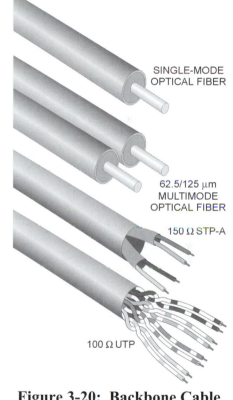

As you already know, data cables such as CAT5 UTP can carry 10baseT, 100baseT, Token Ring, or Ethernet networks, as well as video. Applications can include not only networked data for computers, but also telephone, fax, modem, or even ISDN, T1, or other, special-service data lines furnished by the phone company. The obvious advantage to this is that, with some restrictions placed by the video uses, a building, school, hospital, or other facility wired with only CAT5 UTP is ready for connection to any type of telecommunications service now available. However, there are currently no existing TIA/TIA standards that deal with such **shared-sheath** applications. It is therefore recommended that the user consult with equipment manufacturers, application standards, and system providers for additional information when planning shared-sheath applications on UTP backbone cables.

Installers should be aware that TIA/EIA T568A, while not specifically disallowing multiple uses for any cable, will not certify the cabling installation as being T568A-compliant if such multiple uses are wired. However, the noncompliance of one cable in a bundle does not necessarily mean that the entire bundle is noncompliant.

Figure 3-20: Backbone Cable Types

A sensible approach to this problem would be to accurately designate which cables will be T568A-compliant, and to correctly install them to that standard. Any shared-sheath, nonstandard cables, can then be wired as required.

shared-sheath

Another approach would be to install all cabling as T568A-compliant, and to keep any connections to nonstandard applications external to the applicable wall jacks. In such an installation, all wiring could, at any time, be part of a T568A-compliant network and yet, at the change of a connector, be used for a video feed, telephone, T1 line, or whatever.

Vertical Topology

In general, each backbone should be configured as a star topology. In addition, an individual cabling run can serve a maximum of two floor levels, which means that no more than two hierarchical levels of crossconnects can be used per run. Bridge taps are simply not allowed.

However, the data backbone can be in either a star or ring configuration, depending on the equipment being used. In a star configuration, data backbone cables are usually taken to patch panels in the MC located in the ER. Voice backbone cables are nearly always configured in a star arrangement and are taken to the MC or an IC. The MC or IC is usually a "Krone"-type frame utilizing 10-pair, IDC connection strips.

As shown in Figure 3-21, a star and a ring backbone can be combined to provide resilience (recovery from a system failure).

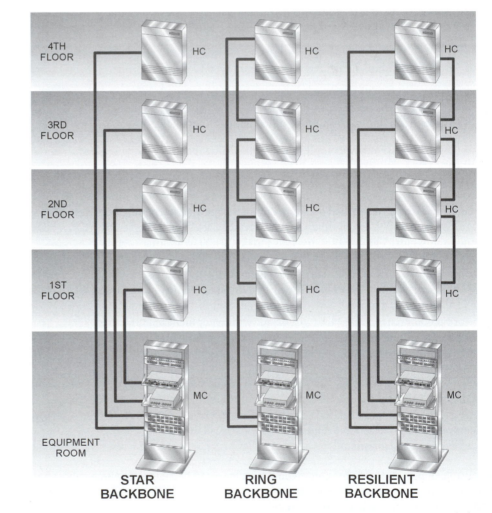

**Figure 3-21:
Backbone
Configurations**

In the event of the failure of any one cable in the backbone, the signal can be rerouted via another available route. This can be carried out automatically by the equipment, or by manually repatching the backbone.

Vertical Cabling Key Points

The important points to remember about vertical backbone cabling systems include:

- Equipment connections to backbone cabling should be made with cable lengths of 30 m (98 feet) or less.

- The backbone cabling shall be configured in a star topology. Each horizontal cross-connect is wired directly to a Main Crossconnect or to an Intermediate Crossconnect, then to a Main Crossconnect.

- The backbone is limited to no more than two hierarchical levels of crossconnects (Main and Intermediate). No more than one crossconnect may exist between a Main and a Horizontal Crossconnect and no more than three crossconnects may exist between any two Horizontal Crossconnects.

- A total maximum backbone distance of 90 m (295 feet) is specified for high bandwidth capability over copper. This distance is for uninterrupted backbone runs. (No Intermediate Crossconnect).

- The distance between the terminations located in the Entrance Facility and the Main Crossconnect shall be documented, and should be made available to the service provider.

- Recognized media may be used individually or in combination, as required by the installation. Quantity of pairs and fibers needed in individual backbone runs depends on the area served. Currently recognized backbone cables consist of:

  - 100-ohm UTP cable.

  - 150-ohm STP-A cable.

  - 62.5/125 micron optical fiber cable (recall that the TIA T568B specification recognizes 50/125 μm multi-mode fiber).

  - single-mode optical fiber.

- Multi-pair cable is allowed, provided that it satisfies the power sum crosstalk requirements.

- The proximity of copper-based backbone cabling to sources of ElectroMagnetic Interference (EMI) shall be taken into account.

- Crossconnects for different cable types must be located in the same facilities.

- Bridged taps are not allowed.

- In ISO/IEC 11801, the equivalent cabling elements to the Main Crossconnect (MC) and Intermediate Crossconnect (IC) are called the Campus Distributor (CD) and Building Distributor (BD), respectively.

- Just as with horizontal cabling, two alternate backbone cabling types are allowed by ISO/IEC (120-ohm, twisted-pair and 50/125-μm, multi-mode optical fiber).

- Again, 50-ohm coaxial cabling is recognized by TIA/EIA T568A, but is not recommended for new installations.

100-OHM UTP REQUIREMENTS

The 100-ohm UTP cabling systems are currently the most flexible, and most economical in the telecommunications industry. This means that the majority of installations now being undertaken are using 100-ohm UTP cable. At the risk of repeating some of the information you have already been given, here in a nutshell are the most important points to keep in mind about 100-ohm UTP installations.

UTP cabling

Although manufacturers are already producing categories of telecommunications cable that outperform them, the following are the currently recognized categories of **UTP cabling** by the TIA/EIA:

- Category 3.

- Category 4.

- Category 5.

The 100-ohm UTP cables used in horizontal cabling systems consist of 24 AWG, thermoplastic insulated solid conductors, formed into four individually-twisted pairs. The maximum diameter of the insulated conductor is 1.22 mm. Under-carpet cables can be used for certain applications. However, only one transition point from round cable to under-carpet cable is allowed. The color code for the 100-ohm UTP cable is shown in Table 3-5.

Table 3-5: Color Code for 100-Ohm UTP

Wire Pair # Lead Function	Banded Colors	Semi-Solid Colors	Twisted CAT5 Solid Colors	8-Pos Jack Pin # (T568A)	8-Pos Jack Pin # (T568B)	6-Pos Jack Solid Colors	6-Pos Jack Pin # (USOC)
1 TIP 1 RING	W/BL BL/W	W/BL BL	W BL	5 4	5 4	GREEN RED	4 3
2 TIP 2 RING	W/O O/W	W/O O	W O	3 6	1 2	BLACK YELLOW	2 5
3 TIP 3 RING	W/G G/W	W/G G	W G	1 2	3 6	WHITE BLUE	1 6
4 TIP 4 RING	W/BR BR/W	W/BR BR	W BR	7 8	7 8		

100-Ohm UTP Work Area Outlets

Usually, if screw-type terminations are utilized for the work area outlets, they will be limited to CAT3 performance. CAT4 and CAT5 terminations are always accomplished with the use of the Insulation Displacement Connection (IDC).

100-Ohm UTP Work Area Outlet Installation

The following is a list of recommendations for the installation of 100-ohm UTP work area outlets.

- All four pairs of a 4-pair cable should be terminated on an 8-position jack.

- All work area outlets should be terminated as either T568A or T568B.

- Both ends of a line should be terminated by the same designation, either as TIA/EIA T568A or TIA/EIA T568B.

- The suggested maximum allowable untwisting for CAT3 cable is 3 inches.

- TIA/EIA T568A specifies that the maximum allowable untwisting of CAT4 cable is 1 inch, and for CAT5 cable, the maximum allowable untwisting is ½ inch.

- It is suggested that the maximum allowable untwisting for any category above CAT5 be less than ½ inch.

- Always leave 1 to 3 feet of service loop for repairs, changes, moves, and additions.

- The bend radius should be no tighter than four times the cable's outside diameter (usually 1 inch).

- The bend radius for cables with more than four cable pairs is ten times the cable's outside diameter.

100-Ohm UTP Cable Installation

The following is a list of specifications for installing 100-ohm UTP cable. Unless these specifications are followed, the system's category rating system will not be maintained.

- Do not exceed 25 pounds of pulling tension on a 4-pair cable.

- Do not cut, or damage, the cable's outer sheath when pulling it.

- Follow the proper color code.

- All of the hardware used should be of the IDC type.

- Installation must be neat, and well organized.

- Document every phase of the installation, including the locations of all components.

- Label cable runs at their beginning, middle, and end for easy identification.

100-Ohm UTP Patch Cords

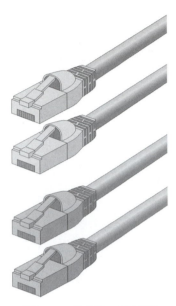

Figure 3-22: CAT5 Patch Cables

We haven't yet gone into much detail about **patch cords**, other than to mention the fact that they are used in TCs and ERs to make the various crossconnections that are required at these locations. Common sense tells us that all of these patch cords should comply with the specified category of the system, but as yet, no general specification exists to ensure such compliance! Often referred to as the installation's missing link, patch cords are the equipment or work area cables that constitute the most overlooked aspect of telecommunications cable systems. Because they do not yet have a standard, patch cords represent the installation's final, unconquered frontier. This is a sad fact, in spite of many years of work by the TIA/EIA.

As has been the case with previous technologies, manufacturers are moving forward with new patch cord products, such as those shown in Figure 3-22, even though the standards-makers have been slow to make much progress. Not surprisingly, manufacturers are creating their own proprietary testing methods, and are fashioning new ways to differentiate between various grades of patch cables. This development will certainly alter the perspective of many installers who formerly considered patch cables as nothing more than commodity items. With the realization that the category compliance of an entire installation may fail due to the use of poor-quality patch cords in the ER or in one of the TCs, the lowly patch cord may finally get the respect it deserves. With the increasing trend towards higher data rates, the signal-quality degradation taking place at the patch panel can no longer be tolerated.

Patch Cord Standard

With the proliferation of standards covering almost every aspect of telecommunications, why has it been so difficult to create a standard for the most inexpensive link in the network? One of the reasons appears to be the lack of agreement as to which testing method should be used. Until this question is resolved, the TIA/EIA cannot release its forthcoming standard. And because it is more difficult to test patch cords than it is data cables, they have become the last part of the cabling system to be defined.

Patch Cord Problems

For being such a vital link in the telecommunications process, there are countless stories about the havoc that has been wreaked on data transmission channels through the use of substandard patch cables. Of course, if there is no "standard" with which to compare an inferior device, can we be so bold as to accuse a suspected patch cable of being substandard? Cable design experts have zeroed in on several chronic problems that have, as their primary cause, bad patch cables being used somewhere along the data path. These include:

- Poor data quality through various crossconnections.

- Communications networks bombarded with increased error rates and junk traffic.

- Severe crosstalk and attenuation problems.

- Generally degraded network performance.

When a defective patch cable causes a poor connection, the system suddenly becomes non-compliant with the CAT5 specifications. This makes the rather large installation investment captive to a lowly, defective patch cord. Some have compared the situation to a Formula 1 racing car hitting the track while outfitted with a set of standard passenger tires.

A patch cord standard that would address these highly publicized problems is what's needed. Most manufacturers would welcome this, because it's generally felt throughout the industry that a good specification would eliminate inferior patch cables, both domestic and imported, from ever showing up on a crossconnect panel. Naturally, most manufacturers claim to have a superior product. However, a solid standard would relegate the judgment of patch cords to their performance.

In lieu of a viable standard, the customer must rely on the manufacturer's reputation or some other subjective criteria. Under normal circumstances, most patch cords have good electrical performance, but the aspects of durability and functionality have become important considerations as well. Functionality relates to the distinctive features—such as color coding—that are now available.

Patch Cord Differentiation

Regardless of the patch cord's past reputation, and even without a clear standard, today's manufacturers are striving to differentiate their patch cable products through:

- Quality testing.
- Performance level warranties.
- Tolerance buffers.
- Strain-relief boots.
- Snagless features.
- Quick turnaround on orders.
- Various colors.
- Icons.
- Cost.
- Manufacturers' labels.
- User-friendliness.

Because many users consider patch cords to be network- and device-independent, the attempt by manufacturers to differentiate between various patch cords is not altogether appreciated. This school of thought considers patch cords to be a generic commodity, with all of them having the same or similar properties.

Various Colors

Most manufacturers now offer CAT5 patch cords in a variety of colors, with the jackets and strain-relief boots matching. In larger jobs, different colors are requested where installers and users assign them to indicate specific functions. Other customers color-code their networks to make them aesthetically pleasing, even going so far as to use only patch cords that match a school's colors.

Installers are not necessarily enamored with colored patch cords. In fact, from a purely installation point of view, colors for patch cables offer no additional value to justify the inventory problems they cause. For smaller jobs, there is simply little need for colored patch cords.

Icons

However, as an alternative to color coding the patch cord itself, some manufacturers have created a colored icon that attaches to the patch cords so that installers can distinguish between network functions. The patch cords, themselves, can always carry the corporate color. Because you can snap on any icon that you need, they are very easy to use. It makes the administration of a network fairly simple by requiring only the replacement of the icon on the boot, and not the wire's color. The icons come in a variety of colors, and use either a computer symbol, or a text symbol.

Strain-Relief Boots

In contrast, both installers and manufacturers alike have reacted enthusiastically to the invention of the strain-relief boot. In fact, it is now a commonplace feature in the industry, and is even considered to be required among installers. A strain-relief boot can prevent an 8-pin modular connector from being pulled off the jack. The boot covers the plug in such a way that when you pull the cable down, the locking tab doesn't hook onto another cable and break off.

Performance Level Warranties

In anticipation of the next series of cabling standards, manufacturers are also designing patch cords with higher performance levels. Not only are the recognized CAT5 specifications being met, but those proposed for enhanced CAT5, and the new CAT6 standard are also being met. CAT5 patch cords meet the CAT5 requirements specified in TIA/EIA T568A, while the enhanced CAT5 and CAT6 patch cords exceed the TIA's specifications in measurable ways.

Tolerance Buffers

Manufacturers are calling the enhanced CAT5 patch cord a product that is specifically "tuned" for CAT5 connecting hardware. They believe that this patch cord tuning helps to guarantee compatibility with all CAT5 modular connectors. They are also confident that any future requirements from the TIA/EIA regarding modular patch cords will be met. To stay ahead of the game, the manufacturers have created a tolerance buffer. They believe that by using the enhanced CAT5 patch cord, installers will feel very comfortable that their installations will pass any future specifications.

Cost

Cost also differentiates between patch cords. When job bids don't request a particular brand (which is normally the case), the installer must decide which brand of patch cords to use. There are occasions when a line in a bid specifies the allowed cost for patch cords, while there are other times when they are figured into the total bid. The best bet is for the installer to shop around for patch cords. You may find a small local manufacturer with a good product, at a cheaper price.

Quality Testing

While manufacturers anxiously wait for the TIA to develop a standard, they continue to find new ways to test and design their products. Many of them are coming to believe that patch cords should be sold as part of a complete system. Being able to control all of the parts that make up a modern telecommunications system may be the only way to meet the required performance levels.

Telecommunications systems in the past were composed mostly of mixed and matched products. But in the future, installers will have a component list that bundles specific products together, in order to guarantee optimal system performance. Under such circumstances, the patch cords would be considered as an integral part of the system, rather than as a generic component.

Testing criteria will dictate the classification of patch cords by performance level, allowing installers to match the patch cord to the system in the same manner that CAT5 jacks are matched to CAT5 cables. The idea that all patch cords have the same or similar properties is no longer valid, because one patch cord will not necessarily work for all systems. Instead, the customer will want to focus on the performance of a channel (including the patch cords through which data is moving), because it will become the more important testing standard.

User-Friendliness

It's difficult to imagine anything easier to use than a simple patch cord, even one that has not been improved upon for category compliance. If ever a manufacturer sinks to producing a patch cable that requires a significant amount of time in order to master the intricacies of using it, it will have created the telecommunications equivalent of the "Edsel."

Regardless of any improvements made to a patch cord, one fact remains constant; patch cords will not reverse their wire positions within the plugs. Pin 1 on the plug at one end of a patch cord will have to connect to pin 1 on the plug at the other end for straight-through wiring. Could anything be more unfriendly than a patch cord wired in any other way?

Lab 15 – Installing and Testing a DB15 Connector

Lab Objective

To understand how to install a DB15 connector on transceiver cable for use with 10base5 Thicknet systems

- Lab 15 work sheet

- Cable, CAT5 STP, 100-ohm, non-plenum (spool or cable fragment)*

- Connector hood/housing, DB15, male (1)

- Connector hood/housing, DB15, female (1)

- Connector shell, DB15, male plug (with hardware) (1)

- Connector shell, DB15, female socket (with hardware) (1)

- Connector pin, DB15, crimp, male (9)

- Connector pin, DB15, crimp, female (9)

- Screwdriver, flat-blade, small

- Insertion/extraction tool

- Crimping tool, pin

- Multimeter, digital

- Wire stripper, CAT5

- Electricians' scissors

- Tension scale/Tape measure

* Because factory standard transceiver cable is expensive, and is unused in any other sections of this course, you may substitute 100-ohm, shielded twisted pair cable for this part. However, if your employer or institution stocks unterminated, factory standard AUI transceiver cable, check into the possibility of using it for this lab. Keep in mind that you won't be using this cable under any actual operating conditions, so if your instructor provides a previously used cable fragment for this procedure, that's OK.

Instructions

Recall the information you read from Chapter 1 about Thicknet systems and the difference between intrusive and non-intrusive taps made to the RG8 cable. In Lab procedure 11 you installed N connectors on a section of RG8 "Thicknet" cable, and in Lab procedure 12, you placed a non-intrusive "vampire" tap on that RG8 cable section.

In order to get the data signals from a tapped RG8 cable, to and from the computer node, an AUI transceiver cable is normally connected between the computer and the transceiver. The transceiver, in turn, interfaces with the RG8 cable via the tap.

In this lab procedure you will create the AUI transceiver cable referred to in Chapter 1 during the discussion on Thicknet computer connections. Factory standard AUI cables contain four or five individually shielded twisted pair wires that are bundled together under both an overall foil shield and a braided shield.

AUI cables run directly between a computer's NIC and the Thicknet transceiver, which itself is either installed in-line on the RG8 backbone, or attached to a tap/transceiver hybrid, as shown in Figure 3-23. Instead of carrying the fully encoded Ethernet signal, AUI cables transfer partially decoded signals such as transmit data, receive data, collision detect, and power. The standard AUI cable comes in lengths of up to 50 meters and must have a characteristic impedance of 78 ohms.

Because of the various layers of shielding in factory standard AUI cable, and the 20AWG diameter of the individual twisted-pair wires, it is as thick as RG8, and difficult to manage. An "office" version has been developed that is thinner and easier to manage in a typical office environment. However, the "office" varieties of these cables suffer from greater signal loss due to higher susceptibility to external noise. Therefore, their maximum usable working distances are much less (10 meters) than that of standard AUI cables.

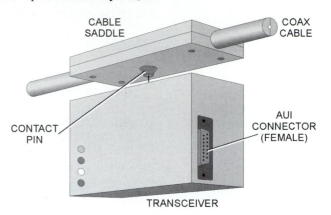

Figure 3-23: Tap/Transceiver Hybrid

As you can see, AUI cables are terminated in 15-Pin D-type connectors. The male end of the cable plugs into the computer, while the female end interfaces with the transceiver. The pinout shown in Table 3-6 shows that AUI cables and interfaces cannot be tested using a "null modem" arrangement, because the AUI cable alone cannot be used to connect two transceivers together. Their operating power must come from the active device to which they are connected (a computer in Figure 3-23).

Table 3-6: AUI Pin Assignments

Pin	Function
1	Control In circuit Shield (Logic Ground)
2	Control In circuit A (Collision +)
9	Control In circuit B (Collision –)
3	Data Out circuit A (Transmit +)
10	Data Out circuit B (Transmit –)
4	Data In circuit Shield (Logic Ground)
11	Data Out circuit Shield (Logic Ground)
5	Data In circuit A (Receive +)
12	Data In circuit B (Receive –)
6	Voltage Common (Logic Ground)
13	Voltage Plus (Power +12VDC)
7	Control Out circuit A (No Connection)
14	Voltage Shield (Logic Ground)
8	Connector Shell/Drain Wire: (Logic/Protective Ground)
15	Control Out circuit B (No Connection)
Shell	Connected to chassis ground

In Table 3-6, the labels in parentheses are functional descriptions for a transceiver connector. Pin 8 could be wired to the grounded shell, and/or to the bare drain wire, in an active circuit operation. Twisted pairs are shaded without a horizontal border between pins.

As is the case with most D-shell connectors, attaching the AUI cable to the computer involves the use of either screws or a sliding-lock mechanism to maintain the connection. Good quality metal must be used to maintain a secure fit with the sliding-lock mechanism.

The weight of a standard AUI cable will bend or break slides composed of thin metal. A more reliable connection can be maintained by using D-shell connectors outfitted with threaded nuts on the equipment and twist screws on the cable.

Before beginning this lab, read through all of the following steps at least once, thoroughly.

1. Use the diagonal cable cutters to cut a 5-foot piece of 100-ohm STP cable from the spool (or use a cable fragment from a previous installation).

2. Begin preparing the cable by first stripping 1 1/4 inches of jacket from each end. Use the tension scale/tape measure for accuracy.

TIP: Before going any further, check with your instructor about whether or not you will be wiring pin 8 to the ground shield and/or the bare drain wire.

3. If you are wiring the shield to pin 8 (along with the bare drain wire), use the electricians' scissors to carefully make a straight, 1-inch cut down the foil shield on each end of the cable, leaving 1/4 inch of it uncut. Then, after sliding the twisted pairs out of the shield, twist the foil tightly (around the bare drain wire, if applicable) until it resembles an uninsulated wire conductor. See Figure 3-24.

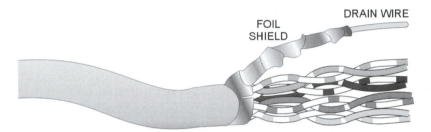

Figure 3-24:
Twisting the Foil
Shield

TIP: If you are not wiring the shield to pin 8, strip one inch of foil shield from each end of the cable, without damaging any of the twisted-pair wires inside the shield. Again, use the tension scale/tape measure for accuracy. This leaves 1/4 inch of foil shielding intact, and 1 inch of the twisted pairs (with bare drain wire, where applicable) exposed at each end of the cable.

4. Strip 3/32 of an inch of insulation from each of the twisted-pair wires at both ends of the cable. Each end of the cable should look like that shown in Figure 3-25.

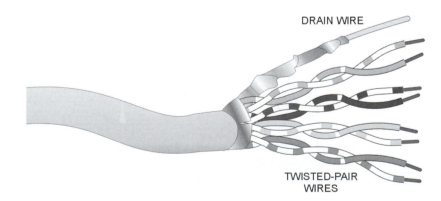

Figure 3-25:
Stripped STP
Cable

TIP: You will begin by installing the male DB15 connector on one end of the cable.

5. Locate the male, DB15 connector pins.

6. Insert the stripped end of one of the twisted-pair wires into a male DB15 connector pin barrel, as shown in Figure 3-26.

7. Locate the various wire sizes on the crimper tool. Then, insert the connector pin/wire combination into the slot on the crimper tool marked "26-28" (the first slot).

TIP: The edge of the crimper tool should line up evenly with the edge of the barrel portion of the pin, as shown in Figure 3-27.

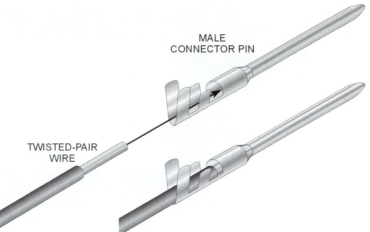

MALE CONNECTOR PIN

TWISTED-PAIR WIRE

Figure 3-26: Inserting a Wire into a Connector Pin

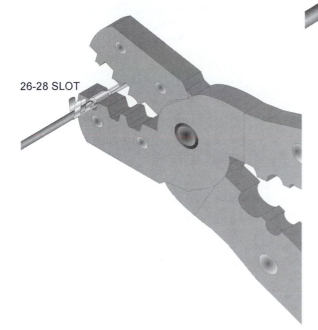

26-28 SLOT

Figure 3-27: Lining Up the Crimper and Pin

10. Examine Figure 3-28 carefully, and orient the tabs on the connector pin within the crimping area above the "22-26" markers, so that their open side is facing up.

11. Wrap the tabs around the insulated portion of the wire by carefully closing the crimper tool completely.

12. For each of the twisted-pair wires (and the twisted foil shield, if applicable), repeat steps 6 through 11.

8. Carefully close the crimper tool completely to make the connection. With the crimper tool fully closed, pull slightly on the wire to ensure that the crimp is tight.

TIP: If the wire is not tight, try rotating the pin a quarter turn, and re-crimp the barrel.

9. When you are convinced that the connection is secure, lift the crimped pin out of the "26-28" slot.

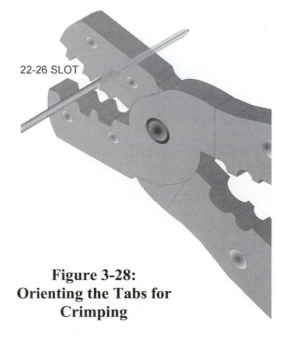

22-26 SLOT

Figure 3-28: Orienting the Tabs for Crimping

13. Locate the DB15 male connector end shell, and orient it so that the pin numbers appear as shown in Figure 3-29.

PIN 8
PIN 15

PIN 1
PIN 9

Figure 3-29: DB15 Male Connector Pin Numbers

TIP: The pin number markings are small, so if you have trouble reading them, Figure 3-29 should be useful as a guide.

14. Carefully compare Table 3-6 with Table 3-7, and note which pin numbers are assigned for each wire of the four twisted pairs.

Table 3-7: DB15 Pin/Wire Assignments

Pin	Pair (Color)
2	Pair 1 (White/Blue)
9	Pair 1 (Blue/White)
3	Pair 2 (White/Orange)
10	Pair 2 (Orange/White)
5	Pair 3 (White/Green)
12	Pair 3 (Green/White)
6	Pair 4 (White/Brown)
13	Pair 4 (Brown/White)
8	Ground (Shield)

TIP: Don't worry about which standard this wiring scheme may or may not resemble. This procedure is designed solely to provide practice.

15. Locate the insertion/extraction tool and fit the back of the pin attached to the white/blue wire (Pair 1 tip) into the tool, as in Figure 3-30. Hold it in place with your thumb.

16. As indicated by Table 3-7, push the pin/wire combination into the hole of the male DB15 connector designated for pin 2. See Figure 3-31.

WHITE/BLUE WIRE
(PAIR 1 TIP)

Figure 3-30: Positioning the Insertion/Extraction Tool

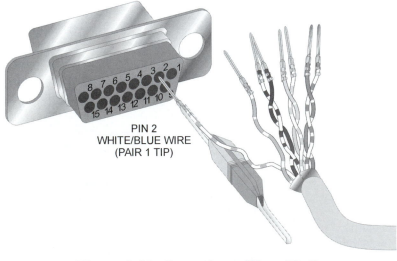

PIN 2
WHITE/BLUE WIRE
(PAIR 1 TIP)

Figure 3-31: Inserting a Pin with the Insertion/Extraction Tool

17. When you feel that it is properly seated in the connector, give it a gentle pull to verify this.

18. Remove the insertion/extraction tool.

19. Consult Table 3-7, and repeat steps 15 through 18 to insert the remaining pins into the DB15 male connector (including the twisted foil shield, if applicable).

20. When you have completed inserting the wires into the DB15 male connector shell, compare your results with Figure 3-32.

TIP: Figure 3-32 includes a wired foil shield to pin 8.

21. Utilizing the multimeter, perform a continuity check on the individual wires of the DB15 male connector you just installed.

22. In addition to the continuity checks between the male connector pins and the corresponding wires (and foil shield, if applicable) at the free end of the STP cable, be sure to check for unexpected short circuits between adjacent pins (or wires). This includes the foil shield.

TIP: The multimeter should show an infinite resistance between any adjacent pins (or wires) with the exception of pin 8 and the foil shield, if applicable, and a short circuit on matching continuity checks. If it does not, the cable is not correctly prepared, and you may have to repeat this procedure. If the cable checks out good so far, continue with the remaining steps.

Figure 3-32: Completed DB15 Male Connector Wiring

23. Locate a DB15 male connector hood housing and its strain relief parts, including the two small screws.

24. Install the two halves of the strain relief on the 1/4 inch of the cable sleeve containing the exposed foil shield, as shown in Figure 3-33.

25. Lay one of the hood halves flat, and position the connector/strain relief combination into the housing, as in Figure 3-34.

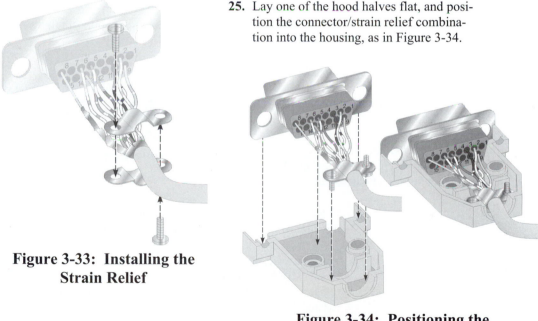

Figure 3-33: Installing the Strain Relief

Figure 3-34: Positioning the Connector/Strain Relief

TIP: Make a preliminary check to be sure that none of the wires are blocking the screw holes.

26. Position the remaining half of the hood housing as shown in Figure 3-35, and use the two long screws and the two nuts to loosely fasten the two halves together.

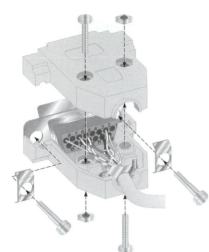

Figure 3-35: Fastening the Housing Halves Together

TIP: You don't want to tighten these housing halves completely until after the next few steps.

27. Locate the two retaining screws (slotted-head) and their mated metal clips.

28. Observe Figure 3-35 again, and thread a metal clip onto one of the retaining screws in the direction shown.

29. Using the small, flat-blade screwdriver, insert the screw/clip combo into one of the mounting holes in the rear of the connector housing, and tighten it until it stops.

30. Then, push on the screw until you see (and feel) it pop through the front of the connector.

31. Repeat steps 28 through 30 for the remaining screw/clip combo.

32. Finally, tighten the two screws from step 26.

TIP: Now it's time to install the female DB15 connector on the free end of the cable.

33. Locate the female, DB15 connector pins.

34. Insert the stripped end of one of the twisted-pair wires into a female DB15 connector pin barrel.

35. For each of the twisted-pair wires (and the twisted foil shield, if applicable), repeat step 34, and then steps 7 through 11.

36. Locate the DB15 female connector end shell, and orient it so that you can read the pin numbers, as shown in Figure 3-36.

Figure 3-36: Reading the Pin Numbers on a Female DB15 Shell

TIP: The pin number markings are small, so if you have trouble reading them, Figure 3-36 should be useful as a guide.

Figure 3-37: Completed DB15 Female Connector Wiring

37. If necessary, check Table 3-6 with Table 3-7 again to see which pin numbers are assigned for each wire of the four twisted pairs.

38. With the exception of the fact that you are now working with a female DB15 connector, repeat steps 15 through 19 to insert the remaining pins into the DB15 female connector (including the twisted foil shield, if applicable).

39. When you have completed inserting the wires into the DB15 female connector shell, compare your results with Figure 3-37.

TIP: Figure 3-37 includes a wired foil shield to pin 8.

40. Utilizing the multimeter, perform a continuity check on the individual wires of the DB15 female connector you just installed.

41. In addition to the continuity checks between the female connector pins and the corresponding wires (and foil shield, if applicable) at the male end of the STP cable, be sure to check for unexpected short circuits between adjacent pins (or wires). This includes the foil shield.

TIP: The multimeter should show an infinite resistance between any adjacent pins (or wires) with the exception of pin 8 and the foil shield, if applicable, and a short circuit on matching continuity checks. If it does not, the cable is not correctly prepared, and you may have to repeat this procedure. Remember that the cable checked good before you mounted the male hood. Any problems should be confined to the newly installed female connector shell.

42. Locate a DB15 female connector hood housing and its strain relief parts, including the two small screws.

43. Install the two halves of the strain relief on the 1/4 inch of the cable sleeve containing the exposed foil shield, just as you did for the male connector.

44. Lay one of the hood halves flat, and position the connector/strain relief combination into the housing.

TIP: Make a preliminary check to be sure that none of the wires are blocking the screw holes.

45. Position the remaining half of the hood housing as shown in Figure 3-38, and use the two long screws and the two nuts to loosely fasten the two halves together.

TIP: You don't want to tighten these housing halves completely until after the next few steps.

46. Locate the two wire slide clips and orient one of them in the direction shown in Figure 3-38.

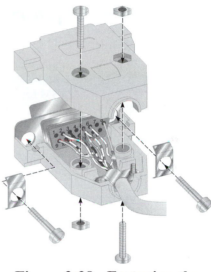

Figure 3-38: Fastening the Housing Halves Together

47. Push the slide clip horizontally until it seats into its permanent slot. Then, repeat the procedure for the remaining slide clip.

48. Finally, tighten the two screws from step 45.

49. Check to be sure that all of the tools, cables, and parts have been returned to their storage area, or previous locations.

50. Once the completed AUI (from STP) cable passes a final continuity check, ask your instructor whether to store the cable or recover the usable parts.

TIP: To recover usable parts, clip the connectors and store the remaining length of cable, and disassemble the hoods, strain reliefs, and end connector shells. Use the insertion/extraction tool to remove any male/female pins from the end connectors. Discard the used pins and twisted-pair fragments and recover all of the short screws, slide clips, metal clips, retaining screws, connector shells, housing hoods, and long screws.

51. Ask your instructor how to store the parts recovered from this procedure.

52. Before leaving your lab work area, check to be sure that you are leaving it in a clean and orderly condition.

Lab 15 Questions

1.   Describe the differences in how pins are numbered between male and female D connectors.

2.   Why should the continuity check be done before installing the hood housings?

3.   When disassembling the D connectors, which parts must be discarded?

150-OHM STP REQUIREMENTS

Although Shielded Twisted-Pair (STP) cable is also manufactured in the 100-ohm variety, similar to UTP, the most common form of STP was originally developed and introduced by IBM for its Token Ring network architecture (IEEE 802.5). It is known simply as 150-ohm STP. As you would expect, its characteristic impedance is engineered for 150 ohms, which normally varies at ± 10 percent over a frequency range of 3 MHz to 20 MHz.

In order to create as dependable a network as possible, IBM developed 150-ohm STP cable to benefit from three different forms of shielding, as shown in Figure 3-39.

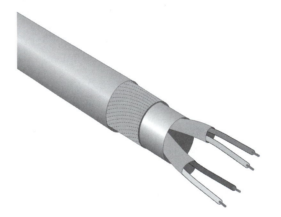

Figure 3-39: IBM Token-Ring 150-Ohm STP Cable

A copper braided shield circles the entire length of the cable just below the outer jacket to help reduce the effects of EMI and RFI. In addition, each of the two twisted pairs is outfitted with its own shield for reduced exposure to crosstalk. Because the pairs are twisted, additional protection from electrically-generated noise is afforded. The braided shielding is grounded at both ends, and not used for any part of the signal path. As you would expect, this extensive shielding means that 150-ohm STP cabling can carry large data rates, while the distortion levels are kept to a minimum.

The downside to all this is that the additional spacing required for all the shielding means that the cable itself must be large, heavy, and expensive. In fact, 150-ohm STP cable has an outer diameter of 0.4 inches (0.98 mm), meaning that wiring closets and conduits must be much larger than those required for a comparable UTP installation. This is in spite of the fact that the copper wire used for the two pairs is only 22 AWG solid strand.

A second and newer variety is the "Extended" 150-ohm STP cable, also called "1A", because of its reference to the extended version designator for the newer Type 1 cable. This extended cable maintains its 150-ohm characteristic impedance up to frequencies of 300 MHz, ± 10 percent, and is not only applicable for Token Ring applications to 100 Mbps, but is also being suggested for broadband video. Extended 150-ohm STP is defined in TSB-53. In TIA/EIA T568A, the extended cabling system replaces the initial version, with extended components being designated by an "A" following the type designation on the cables.

Table 3-8 lists the criteria used for rating the maximum acceptable Balanced Mode Attenuation/NEXT signal loss for the worst pair in both horizontal and backbone STP-A cable. The conditions stipulated for these values are through 100 meters (328 feet) of cable at a temperature of 25 degrees Celsius.

Table 3-8: STP-A Balanced Mode Attenuation/NEXT Signal Loss

Frequency (MHz)	Attenuation (dB)	NEXT (dB)
4	2.2	58.0
8	3.1	54.9
10	3.6	53.5
16	4.4	50.4
20	4.9	49.0
25	6.2	47.5
31.25	6.9	46.1
62.50	9.8	41.5
100	12.3	38.5
300	21.4	31.3

The physical design of 150-ohm STP-A cable should meet the following specifications:

- The insulated conductor shall not exceed the diameter of .26 mm (0.102 inches) maximum.

- The cable shall be restricted to 2-pair only.

- The color code for 150-ohm STP cable is as follows:
    - pair 1: green is tip, and red is ring.
    - pair 2: black is tip, and orange is ring.

150-Ohm STP Closet and Connecting Hardware

Patch panels and passive or **electronic hubs** are the most common crossconnect hardware, but crossconnect blocks are rarely used, and are not recommended. Horizontal and backbone cables are most commonly terminated to a IDC on the patch panel, or to a 150-ohm, STP media interface connector.

Passive or active hubs usually are connected via the 150-ohm STP media interface connectors, and patch cords, to the horizontal cabling. Backbone cables may be optical fiber, or 150-ohm, STP cables, and are usually outfitted with proprietary connectors that interface directly to the hub, minimizing the number of necessary connections. In either case, an installer should follow the recommendations for 150-ohm STP patch panels.

Patch panels

electronic hubs

Everything that was presented in the telecommunications closet section earlier is valid also for the TCs used to service 150-ohm systems. However, the size of the closets should be larger in order to accommodate the larger-sized cables and conduit being used, and the increased amount of equipment that would be needed to service them.

150-Ohm Crossconnect Panels

crossconnect panels

Two types of **crossconnect panels** are generally used for STP patching:

- A panel with IDCs for termination of the building cables.

- An open panel with lock openings for 150-ohm, STP snap-type media interface connectors.

In either case, the installer needs to follow the manufacturer's recommendations for making the terminations. Other recommendations should include:

- Allow 1 to 3 feet (1/3 to 1 meter) of service loop for future adds, moves, and changes.

- For 19-inch (483-mm) rack-mounted crossconnect panel installations, allow room on the rack for possible telecommunications equipment associated with the 150-ohm STP cable.

- Racks should have at least the following clearances for access and cable dressing space:
    - 30 inches (762 mm) in the rear.
    - 36 inches (915 mm) in the front.
    - 14 inches (356 mm) on the side.

hermaphroditic

The telecommunications connector to be used for terminating the 150-ohm STP cable shall be that specified by IEEE 802.5 for the media interface connector. This connector is **hermaphroditic** in design (having both male and female connector elements), so that the two identical units will mate when oriented 180 degrees with respect to each other.

The new extended version (specified in TSB-53) will mate with the old version. It is recommended that the extended version be used in all new installations. This connector is installed directly on the horizontal cable at the work area, or in the telecommunications closet.

Although it is suggested that a 1 to 3 foot (1/3 to 1 meter) service loop be added at both locations for adds, moves, and changes, you should keep in mind that this may be difficult with the cable, due to its large size.

150-Ohm STP Patch and Equipment Cords

Patch cords for 150-ohm STP systems are usually purchased items and are not normally constructed in the field. However, if field construction becomes necessary, follow the patch panel or hub vendor's recommendations. Patch cord lengths are limited to 23 feet (7 meters).

Just as for patch cords, the equipment cords for 150-ohm STP systems are usually provided by the equipment vendor, and are not normally constructed in the field. Follow the equipment vendor's recommendations if and when field construction of any equipment cords becomes necessary. Keep in mind that the maximum lengths for equipment cords should be limited to 10 feet (3 meters).

General Rules for Installation of 150-Ohm STP

By way of repetition, the following list summarizes the most important recommendations for the installation of 150-ohm STP cables:

- Always follow the manufacturer's recommended termination methods.

- Maintain at least 1 meter (approximately 3 feet) of cable for a service loop.

- Telecommunications closet design must follow TIA/EIA 569.

- Patch cord lengths should be limited to 7 meters (approximately 23 feet).

- Equipment cord lengths should be limited to 3 meters (approximately 10 feet).

CABLING MANAGEMENT

The TC will have conduits or raceways entering the TC for backbone and horizontal cables. To facilitate ongoing cabling system management and changes, wiring installations should be documented per the requirements of TIA-606. In fact, proper wiring administration is a requirement of TIA-T568A.

Get used to the fact that a neat and accurate record of the installation is a must. Always keep an updated version of all the wiring runs on file for ease of troubleshooting, for planning additions and moves, and to facilitate the making of repairs. Realize that the orderly management of the telecommunications closet is a vital link to a smooth operation.

For example, if and when any wires are found disconnected at the telecommunications closet, they should be removed. It will improve the appearance of the closet, and it will make it easier not only to run new cable, but to troubleshoot existing cable. By doing this simple step, a technician could save hours of frustration.

In order to maintain confidence between the employer and the contracting company, always clean up the work area once an installation is completed. This will certainly make it easier to do future installations, as well as troubleshooting, because there will be no wire scraps or unused equipment to work around.

Cabling management involves both the planning and documentation of each installation. As time passes, changes to even the most well-planned installation can (and often do) degrade a closet into an impossible situation. As "quick fixes," leftover disconnected equipment, and nonstandard installation changes accumulate in TCs, the resulting reality is that even simple changes are more difficult and time-consuming to make.

Good management is every bit as important as making good connections to wires. The following points constitute a basic checklist of administration practices. If additional or alternate local practices have been established, these should be discussed with the building owner or manager and adhered to carefully.

To maintain proper cable management techniques, you should:

- Plan for those inevitable future changes by allowing sufficient space in closets, and elsewhere, to make them. Assure that all connections include the necessary cabling service loops to effect these changes, and to maintain category compliance.

- Include the following items in your worksheet documentation for each installation:

    ◆ Indicate the length of cable runs to each room.

    ◆ In order to save hours of future troubleshooting, document all of your wiring and test results.

    ◆ Also document any problems that may have occurred, along with any solutions that were found.

    ◆ Differentiate between all primary and secondary outlet locations in your documentation, and include the station phone numbers for each access line.

    ◆ Be sure to document the cabling color code, the wire color combinations for each line, and the relationships between the wire colors. This is very important in situations where colors have been converted at a distribution device.

    ◆ Maintain the relationship of pairs and lines at the distribution device, and clearly label the lines if the distribution device does not have a valid marking scheme.

    ◆ Note any special circumstances.

    ◆ For future reference, remember to provide a copy of the documentation at the wiring location (near a distribution device) and to keep a copy for your records.

- Be certain that all your cable runs are labeled in the middle, and at both ends, and clearly indicate the marking scheme on your worksheet.

- Logically plan the system crossconnect points so that each individual channel can be tested without making additional or unnecessary connections, which would degrade system performance. This is also an important point for Category compliance.

- Avoid making quick fixes to any installation. Instead, use standard wiring practices to ensure that all connections can be easily found and identified, now and later. Crossconnections made in ceilings, for example, are not standard, and will probably be forgotten about and difficult to manage.

- Make your wiring plan consistent throughout your documentation. Regardless whether the installation calls for a simple or complex wiring plan, it must:

    ◆ Be complete and orderly.

    ◆ Be scrupulously maintained.

    ◆ Be readily accessible.

    ◆ Be decipherable by other installers.

- Neatness is an essential ingredient of any installation. Therefore:

    ◆ All wiring should be laid out methodically and consistently.

    ◆ Closets should be kept clean, with adequate room to work.

♦ Unused equipment, leftover materials, and miscellaneous items should be removed from all ERs and TCs to provide full access to the cabling and the equipment currently in use at these locations.

CHAPTER SUMMARY

This chapter introduced you to the philosophy behind the existence of cabling standards in general, and described the concept of structured cabling. The history leading up to the creation of the TIA/EIA T568A series of telecommunications cabling standards was also included, as well as how these standards were specifically formulated for commercial buildings.

Following an explanation of the benefits in using structured cabling, you learned what horizontal cabling is, and the definitions for patch cords and jumpers. You read about the differences between a telecommunications closet and an equipment room, and about some specific attributes of each. You became familiar with the various types of crossconnects, their cabling color schemes, and the differences between them. Various cable lengths were also discussed, with particular attention being given to maximum distances.

You examined the subject of telecommunications outlets and connectors, how they interface with work area cabling and components, and why the horizontal topology is normally configured as a star. You gained experience in Lab procedure 13 with installing a punch-down modular telecommunications outlet, as well as testing a modular jack and its associated 100-ohm, UTP cabling. You also gained an opportunity to fit the outlet into a wall plate. For Lab procedure 14, you installed and tested an edge-connector modular outlet using 100-ohm, UTP cable.

You learned what backbone cabling is, and how it relates to the TIA/EIA 569 specification. You read about how the entrance facilities relate to the equipment room, and how grounding and bonding issues are covered by the TIA/EIA 607 specification and the NEC. The ER/EF sizing requirements were discussed, along with various examples.

The way in which backbone cabling interfaces with different types of crossconnects was examined, as were the maximum distances allowed for various types of vertical cables, jumpers, and patch cords.

You learned how the star topology, although normally used for vertical cabling systems, can be combined with a ring topology to provide resilience in a cabling system. In addition to contrasting the differences between 100-ohm UTP and 150-ohm STP systems, you learned why so much attention is now being paid to the patch cords that make the various crossconnections between ERs, TCs, and the equipment/components.

You performed Lab procedure 15, where you installed and tested both male and female DB15 connectors according to the AUI cabling specification (probably using 100-ohm STP rather than standard AUI twisted-pair).

You then examined the particular requirements and specifications for 150-ohm, STP cabling, as well as the general rules for its installation. Finally, you learned why proper cabling management practices are so important to maintaining a robust telecommunications installation.

REVIEW QUESTIONS

The following questions test your knowledge of the material presented in this chapter:

1. How much extra cabling should be allowed for a Service Loop?

2. How large should the telecommunications closet be for an office space of 10,000 sq. ft?

3. Where should cabling be labeled?

4. What is the maximum recommended pulling tension on a 4-pair cable?

5. Describe a telecommunications equipment room.

6. What size door is required on a telecommunications closet?

7. State the color code for an 150-ohm, STP cable.

8. How much of a CAT3 cable can be untwisted? CAT4? CAT5?

9. What type of topology is horizontal cabling?

10. Define horizontal cabling.

CHAPTER
4

TESTING AND TROUBLESHOOTING

Upon completion of this chapter and its related lab procedures, you should be able to perform these tasks:

1. Explain the benefits of certifying cable installations.

2. Differentiate between a channel and a basic link.

3. Identify standard tests that should be performed on CAT5 cable runs to certify them.

4. Describe the functions typically associated with cable testing instruments.

5. Describe the steps associated with testing and certifying basic links and channels.

Testing and Troubleshooting

INTRODUCTION

Any individual can install telecommunications cabling inside a given building. However, it requires a skilled craftsman to be able to install the cable runs according to the applicable standards. Specific skills are also required when it comes time to test the runs according to the criteria contained in the standards. It is not only important to know which tests must be performed, but also when to conduct these tests.

COMMERCIAL BUILDING—CAT5 INSTALLATION

In this chapter, the focus narrows primarily on the installation of CAT5 cable runs for a commercial building or educational facility.

Installer and End-User Benefits

Following their installation, cabling systems that have been certified to meet the CAT5 TIA/EIA standards will help to ensure that any end-user performance requirements will be met. Also, the performance of a cabling system certification verifies that there are no faults due to the installation itself. Testing the newly installed cable runs will make certain that later problems, which may occur, are not the result of error during the original installation.

An important point to remember is that when the network equipment is connected to the cabling, the behavior of the network can be significantly altered, in spite of the fact that all of the cable runs have passed the CAT5 testing specifications, as documented in the TIA/EIA T568A standard. Although TSB 67 allows for the testing of all, or portions, of the cable runs in order to meet the required standards, it does not guarantee how the network equipment will affect the various high-speed data transmission parameters.

Channel Vs. Basic Link

It is important to understand the differences in terminology between a **channel** and a **basic link**. Both of these terms are used when describing specific CAT5 certification tests, and both of these tests are concerned with the testing of a horizontal cabling run. The main difference between the two is that a channel certification would also include the patching and/or equipment cords being used, while a basic link certification would not.

channel

basic link

Testing the Basic Link and Channel

contractor link

contractor model

user model

The basic link, also known as the **contractor link**, or **contractor model**, is the permanent part of the cable run. It includes two 90-meter horizontal cables, a telecommunications outlet, an optional transition connection close to the work area.

The channel, also known as a **user model**, provides end-to-end testing of the run and is performed after equipment and patch cords are in place. The total maximum length of equipment cords, patch cords (including work area), and jumpers is 10 meters, as shown in Figure 4-1.

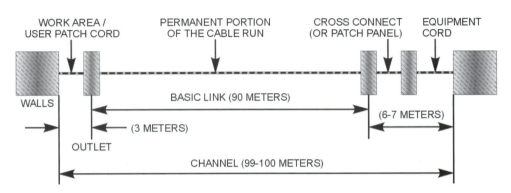

Figure 4-1: Basic Link and Channel Lengths

It is important to test both the basic link and the channel in order to ensure the category performance. To the installer, it should be obvious that separate testing and documentation of the basic link and the channel, immediately after installation has been completed, will become an important reference source for any future testing or troubleshooting. Thorough testing of all cable runs, before any equipment is connected to a network, and before the network is up and running, will rule out the cabling as the source of any problems that may later arise.

Maximum Cable Lengths for Horizontal Cable

Table 4-1 indicates the maximum cable lengths allowed for the individual members of a horizontal cable run.

Table 4-1: Maximum Horizontal Cable Lengths

Run	Length
Basic Link	90 meters
Channel	100 meters
Equipment Cord	7 meters (no one cord longer than 6 meters)
Patch Cord	3 meters (recommended)

Quality Workmanship and Components

The type of components chosen can greatly affect the channel performance, especially with respect to Near-End CrossTalk (NEXT). If quality components are selected, and skilled workmanship goes into a cabling system, the performance of the system will be preserved for a long period of time. The negative impact on the channel's performance, brought about by various conditions of workmanship, is shown in Table 4-2.

Table 4-2: Workmanship's Impact on Channel Performance

Condition	NEXT Performance Loss
Properly Installed, Full Channel	Benchmark
Cable flexed 1,000 times	No change
2 feet of CAT5 patch cord replaced with 2 feet of CAT3	8.0 dB
2 feet of CAT5 patch cord replaced with 20 feet of CAT3	13.0 dB
Cable coiled in 6-foot circle, 2-inch diameter cross section	No change
Cable bundled and secured with ties (20 pounds)	No change
1 inch of cable sheath at workstation end removed	1.2 dB
1 foot of cable sheath at workstation end removed	2.0 dB
½ inch of untwisted pairs at workstation end	1.5 dB
2 inches of untwisted pairs at workstation end	3.8 dB
6 inches of untwisted pairs at workstation end	11.6 dB
Bent cable around a 3-inch diameter	1.9 dB
Bent cable around a 1-inch diameter	2.1 dB
Kinked cable	2.4 dB
Cable run in aluminum conduit	No change

Labeling

As mentioned in a previous chapter, it is very important to take the time to label all of the CAT5 cabling runs before beginning the installation. If any of the runs are not labeled, or are improperly marked, it will be difficult, if not impossible, to identify them later. Depending on how large the overall installation is, trying to identify an unlabeled cable run by using a tone from a **tone generator** might be like trying to find that proverbial needle in a haystack, especially if that cable is located in the middle of a large CAT5 sheath. Tone generation is a subject more suitable to the context of test tools, and will be examined later.

tone generator

It was mentioned in a previous section that many installers have learned that using a color code for various types of cabling can aid in the labeling process. For example, the horizontal runs can be outfitted with blue labels or tags, while the backbone runs can be green, and the equipment cords can be labeled with gray. More controversial are the cables that have been manufactured in various colors, even before the thought of labeling has taken place. There are certain installers who absolutely hate this idea, even though they may accept the thought of color-coded labels.

Required Field Tests

Key to the rapid growth in the use of CAT5 cabling is the adoption of **Technical Systems Bulletin (TSB 67)** from the TIA/EIA. This specification, called the **Transmission Performance Specifications for Field Testing of Unshielded Twisted Pair Cabling Systems**, defines the required test functions, test configurations, and minimum tester accuracy for certifying CAT5 cable installations in the field. Since the emergence of solid guidelines for CAT5 testing and certification, there has been an escalating demand for certification services from the customers of cable installers and contractors.

TIA/EIA specifications, as outlined in TSB 67, require that four different tests be performed on CAT5 runs immediately after the cable is pulled, and terminated, in order to meet the certification guidelines and to avoid having to troubleshoot certain types of problems later. The four specified tests are:

- the Wire Map test.
- the Length test.
- the Attenuation test.
- the NEXT test.

Several additional tests that may be required under certain circumstances include:

- the Power Sum NEXT test.
- the ELFEXT test.
- the Power Sum ELFEXT test.
- the Structural Return Loss (SRL) test.
- the Attenuation-to-Crosstalk Ratio (ACR) test.

The **Wire Map test** will identify any incorrect pairing that was done during the installation. For example, if one end of a cable run was terminated according to the T568A spec, and the other end was terminated according to the T568B, the wrong pairs will be identified. A successful test would indicate a pin-to-pin continuity from end-to-end.

The **Length test** verifies that each cable link falls within the standard's specified maximum physical lengths, and on some testers, it can be used to pinpoint the distance to a fault. The maximum length of the basic link is 94 meters (including test equipment cords). The maximum length of a channel is 100 meters (including equipment and patch cords).

The **Attenuation test** measures how much lower in volume the received signal is when compared to the transmitted signal, and is performed from one end of the link to the other. Attenuation is the loss of signal power along the length of a cable, and is directly related to the length of the cable. Expressed in decibels, attenuation increases as the signal frequency increases. It indicates the ratio of the strength of the original transmitted signal to the strength of the received signal. A Level I tester provides an accuracy of ± 1.3 dB for attenuation. A Level II tester's accuracy is ± 1.0 dB for attenuation.

Attenuation test

A high-speed LAN device may transmit and receive simultaneously. **Near-End CrossTalk (NEXT)** is the unwanted signal coupling between the transmit wire pair and receive wire pair, which adversely affects the quality of the received signal, and is perhaps the most important measurement used in evaluating cabling performance. NEXT measurements are expressed in decibels (dB), which indicate the ratio between the transmit signal and the crosstalk. You may see charts that show NEXT (expressed as negative numbers) or NEXT loss (expressed as positive numbers). In either case, the larger the number, the lower the crosstalk (for example, 40 dB is better than 30 dB and –40 dB is better than –30 dB).

Near-End CrossTalk (NEXT)

For new installations, the **NEXT test** should be performed to determine exactly how much signal coupling is occurring from one pair, to another, within a UTP cable link. In situations where two signals are transmitted in opposite directions over one pair of wires, and the stronger signal overrides the weaker signal, the NEXT test must be individually performed at both ends of the questionable run. Alternately, some testing equipment will allow the installing technician to configure the test so as to evaluate both ends of a cable simultaneously. The test is designed to measure the portion of the transmitted signal that is electromagnetically coupled back into the received signal. A Level I tester provides an accuracy of ± 3.4 dB for NEXT. A Level II tester's accuracy is ± 1.6 dB for NEXT.

NEXT test

Standard NEXT measurements (pair-to-pair) reflect the common application of one device using one pair to transmit and one pair to receive. That's OK for 10BaseT and Token Ring, even 100BaseT and 155Mbps ATM. However, sometimes it's useful and necessary to use those other two pairs for another workstation. It's also probable that faster LANs, such as 622Mbps ATM and 1000BaseT, will utilize all four pairs for transmit and receive.

Using more than one pair in a cable for transmission increases the crosstalk level in that cable, but existing 4-pair CAT5 requirements do not take this into consideration. The **Power Sum NEXT test** involves a mathematical process of combining the NEXT generated from multiple transmit pairs. If a cabling system can still provide CAT5 NEXT performance at the power sum level, it should be able to handle anything from shared-sheath applications to the fastest LAN applications of the future. On a 90-meter link, some manufacturers are already producing Enhanced CAT5 cable, connecting jacks and plugs and patch panels, which provides as much as 8.3 dB margin over current CAT5 NEXT requirements, and 6.6 dB margin over the CAT5 Power Sum NEXT requirements.

Power Sum NEXT test

The **FEXT test** by itself is not a very useful measurement. **Far-End CrossTalk (FEXT)** is the interference of two signals traveling in the same direction in a cable, where the stronger signal drowns out or distorts the weaker signal. It is highly influenced by the length of the cable, because the strength of the crosstalk-inducing signal is affected by how far it has been transmitted from its source. For this reason, Equal Level Far-End CrossTalk (ELFEXT) is measured instead.

FEXT test

Far-End CrossTalk (NEXT)

The **ELFEXT test** subtracts the attenuation from the FEXT result, so that the ELFEXT result is normalized for attenuation. Simply put, ELFEXT = FEXT – ATTENUATION. The **Power Sum ELFEXT test** is the calculation of the individual FEXT effects on each pair by the other three pairs in the same cable, and is becoming a critical testing parameter given the certainty that higher-speed networks will employ multi-pair transmission properties.

ELFEXT test

Power Sum ELNEXT test

**Structural Return
Loss test**

The **Structural Return Loss test** is a measure of the overall uniformity of a link's impedance (resistance to the flow of alternating current in a circuit). Any variations in this impedance will cause return reflection, which causes noise to occur as a portion of the signal's energy is reflected back toward the transmitter. These variations in impedance are caused by structural irregularities in the cable, and SRL is a measure of the reflected energy set up by these imperfections. TIA/EIA T568A requires a SRL of at least 16 dB at 100 MHz. Some manufacturers are producing cable that exceeds this spec. Any advantage of this sort will mean superior structural uniformity in the cable and less reflected energy. This, in turn, will mean greater signal integrity and less noise on the cable.

**Attenuation-to-
Crosstalk Ratio test**

The **Attenuation-to-Crosstalk Ratio test** is a qualitative measure of performance incorporating both signal attenuation and crosstalk parameters. ACR is the difference between the NEXT and attenuation measurements, at a given frequency. While not a requirement of the TIA/EIA T568A standard, ACR is still a useful measure of performance, because it expresses the relationship between the signal level at the device and the noise level generated from crosstalk. This makes it the only measurable aspect of the system's **Signal-to-Noise Ratio** (**SNR**), which is the determining factor in network performance. Encompassing both attenuation and crosstalk, the ACR is also the true indicator of performance headroom (performance margins above and beyond the requirements of CAT5).

**Signal-to-Noise Ratio
(SNR)**

The ISO/IEC 11801 specification requires 4 dB, at 100 MHz. The derived TIA/EIA T568A ACR is 7.7 dB, at 100 MHz. Vendors of network equipment mostly recommend a minimum of 5 dB, with 10 dB being the preferred capability. Additional headroom in the ACR means a potentially better SNR over the communications path. A better SNR, in turn, means an improved **Bit-Error Rate** (**BER**). A lousy SNR raises the BER and lowers the performance of a network by causing excessive re-transmissions due to errors.

Bit-Error Rate (BER)

Delay Skew

Delay Skew

A rather new and unfortunate problem that has been recently discovered is the **Delay Skew** parameter, unknown prior to the introduction of cabling containing non-Teflon coatings on one or two of the four pairs. Many cable manufacturers recently began using such coatings in order to overcome a shortage of Teflon, due to the increased demand for it. These new products meet CAT5 crosstalk and attenuation requirements, as well as the fire rating required for plenum cable. Known in the industry as "3 + 1" or "2 + 2" cables, their new insulating materials create a difference or "skew" in propagation delay for each pair, much to the surprise of both cable manufacturers and users.

Skew Problems

New and popular high-speed LAN technologies, such as 100VG AnyLAN and 100BaseT4, transmit over all four pairs to maximize pair bandwidth and to minimize emissions. In order to achieve these results, these topologies basically break the signal into several components, send the components down the cable on separate pairs, and then reassemble them into the original signal at the far end. Compensation for small differences in propagation delay is accomplished by using built-in buffers, and assuming that the signal will arrive at the end of each twisted pair at about the same time. Using a 100BaseT4 network architecture, signals arriving more than 50 nanoseconds apart will cause the buffers to overflow and information will be lost. For the 100VG-AnyLAN architecture, a maximum delay skew limit of 67 nanoseconds is specified.

When customers began calling the manufacturers about their testers giving inaccurate readings, concern about the problem began to grow. This situation existed because of the way in which most cable testers measure the length of a cable. Normally, the signal-propagation delay is factored with a parameter called the **Nominal Velocity of Propagation** (**NVP**), in order to calculate the length of a given cable. The NVP is an expression of the speed with which signals travel in the cables, and the cable tester assumes that the NVP is relatively constant from one pair to another.

This assumption changes however, when the insulating material is altered; the NVP from pair to pair differs significantly. Installers were recording readings indicating that wire pairs within the same sheath were different lengths, simply because the NVP readings for one or more pairs were not identical.

After testing cables from a number of manufacturers, tester manufacturers found that some failed the skew requirement for 100BaseT Fast Ethernet. Installers could, in effect, buy CAT5 cable, use CAT5 connectors, install them as recommended by TSB 67 and TIA/EIA-T568A, test the link with an approved Level II tester, meet all of the known CAT5 requirements, but fail to run high-speed network topologies reliably.

Skew Compensation

Obviously, as network architectures attempt to squeeze increasingly more cable bandwidth by transmitting in parallel over multiple pairs, this propagation delay-skew issue can only gain more attention. Cable tester vendors are already adding skew capability to their products to compensate for this problem. For example, the WireScope 155 from Scope Communications, shown in Figure 4-2, offers a skew reading that will identify cables that could pose a problem to network applications. The unit independently measures one-way propagation delay for each of the four pairs, and reports total delay and delay skew between pairs. The installer can then add a delay-skew specification into the overall pass/fail requirements via an option in the unit's autotest menu.

In spite of the rapid adjustment to this issue, industry observers agree that the ultimate solution to the problem lies with cable manufacturers. The TIA has already drafted a proposed change to the TIA/EIA T568A standard that would strengthen the current propagation-delay guidelines, and add delay skew requirements for 4-pair cables covering all recognized cabling categories. The new requirements will apply to qualifying cables in a laboratory environment, using a network analyzer, rather than to field testing. Table 4-3 compares several test parameters for CAT5, CAT5E, and CAT6 worst-case channel performance @ 100 Mhz.

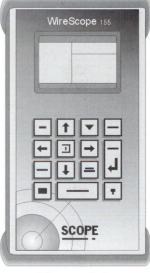

**Figure 4-2:
WireScope 155**

Table 4-3: Industry Standards Performance Comparison
(Worst-Case Channel Performance @ 100 MHz)

Parameter	CAT5	CAT5E	CAT6
Specified MHz Range	1-100 MHz	1-100 MHz	1-250 MHz
Attenuation	24 dB	24 dB	21.7 dB
NEXT	27.1 dB	30.1 dB	39.9 dB
Power Sum NEXT	N/A	27.1 dB	37.1 dB
ACR	3.1 dB	6.1 dB	18.2 dB
Power Sum ACR	N/A	3.1 dB	15.4 dB
ELFEXT	17 dB *	17.4 dB	23.2 dB
Power Sum ELFEXT	14.4 dB *	14.4 dB	20.2 dB
Return Loss	8 dB *	10 dB	12 dB
Propagation Delay	548 ns	548 ns	548 ns
Delay Skew	50 ns	50 ns	50 ns

* New requirements addressed
in TSB95

And regardless of which version of TIA/EIA T568A is currently in effect, remember that all of these test results must be documented immediately after being performed, if possible. Any cable that barely passes a test must be duly noted, including a warning stating the true situation. The TIA/EIA T568A-5 standard that covers CAT5E cabling is in circulation, and was due to be published in late 1999. The CAT6 specification was expected to be released for an industry-wide ballot during the fourth quarter of 1999.

Simple Rules for Testing

- NEXT test must be performed at both ends of the run.

- Documentation of test results must be included.

- Cabling/components must not be moved during testing, or retesting will be required.

- End user patch cords must be tested in place. It is suggested that patch cords be three meters, but this subject does not as yet fall into the scope of the TIA/EIA T568A standards. It is expected that future releases will deal with this issue.

Test Tools

Because new cabling installations now almost always require certification, the cable installer will need to include cable testing equipment among the tools at his/her disposal. For certain tests described earlier, this equipment is both sophisticated and expensive.

While simple continuity testers can be purchased for well under $100, more expensive testers are required to fully certify a new installation for the required NEXT parameters. Units running several hundreds of dollars can pinpoint shorts, opens, reversed or crossed pairs, split pairs, and cable lengths and distances to a fault, or provide test tones for cable identification purposes. More sophisticated units, designed for testing NEXT at frequencies up to 350 MHz, can run several thousands of dollars.

Low-End Testers

Although sophisticated and expensive cable certification tools are capable of graphically displaying the pin-to-pin configuration of a cable, including all wiring faults, an experienced cable installer can use a simple continuity-type testing device to determine the wire map of a new installation. Figure 4-3 depicts such a tester, along with its corresponding remote terminator unit. It comes equipped with RJ11, RJ45, and Mini-DIN jacks (for AppleTalk systems), as well as a BNC connector for testing Ethernet terminator values.

Figure 4-3: Simple Cable Tester

The various features for this tester include:

- Before & after installation checkout.

- Offers easy one-step operation.

- Has 14 LEDs to confirm testing status.

- Shuts off automatically to maximize power saving.

- Tests for shorts, connections, non-parallel wiring, and 25/50-ohm BNC terminations.

- Requires a 9-volt battery.

While a simple cable tester would not be appropriate for many types of cabling tests, it can be used as a logical companion to the high-priced cable testers. Given the large expense of full-featured cable testers ($3,000 to $4,000), it's not always possible to send one out to the field with every installation crew. For example, a high-end tester may indicate that a wire pair has been punched down incorrectly. An installation team can be sent out to repunch the cable, and check it again with simply a continuity-testing device.

High-End Testers

The certification requirements for CAT5 wiring have created a situation in which continuity tests, using a simple cable tester as described above, cannot hope to adequately document a successful cabling installation. In order to perform the required frequency-dependent tests for CAT5 cabling, the tester being used must be able to provide signaling frequencies up to 100 MHz. Therefore, such a tester must be specified as being suitable for CAT5 operation. Just as with the cabling manufacturers, companies that produce cable certification tools have been developing features for their products designed to meet the current (CAT5) and future specifications (CAT5E, CAT6, and CAT7). In fact, there are testers already on the market that operate at frequencies up to 350 MHz.

There currently are five vendors offering products that test for CAT5 conformance. Vendors for these high-end cable testers include: Datacom, Fluke, Microtest, Agilent Technologies, and Wavetek. All of the devices are handheld, but some might require a firmer grip than others. Those from Fluke Corporation (Everett, WA), shown in Figure 4-4, and Wavetek Corporation (San Diego), shown in Figure 4-5, tip the scales at more than 3 pounds.

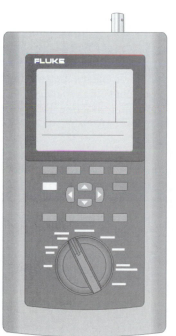

Figure 4-4: Fluke CAT5 Tester

Figure 4-5: Wavetek CAT5 Tester

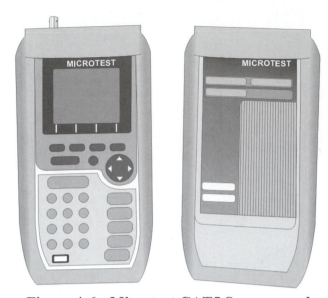

Figure 4-6: Microtest CAT5 Scanner and 2-Way Injector

Testers from Microtest Inc. (Phoenix), shown in Figure 4-6, Datacom Technologies Inc. (Everett), and Agilent Technologies (Northborough, MA), weigh in at under 3 pounds (both the main and remote units). TSB 67 specifies the electrical characteristics of field testers, test methods, and minimum transmission requirements for UTP cabling. The purpose of this bulletin is to specify transmission performance requirements for UTP cabling links consistent with all three categories of UTP cable, and connecting hardware, specified in TIA/EIA T568A.

The field testing methods, as well as the interpretation of the test data, provide installers with the necessary guidelines to determine the Pass/Fail criteria, and to verify that the installed cabling will pass certification. Laboratory procedures and test setups to measure transmission performance are described, allowing the comparison of results between field testers and laboratory equipment. The bulletin also defines channels and basic links, levels of accuracy (Level I and Level II), and the tests necessary to qualify installed cabling. For further details about these levels of accuracy, it may be necessary for the cable installer to acquire a copy of TSB 67, and to study it carefully.

Level I accuracy includes the capabilities of testers that were available prior to the TSB 67 publication, and they are not as stringent as those defined for Level II testers. Level II testers are much closer to laboratory grade in quality and much more accurate. They are designed to fail fewer good cable runs, and pass fewer bad ones. Obviously, testing devices that meet Level II specifications can determine with more certainty whether or not a cabling installation passes the existing standards. In addition, the near future will see Level IIE equipment being produced to verify the CAT5E performance requirements, including EL-FEXT, Return Loss, Delay, and Delay Skew as per the requirements documented in TIA/EIA TSB 95 and TIA/EIA T568A-5.

CAT5 testing equipment must be able to measure all of the data for the required tests, as previously defined. The unit should also come with manufacturer's information on how to interpret data with the device. In addition, a 310-foot meter range is desirable to allow for approximate measurements of available cable on spools. This also helps the installer to determine if there is enough cable left on a spool to run a maximum-length segment.

Lab 16 – Calibrating and Self-Testing Level II Testers

Lab Objective

To understand the procedure for properly calibrating the Level II testers, performing the self test, and documenting the test results

Materials Needed

- Lab 16 work sheet

- Cable, UTP CAT5, 2-meters

- Cable, UTP CAT5, 55 feet

- Level II testers, w/case & cables (control unit and remote)*

- Users Manual for Level II testers

* It's a good idea to prepare ahead of time for any lab procedures that involve the use of the Level II testers. Charge the Nickel-Cadmium battery packs in both the control unit and the smart remote for about 3 hours before use. In an emergency situation, the units will operate during a charge operation, but they will have to remain near an ac outlet during this time.

Instructions

Before beginning this lab, read all of the following steps thoroughly, at least once.

1. Locate the Users manual for the Level II meters and browse through the first 2 chapters, plus Chapter 7. Pay particular attention to the list of features, the various standard accessories that come with the meters, and the safety precautions that must be observed during their operation.

TIP: Several warnings are listed in the manual that indicate the possibility of causing the meters to give erroneous test results. Become familiar with these error-causing situations.

2. Begin the calibration of the Level II testers by locating the supplied 8.25-inch calibration cable.

3. Turn the rotary switch to the Level II tester remote unit on first, and then the control unit.

4. Rotate the function selector on the control unit clockwise, as shown in Figure 4-7, to the SPECIAL FUNCTIONS menu.

5. Use the arrow keys to highlight the Self Calibration function.

6. Press the ENTER button on the front panel of the control unit.

7. Check to be sure that no cable is connected to the control unit's BNC connector. If a cable is connected, remove it now.

8. Read the instructions on the control unit's display screen, and compare them with those shown in Figure 4-8.

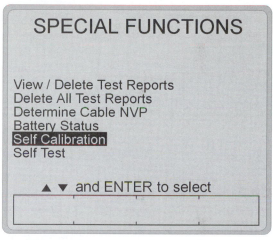

Figure 4-7: Locating the Special Functions Menu

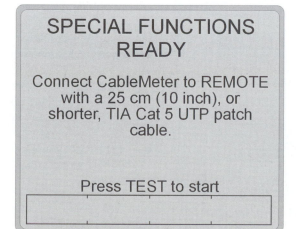

Figure 4-8: Self Calibration Ready

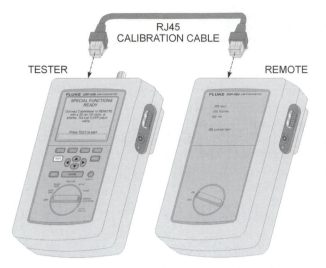

Figure 4-9: Connecting the Calibration Cable

9. As per the instructions, plug the 8.25-inch calibration cable into the Level II test control unit first, and then into the remote unit. See Figure 4-9.

10. To begin the self-calibration procedure, press the control unit's TEST button.

TIP: After a slight delay, during which time the control unit reports that the self-calibration procedure is in progress (and various clicks and beeps can be heard), the control unit will display the "SELF CALIBRATION COMPLETE" message, as shown in Figure 4-10.

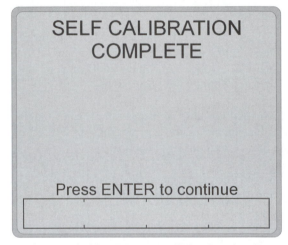

Figure 4-10: Message Display in the Control Unit

11. Press the ENTER button to continue.

TIP: The screen menu returns to the SPECIAL FUNCTIONS display. Always remember to calibrate the Level II testers before using them.

12. Unplug the 8.25-inch calibration cable, first from the remote unit, and then from the control unit.

TIP: The Self Test function of the Level II testers is used to check the operation of both the control unit and the remote unit. It's a good idea to run this test periodically, as well.

Figure 4-11: Self Test Ready

13. Use the arrow keys to highlight the Self Test function, and press the ENTER button on the front panel of the control unit to select it.

14. Read the instructions on the control unit's display screen, and compare them with Figure 4-11.

15. As per the instructions, plug the 2-meter CAT5 cable, with shielded RJ45 plugs, into the Level II test control unit first, and then into the remote unit.

16. At the control unit, press the TEST button to begin the self-test procedure.

TIP: After a slight delay, during which time the control unit reports that the self-test procedure is in progress (and various clicks and beeps can be heard), the control unit will display the "SELF TEST PASS" message, as shown in Figure 4-12.

17. Press the ENTER button to continue.

18. Unplug the 2-meter CAT5 cable, with shielded RJ45 plugs, first from the remote unit, and then from the control unit.

19. Turn the rotary switch on the remote unit to the OFF position.

TIP: While the control unit is still displaying the SPECIAL FUNCTIONS screen, it would be a good idea to check the Battery Status function. Because the control unit was charged just prior to performing this procedure, it should indicate that fact during a check.

20. Use the arrow keys to highlight the Battery Status function.

21. Press the ENTER button on the front panel of the control unit to select it.

22. Read the instructions on the control unit's display screen, and compare them with those shown in Figure 4-13.

TIP: If the status of the battery ever drops to a value such that the Level II testers would not operate properly, there will be a specific warning message displayed on the screen. See Chapter 2 in the Users Manual for more details.

Nominal Velocity of propagation (NVP) is the speed at which the signal travels through a cable. It is expressed as a percentage of the speed of light. Typically, the speed of a signal through a cable is between 60% and 80% of the speed of light.

NVP values affect the limits of cable length for Ethernet systems because Ethernet operation depends on the systems ability to detect collisions in a specified amount of time. If a cable's NVP is too low or the cable is too long, signals are delayed and the system cannot detect collisions soon enough to prevent serious problems in the network.

Length measurements depend directly on the NVP value entered for the selected cable type. To measure length, the test tool first measures the time it takes for a test pulse to travel the length of the cable. The test tool then calculates cable length by multiplying the travel time by the signal speed in the cable.

Because the test tool uses the length measurement to determine cable resistance limits, the NVP value also affects the accuracy of resistance measurement.

23. Make a cable of exactly 55 feet of length with the T568A standard as we did in Lab 6.

24. Plug the cable into the Fluke meter and the Fluke smart remote.

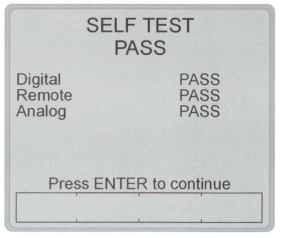

Figure 4-12: Display Screen for SELF TEST PASS

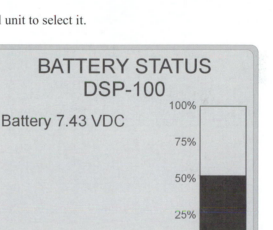

Figure 4-13: Display Screen for BATTERY STATUS

```
CABLE NVP

CABLE TYPE:
UTP 100 Ohm Cat 5
NVP: 50.0%
Default NVP = 69.0%

Connect 15m (50ft), or
longer,
of cable type:
UTP 100 ohm Cat5
cable to the CableMeter.
Press TEST to start
┌────────┬────────┬─────────┐
│ INC    │ DEC    │ Set to  │
│ + .1   │ - .1   │ Default │
└────────┴────────┴─────────┘
```

Figure 4-14: Changing the Display to Read NVP=50%

```
CABLE NVP

CABLE TYPE:
UTP 100 Ohm Cat 5
NVP: 50.0%

█LENGTH:        40 ft█

Use  ▲ ▼  Keys to adjust length


Press SAVE when done
```

Figure 4-15: Cable NVP Test Results Screen

```
CABLE NVP

CABLE TYPE:
UTP 100 Ohm Cat 5
NVP: 99.9%
Default NVP = 69.0%

Connect 15m (50ft), or
longer,
of cable type:
UTP 100 ohm Cat5
cable to the CableMeter.
Press TEST to start
┌────────┬────────┬─────────┐
│ INC    │ DEC    │ Set to  │
│ + .1   │ - .1   │ Default │
└────────┴────────┴─────────┘
```

Figure 4-17: Changing the Display to Read NVP=99.9%

25. Select the special functions setting on the Fluke meter.

26. Arrow down to Determine Cable NVP on the display.

27. Press ENTER.

28. Change the display to read NVP=50% by pressing and holding the down arrow as shown in Figure 4-14. (50% is the lowest setting possible, it will beep when it reaches it.)

29. Press the TEST button on the Fluke meter.

30. It will test the cable and give you a reading in feet that will be much lower than the actual length of the meter and the control unit will display a message similar to that shown in Figure 4-15.

31. Press Save (without changing the length).

32. Switch to the single test function on the Fluke meter.

33. Arrow down to length and press ENTER.

34. It will give you the length of each tested twisted pair and the control unit will display a message similar to that shown in Figure 4-16.

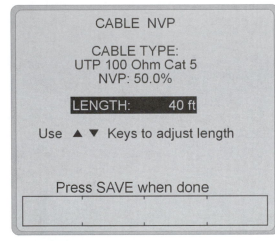

```
                    LENGTH

          Length   Limit
Pair      (ft)     (ft)      Result
1,2       41       328       PASS
3,6       41       328       PASS
4,5       40       328       PASS
7,8       40       328       PASS
```

Figure 4-16: Cable Length Test Results Screen

35. Switch the selector to special functions again.

36. Select Determine Cable NVP and press ENTER.

37. Now press and hold the arrow up key until we get to 99.9%, as shown in Figure 4-17.

38. Press the Test button.

39. Press Save.

40. Change the selector to Single Test.

41. Arrow down and select length.

42. Now go back to Special Functions.

43. Select Determine Cable NVP.

44. Press enter.

45. Press Test.

46. Now change the length to 55.

47. Rotate the function selector on the control unit counterclockwise, to the OFF position.

48. Document all of your test results on the Lab 16 work sheet.

49. Return the Level II testers (and charge units, if necessary) to their appropriate storage bags, along with any of the accessory cables you used.

TIP: If any of the cables were organized using tie wraps, go ahead and use the tie wraps again before packing the various cables away.

50. Before leaving your lab work area, check to be sure that you are leaving it in a clean and orderly condition.

Lab 16 Questions

1. When should the various cables be plugged into the Level II testers?

2. How long should the Level II testers be charged, according to the Users Manual?

3. What three measurements are taken during a self test between the control unit and the remote unit?

4. When plugging a cable into the Level II testers, which unit (control or remote) should be connected first?

5. How long is the cable that is used to calibrate the Level II testers?

6. Under what circumstances should the Level II testers be calibrated?

7. How long does the self-calibration test take to complete?

Lab 17 – Setting Up Level II Testers

<u>Lab Objective</u>

To understand the procedure for properly setting up the Level II testers, and documenting the results

<u>Materials Needed</u>

- Lab 17 work sheet

- Level II testers, w/case & cables (control unit and remote)*

- Users Manual for Level II testers

* Prepare ahead of time for any lab procedures that involve the use of the Level II testers by charging the Nickel-Cadmium battery packs in both the control unit, and the smart remote. In an emergency situation, the units will operate during a charge operation, but they will have to remain near an ac outlet during this time.

<u>Instructions</u>

Before beginning this lab, read all of the following steps thoroughly, at least once.

1. Locate the Users manual for the Level II meters, and review the list of activities that can be performed from the SETUP menu of the control unit.

TIP: Not all of the adjustable parameters displayed need to be tinkered with. Once some settings have been selected, they can normally be ignored from then on. Settings that may require periodic changes involve the particular test standard to be run and the type of cable to be tested.

2. Begin this procedure by turning the rotary function selector on the control unit clockwise to the SETUP menu position.

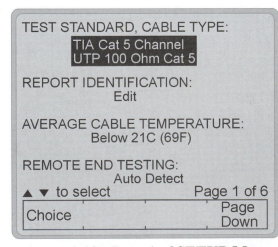

Figure 4-18: Page 1 of SETUP Menu

3. Check the screen on the control unit for a display on its Page 1, similar to that shown in Figure 4-18.

TIP: Notice that the very first parameter displayed is the TEST STANDARD, CABLE TYPE, because it is the parameter that most often needs to be adjusted.

4. Press the Choice button (1) on the control unit.

TIP: Notice that the bottom of the display indicates which of the four numbered buttons to press for the desired screen. You should now see the SELECT TEST STANDARD screen.

5. Press the Choice button (4) to get to page 5 of 5. Then, use the arrow keys to highlight the All Tests selection, and press the ENTER button.

TIP: You should now be looking at the SELECT CABLE TYPE screen.

6. Use the arrow keys to highlight the UTP 100-ohm CAT5 selection, and press the ENTER key.

TIP: The control unit returns to the Page 1 display with the selections you just made highlighted. Any tests conducted at this point would expect to see CAT5 100-ohm UTP cable connected to the Level II testers.

7. Use the arrow keys to highlight the REPORT IDENTIFICATION parameter.

TIP: This parameter allows the user to edit the identification information that will appear in any reports that are generated by the Level II tester.

8. With the Edit selection highlighted, press the EN-TER button or CHOICE button (1) to get to the REPORT IDENTIFICATION editing screen, as shown in Figure 4-19.

TIP: Report identification parameters that can be edited include the CUSTOM HEADER, the OPERATOR, and the testing SITE. Normally, there will be some existing text that was input from another technician or student.

9. With the text directly below the CUSTOM HEADER highlighted, press the ENTER key.

TIP: An alphanumeric listing appears, along with an edit box for the CUSTOM HEADER, containing the previously input text, as shown in Figure 4-20. In order to input the desired text in the editing box, you can use the arrow keys to select a character from the alphanumeric display. Then, use the direction arrow button (1) to select where the character will appear and press the ENTER key. Alternately, you can select an existing character position, and use the INC or DEC buttons to move up or down the alphanumeric list to the desired character. The Delete button (4) will delete the character immediately to the left of the cursor.

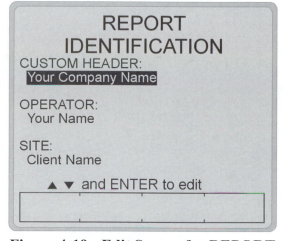

Figure 4-19: Edit Screen for REPORT IDENTIFICATION

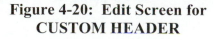

Figure 4-20: Edit Screen for CUSTOM HEADER

10. Using the input method of your choice, create a custom header for your report identification. You can use the name of your school or company.

11. When the header looks the way you want it, press the SAVE button.

TIP: The control unit returns to the REPORT IDENTIFICATION screen.

12. Using the same techniques you just learned, edit the OPERATOR and SITE parameters under the REPORT IDENTIFICATION heading. Then save your edits.

TIP: If you receive any message regarding the fact that the selected parameter has been used to generate a report, and you don't wish to delete the existing information, simply press the NEW button (2) to add another line of text.

13. Once you have edited the report identification parameters, press the EXIT button twice to return to the opening screen.

14. Use the arrow keys to highlight the AVERAGE CABLE TEMPERATURE setting, which in the case of CAT5 100-ohm UTP displays N/A (Non-Applicable).

TIP: This means that there are no temperature-dependent limits that would affect the attenuation parameters using this cable. If temperature was an important factor in the use of the selected cable, you could use the CHOICE button (1) to select an average cable temperature that accurately reflected the actual conditions under which the this cable was required to operate. Notice that if you press the CHOICE button (1) at this time, nothing happens!

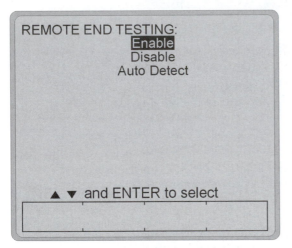

15. Use the arrow keys to highlight the REMOTE END TESTING setting. Then, press the CHOICE button to get to the screen shown in Figure 4-21.

16. Use the arrow keys to highlight the Enable setting. Then, press the ENTER key to select it.

TIP: This setting will allow you to use the remote unit to measure NEXT and ACR parameters from the far end of the cable under test, provided the test standard selected requires them. Keep in mind that the only time the remote unit is used is when testing twisted-pair cable.

17. Press the Page Down (4) key to move to the next page of settings (2 of 6) on the control unit. The screen shown in Figure 4-22 appears.

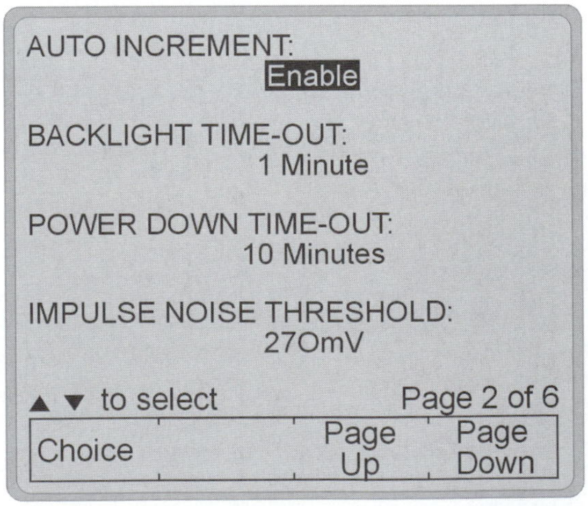

Figure 4-21: Edit Screen for REMOTE END TESTING

18. With the AUTO INCREMENT setting highlighted, press the ENTER key.

TIP: Normally, when the Level II testers are used in the field, the results of the various tests that are conducted need to be recorded, and printed, in order to provide the customer and the installer with copies.

This information serves as proof of the condition of a CAT5 cabling installation immediately following its completion. For this course, you should set this parameter to Disable.

Figure 4-22: Screen for Page 2 of SETUP

19. Set the AUTO INCREMENT parameter to Disable, and press the ENTER key.

20. Move down to the BACKLIGHT TIME-OUT setting and press the ENTER key. The screen displays the choices shown in Figure 4-23.

TIP: This screen allows you to specify how long the control unit will allow the backlight to operate before shutting it off. Unless you were a cable installer doing work in a dark wiring closet, there would be no reason to operate the backlight during normal daylight. To conserve the charged batteries, make sure the setting shows 1 Minute. This will turn the backlight off after the control unit detects 1 minute of inactivity. Better yet, conserve even more power by not pressing the WAKE UP button at all, and keeping the backlight off.

21. If necessary, highlight the 1 Minute setting on the BACK-LIGHT TIME-OUT screen, and press the ENTER button.

22. Move down to the POWER DOWN TIME-OUT setting and press the ENTER key. The screen displays the choices shown in Figure 4-24.

TIP: Again, you have several timing choices displayed. This parameter has to do with powering down the display altogether when the control unit detects inactivity. This is another way that the meter conserves its limited battery power. When the meter shuts down the display, it broadcasts an audible signal. The control unit does not forget where it is when it shuts down. To activate the unit once it shuts down, simply press its ENTER key.

23. If necessary, set the AUTO INCREMENT parameter to 10 Minutes, and press ENTER.

24. Move down to the IMPULSE NOISE THRESHOLD setting and press ENTER.

TIP: This measurement allows the monitoring of electrical noise, on lines 3 and 6 (pair 2 for T568A), that may be present on a twisted-pair cable. The control unit will sample the voltage on the test pair once each second, and records any noise that exceeds this setting as a hit. For 10baseT purposes, 2 hits within a 10-second time span is interpreted as a failure. The default threshold for the control unit is 270 mV, and can be adjusted in 10mV increments.

25. If necessary, use the DEC (3) or INC (4) buttons to set the IMPULSE NOISE THRESHOLD at 270 mV. Alternately, you can use the up or down arrow keys. Press ENTER when the setting is correct.

26. Press the Page Down (4) key to move to the next page of settings (3 of 6) on the control unit. The screen shown in Figure 4-25 appears.

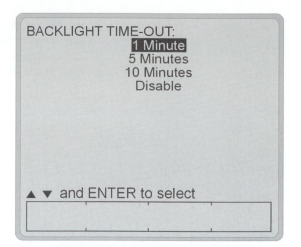

Figure 4-23: Edit Screen for BACKLIGHT TIME-OUT

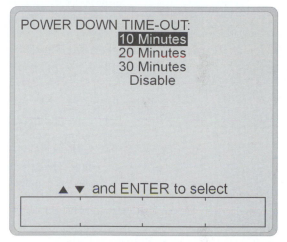

Figure 4-24: Edit Screen for POWER DOWN TIME-OUT

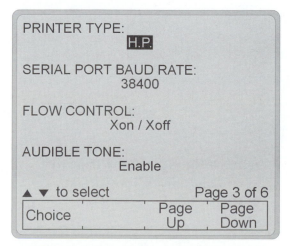

Figure 4-25: Screen for Page 3 of SETUP

TIP: You will see settings for handling the serial printing functions available with the control unit. These include PRINTER TYPE, SERIAL PORT BAUD RATE, and FLOW CONTROL. Unless your instructor is planning on printing out reports for your class activities, do not change these settings. The remaining setting on this page is for the AUDIBLE TONE.

DATE:
        5/03/02

DATE FORMAT:
        01/31/00

TIME:
        10:56:28 am

TIME FORMAT:
        12:00:00 am

▲ ▼ to select   Page 4 of 6

| Choice | Page Up | Page Down |

Figure 4-26: Screen for Page 4 of SETUP

27. Highlight the AUDIBLE TONE function, and if necessary press the ENTER key to set this parameter to Enable. Then page down to Page 4 of 6 for the display shown in Figure 4-26.

28. Using what you have already learned, use the arrow keys to set and enter the correct DATE: and TIME: information in the control unit.

29. Press the Page Down (4) key to get to the screen shown in Figure 4-27. Using what you have already learned, adjust the following settings, if necessary.

- LENGTH UNITS: Feet (ft)

- NUMERIC FORMAT: 00.0

- LANGUAGE: English

- POWER LINE FREQUENCY: 60 Hz

30. Press the Page Down (4) key to move to the next page of settings (6 of 6) on the control unit. The screen shown in Figure 4-28 appears.

LENGTH UNITS:
        Feet (ft)

NUMERIC FORMAT:
        00.0

LANGUAGE:
        English

POWER LINE FREQUENCY:
        60 Hz

▲ ▼ to select   Page 5 of 6

| Choice | Page Up | Page Down |

Figure 4-27: Screen for Page 5 of SETUP

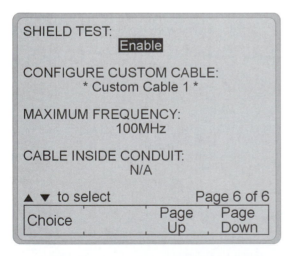

SHIELD TEST:
        Enable

CONFIGURE CUSTOM CABLE:
        * Custom Cable 1 *

MAXIMUM FREQUENCY:
        100MHz

CABLE INSIDE CONDUIT:
        N/A

▲ ▼ to select   Page 6 of 6

| Choice | Page Up | Page Down |

Figure 4-28: Screen for Page 6 of SETUP

31. Using what you have already learned, adjust the following settings, if necessary.

- SHIELD TEST: Enable

- CONFIGURE CUSTOM CABLE: * Custom Cable 1 *

- MAXIMUM FREQUENCY: 100MHz

- CABLE INSIDE CONDUIT: N/A

TIP: The CONFIGURE CUSTOM CABLE setting allows the user to set up the control unit for testing cables that possess other than standard parameters. You may look through the various screens that are used to define a custom cable, but there is no need to make any changes to the information you see.

32. Once all of the required settings have been completed, turn the control unit's rotary switch to OFF.

33. Be sure to document all of your settings on the Lab 16 work sheet.

34. Return the Level II testers (and charge units, if necessary) to their appropriate storage bags, along with any of the accessory cables you used.

35. Before leaving your lab work area, check to be sure that you are leaving it in a clean and orderly condition.

Lab 17 Questions

1. When does the audible tone sound on the Level II testers?

2. What type of cable would you use the remote unit to test?

3. What would the primary use of the control unit's backlight feature be?

4. How do you reactivate a control unit that has powered down in order to conserve battery power?

5. Why would it be necessary for the results of the various tests that are conducted by the Level II testers to be recorded and printed?

Lab 18 – Autotesting Patch Cables on Level II Testers

Lab Objective

To understand the procedure for autotesting previously created patch cables using the Level II testers, and documenting the results

Materials Needed

- Lab 18 work sheet

- Level II testers, w/case & cables (control unit and remote)*

- Users Manual for Level II testers

- RJ45 patch cable, greater than or equal to 2 meters in length (2)

- Cable, patch, CAT5 from Lab 4 (terminated with RJ45/T568A jacks)

- Cable, patch, CAT5 from Lab 5 (terminated with RJ45/T568B jacks)

- Cable, patch, CAT5 from Lab 6 (terminated with RJ45/T568A plugs)

- Cable, patch, CAT5 from Lab 7 (terminated with RJ45/T568B plugs)

- Cable, RJ45/T568A Jack/Wallplate combination

- Cable, RJ45/T568A jack/edge connector combination

- Cable, RG58 coax from Lab 9, 5-meter, (terminated with BNC connectors)

- Cable, RG8 coax from Lab 11, 5-meter, (terminated with N connectors)

- Adapter, RG8/BNC (2)

- Terminator, 50-ohm, BNC

- Straight-through barrel, BNC

- Label markers

* Prepare ahead of time for any lab procedures that involve the use of the Level II testers by charging the Nickel-Cadmium battery packs in both the control unit, and the smart remote. In an emergency situation, the units will operate during a charge operation, but they will have to remain near an ac outlet during this time.

Instructions

Before beginning this lab, read all of the following steps thoroughly, at least once. Be advised that this is a multi-part lab that will require several sessions to complete!

1. Locate the Users manual for the Level II meters, and review the list of activities that can be performed from the AUTOTEST menu of the control unit.

TIP: Notice that the information provided for the Autotest functions is divided between those designed for twisted pair cable and those designed for coaxial cables. Although the tests conducted for unshielded and shielded twisted pair are identical, the shielded variety is also tested for continuity along the shield.

Cable Test with RJ45/T568A Jack Terminations

2. Begin this procedure by closely examining Figure 4-29. Then, locate the CAT5 patch cable you created in Lab 4 (terminated with RJ45/T568A modular jacks).

TIP: Figure 4-29 displays a near minimum configuration. A minimum configuration would consist of only one cable. Keep in mind that the cables could be shown routed through a patch panel, wall plate, or various wiring closets. One thing to realize is that the Level II testers will not run an Autotest on twisted pair cable, unless a remote unit (or control unit operating in SMART REMOTE mode) is connected at one end of the wire run.

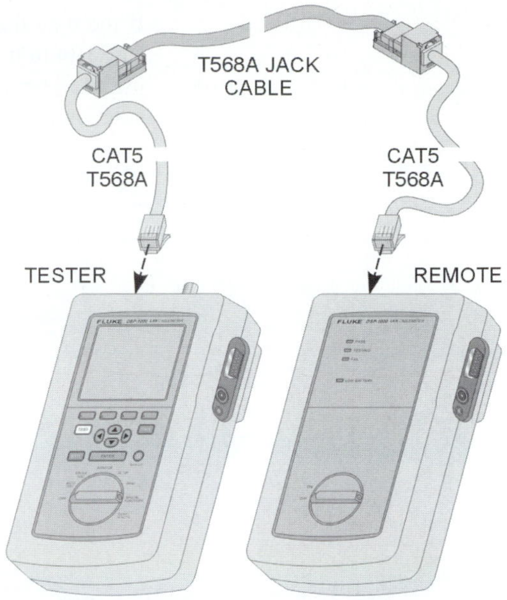

3. Create an identification label for the CAT5 patch cable you created in Lab 4 (terminated with RJ45/T568A modular jacks), and attach it.

4. Create a single cable by connecting a known good RJ45 patch cable (of 2 meters or more in length) to each end of your CAT5 patch cable (with T568A termination jacks) from Lab 4. Use the cables supplied with the Level II meters, if necessary, to make the cable combination.

TIP: The idea here is to examine the wiring condition of the cable you created in Lab 4. Because the other two cables are known to be good, any problems that occur should be traced to the cable under test.

Figure 4-29: Autotest Wiring Diagram for Jack Terminations

5. Turn the rotary function selector on the smart remote unit to the ON position.

TIP: The remote unit will emit several audible tones.

6. Connect the end of the combination cable that represents the far end of the network to the smart remote unit (using the RJ45 jack located in the unit's top).

TIP: Always check to be sure that there are no BNC cables attached to the control unit when testing twisted pair cabling.

7. Turn the rotary function selector on the control unit clockwise to the AUTOTEST menu position.

TIP: The control unit will emit several audible tones.

8. Check the settings displayed on the control unit's screen for accuracy, to be sure that the test being conducted is for the cable type under test. Then, if necessary, make any required adjustments from the control unit's SETUP menu.

9. Connect the end of the combination cable that represents the near end of the network to the control unit (using the RJ45 jack located in the unit's top).

10. To begin running the tests on the combination cable, press the TEST button on the control unit.

TIP: Depending on the wiring setup and the battery condition of the Level II testers, several things may happen. If for some reason the connection to the remote unit is not detected, or if the remote's batteries are low, you may receive a continuous Scanning for Remote message, rather than a very short one. In this case the Autotest will not be run until a properly charged remote unit is detected. Or, if a calibration message appears on the control unit's screen, perform the calibration procedure presented in Lab 16.

If the data that was generated from a previous test procedure was not saved, pressing the TEST button will give the control unit an opportunity to warn the user about this before going on. The user can choose to save the previous results, or simply discard the previous results by pressing the TEST button once again.

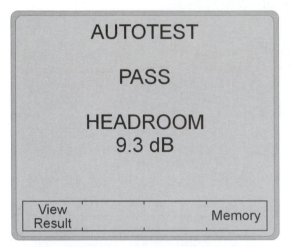

Figure 4-30: Autotest PASS and HEADROOM Report Screen

After several clicks and tones, the control unit will display a message similar to that shown in Figure 4-30, if all tests are passed. The HEADROOM reading shown will represent the worst case measurement of the four pairs under test. In the case where the wire map test fails, the word "Fail" will appear on the wire map screen, and the rest of the testing will be suspended until the user presses the Continue Test button (4).

11. To view the results of the current test, as shown in Figure 4-31, press the View Result button (1).

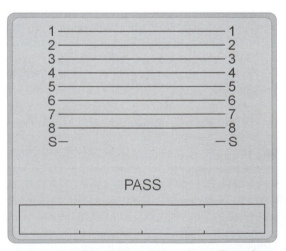

Figure 4-31: Autotest Test Results Screen

12. Record the test results in your Lab 18 work sheet for the CAT5 patch cable you created in Lab 4 (terminated with RJ45/T568A modular jacks).

13. Highlight the Wire Map parameter from the results listing, and press the ENTER button to view the information.

TIP: All four pairs of wires are displayed, with the various conditions of each wire depicted. Notice that for UTP cable, the shield wire is displayed as an open. If there are no problems detected, the display will appear as shown in Figure 4-32.

14. Check the Users Manual for examples of how various wire map problems will be depicted on the control unit's display.

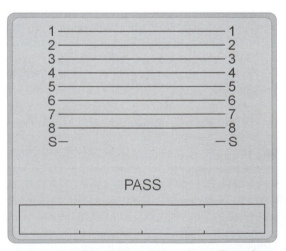

Figure 4-32: Wire Map Test Results Screen

15. Record the wire map test results on your work sheet, and if you detect a problem, be sure to clearly describe it, and mark the label on the cable accordingly.

TIP: If the cable under test is shown to have some type of problem, you should describe the problem as accurately as possible on your work sheet, and then continue checking the rest of the test results. Once you have finished checking the test results, ask your instructor about correcting any problems that were discovered with the current cable under test.

16. Press the EXIT button to return to the listing of test results, and highlight the Resistance parameter from the listing.

17. View the Resistance parameter results by pressing the ENTER button.

TIP: As shown in Figure 4-33, a typical reading should be far below the 40-ohm limit. The values shown (around 0.5 ohms per pair) are fine for fairly short cable runs. Long runs will show readings above those values, but should still remain below the limit.

18. Record the resistance test results on your work sheet, and if you detect a problem, be sure to clearly describe it, and mark the label on the cable accordingly.

19. Press the EXIT button to return to the listing of test results, and highlight the Length parameter from the listing.

20. View the Length parameter results by pressing the ENTER button.

TIP: The length is displayed according to the units you selected during the setup procedure. For longer lengths, a slight difference would be normal between the various pairs. A typical display for length is shown in Figure 4-34. If a measured length exceeds the limit, a warning is displayed.

21. Record the length test results on your work sheet, and if you detect a problem, be sure to clearly describe it, and mark the label on the cable accordingly.

22. Press the EXIT button to return to the listing of test results, and highlight the Propagation Delay parameter from the listing.

RESISTANCE

Pair	Resistance (Ω)
1,2	0.5
3,6	0.5
4,5	0.5
7,8	0.5

Figure 4-33: Resistance Test Results Screen

LENGTH

Pair	Length (ft)	Limit (ft)	Result
1,2	5	328	PASS
3,6	4	328	PASS
4,5	4	328	PASS
7,8	4	328	PASS

Figure 4-34: Length Test Results Screen

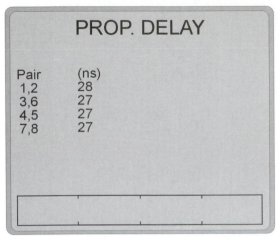

PROP. DELAY

Pair	(ns)
1,2	28
3,6	27
4,5	27
7,8	27

Figure 4-35: Propagation Delay Test Results Screen

23. View the Propagation Delay parameter results, an example of which is shown in Figure 4-35, by pressing the ENTER button.

TIP: The delay time (in nanoseconds) is given for each individual pair. In cases where a propagation delay is detected, the display includes the longest delay time detected of all four pairs.

24. Record the propagation delay test results on your work sheet, and if you detect a problem, be sure to clearly describe it, and mark the label on the cable accordingly.

25. Press the EXIT button to return to the listing of test results, and highlight the Delay Skew parameter from the listing.

26. View the Delay Skew parameter results, an example of which is shown in Figure 4-36, by pressing the ENTER button.

TIP: An important electrical characteristic of cables with regard to high-speed computer network performance is delay skew. This characteristic involves the movement of electrical signals traveling down a length of a multi-pair CAT5 cable, and arriving at the far end of the cable at different times, depending upon which twisted pair the signals used. The difference in time taken by these electrical signals between the fastest and slowest wire pairs is the measurement called the delay skew. If a computer system's operation is dependent on simultaneous arrival of these signals, problems will arise at some point.

DELAY SKEW

Pair	(ns)	Limit (ns)	Result
1,2	1	50	PASS
3,6	0	50	PASS
4,5	0	50	PASS
7,8	0	50	PASS

Figure 4-36: Delay Skew Test Results Screen

These problems seem to become more pronounced when values of delay skew approach 45 to 50 nanoseconds per 100 meters of CAT5 cable. As Figure 4-36 indicates, the Level II meters will fail any reading that goes above 50 ns, while recent industrial tests have suggested that delay skew values anywhere over 45 ns can be problematic. Keeping a CAT5 cable below a 45ns delay skew will result in optimized performance.

One of the main causes of CAT5 twisted pair cables exceeding this limit seems to be the production of 4-pair cable in which 2 of the pairs use a different type of dielectric than the remaining 2 pairs.

27. Record the delay skew test results on your work sheet, and if you detect a problem, be sure to clearly describe it, and mark the label on the cable accordingly.

28. Press the EXIT button to return to the listing of test results, and highlight the Impedance parameter from the listing.

29. View the Impedance parameter results, an example of which is shown in Figure 4-37, by pressing the ENTER button.

TIP: The characteristic impedance test will require the testing of a cable of at least 16 feet (5 meters) in length. A shorter length than this will automatically pass the test. If the cable under test exceeds the impedance limit specified by the test, the Result column will display the FAIL message. Any pair meeting the specified limit will be given a PASS result.

If a selected test standard does not require an impedance test, measurements that exceed the normal test limits will be issued a Warning as a test result. If some type of catastrophic failure (called an anomaly) is detected somewhere along the length of the cable under test, the distance to the failure will be displayed (in feet or meters), and the test result will be given a FAIL result. In cases where more than one catastrophic failure is detected, only the distance to the largest one is indicated.

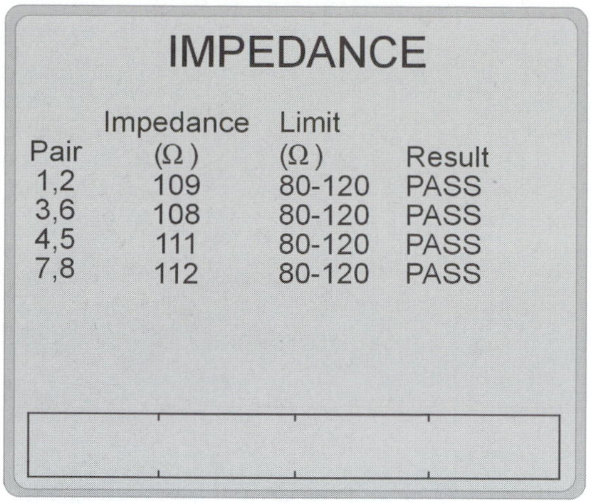

30. Record the impedance test results on your work sheet. If you detect a problem, be sure to clearly describe it and mark the label on the cable accordingly.

31. Press the EXIT button to return to the listing of test results, and highlight the NEXT parameter from the listing.

32. View the NEXT parameter results, an example of which is shown in Figure 4-38, by pressing the ENTER button.

TIP: This test is concerned with the levels of crosstalk that exist between the various wire pairs on the cable being checked. The measurement itself is expressed as the amplitude difference between the test signal and the signal being detected on another pair. The first wire pair listed is the pair into which the test signal is being injected, while the second wire pair listed is the pair from which the crosstalk is being measured. The measurement takes place at the end of the cable attached to the control unit. The screen shown in Figure 4-34 shows the test results for each pair listed.

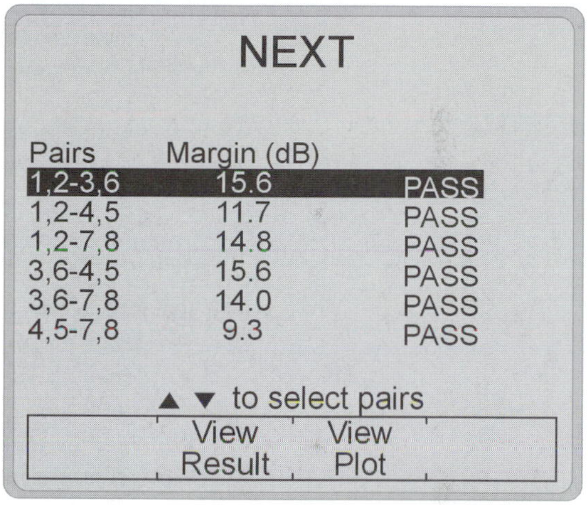

33. Highlight a specific wire pair, and press the View Result button (2) to see a detailed presentation of the test results for that pair, as shown in Figure 4-39.

TIP: The specific parameters reported for the selected pair include:

- **Pairs**–The specific pairs for which the displayed data is relevant.

- **Result**–The final determination as to whether the tested wire pairs meet the NEXT requirements for the selected test standard. If the result shown is PASS, then the NEXT between the tested pairs is higher than the standard specifications. A result that reports a FAIL indicates that the NEXT is lower than the standard.

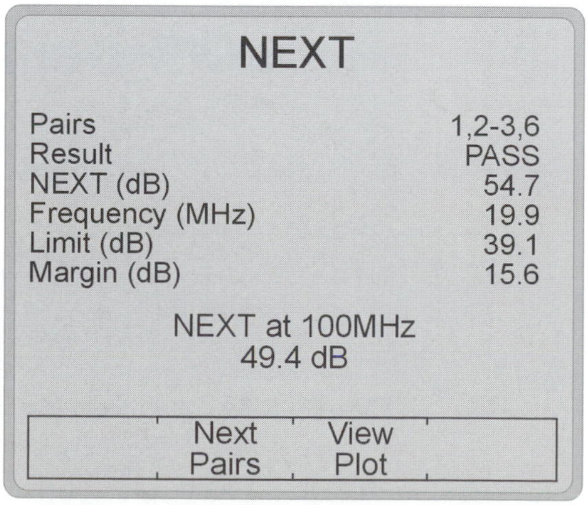

IMPEDANCE

Pair	Impedance (Ω)	Limit (Ω)	Result
1,2	109	80-120	PASS
3,6	108	80-120	PASS
4,5	111	80-120	PASS
7,8	112	80-120	PASS

Figure 4-37: Impedance Test Results Screen

NEXT

Pairs	Margin (dB)	
1,2-3,6	15.6	PASS
1,2-4,5	11.7	PASS
1,2-7,8	14.8	PASS
3,6-4,5	15.6	PASS
3,6-7,8	14.0	PASS
4,5-7,8	9.3	PASS

▲ ▼ to select pairs

| View Result | View Plot |

Figure 4-38: NEXT Test Results Screen

NEXT

Pairs	1,2-3,6
Result	PASS
NEXT (dB)	54.7
Frequency (MHz)	19.9
Limit (dB)	39.1
Margin (dB)	15.6

NEXT at 100MHz
49.4 dB

| Next Pairs | View Plot |

Figure 4-39: Detailed NEXT Test Results Screen

- **NEXT**–The NEXT value that is considered to be the worst-case measurement, or the value nearest to failure. If the value displayed does fall below the specification, it will be the one that is the farthest below.

- **Frequency**–The frequency at which the worst-case NEXT value is measured.

- **Limit**–For the worst-case frequency, the lowest acceptable NEXT value.

- **Margin**–The difference between the limit, and the measured NEXT value. If the measured NEXT value is higher than the limit, the number will be positive. If the measured NEXT is lower than the limit, the number will be negative.

34. Record the NEXT test results for this pair on your work sheet. If you detect a problem, be sure to clearly describe it and mark the label on the cable accordingly.

35. View the detailed NEXT results for the remaining pairs tested, record the results, and then return to the main AUTOTEST screen listing.

36. From the test results list, highlight the NEXT @ REMOTE parameter.

TIP: Recall that the NEXT @ REMOTE measurements are taken from the far end of the cable, and are sent to the control unit by the remote.

37. View the NEXT @ REMOTE parameter results, by pressing the ENTER button.

TIP: These test results are similar to the NEXT data except that the measurements are taken from the remote cable end.

38. Record the NEXT @ REMOTE test results for this pair on your work sheet. If you detect a problem, be sure to clearly describe it, and mark the label on the cable accordingly.

39. View the detailed NEXT @ REMOTE results for the remaining pairs tested, record the results, and then return to the main AUTOTEST screen listing.

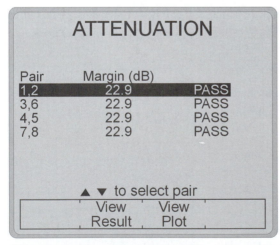

Figure 4-40: Attenuation Test Results Screen

40. From the test results list, highlight the Attenuation parameter, and press the ENTER key. The screen shown in Figure 4-40 appears.

TIP: The screen shown in Figure 4-40 provides a basic PASS or FAIL report for each pair. The test for attenuation is concerned primarily with how much signal strength is lost over the length of the cable. It is important to use the correct conduit or temperature settings for this measurement in order to avoid getting erroneous results. These settings were initially considered in Lab 17. If they need to be changed, they can be reached from the SETUP menu.

41. Highlight a specific wire pair, and press the View Result button (2) to see a detailed presentation of the attenuation test results for that pair, as shown in Figure 4-41.

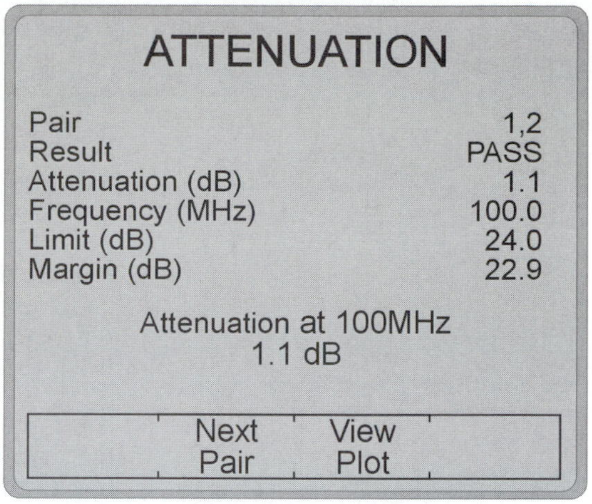

TIP: The specific parameters reported for the selected pair include:

- **Pair**–The specific pair for which the displayed data is relevant.

- **Result**–The final determination as to whether the tested wire pairs meet the Attenuation requirements for the selected test standard. If the result shown is PASS, then the attenuation for the tested pairs is lower than the standard specifications. A result that reports a FAIL indicates that the attenuation is higher than the standard.

- **Attenuation (dB)**–The attenuation value considered to be the worst-case percent margin, at its occurring frequency.

- **Frequency**–The frequency at which the worst-case percent margin occurred.

- **Limit**–For the worst-case frequency, the highest acceptable attenuation value.

- **Margin**–The difference between the limit, and the worst-case attenuation. If the measured attenuation value is lower than the limit, the number will be positive. If the measured attenuation is higher than the limit, the number will be negative.

42. Record the attenuation test results for this pair on your work sheet. If you detect a problem, be sure to clearly describe it and mark the label on the cable accordingly.

43. View the detailed attenuation results for the remaining pairs tested, record the results, and then return to the main AUTOTEST screen listing.

44. From the test results list, highlight the ACR parameter, and press the ENTER key. The screen shown in Figure 4-42 appears.

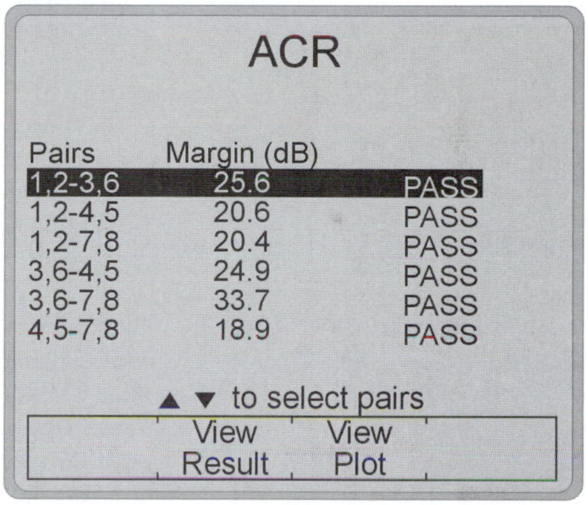

TIP: The screen shown in Figure 4-42 provides a basic PASS or FAIL report for each pair. The Attenuation to Crosstalk Ratio (ACR) is the difference between the measured values of NEXT, and the measured values of attenuation, for each wire pair combination. Obviously, the NEXT and attenuation values (in dBs) must already have been obtained before this parameter can be calculated.

45. Highlight a specific wire pair, and press the View Result button (2) to see a detailed presentation of the ACR test results for that pair, as shown in Figure 4-43.

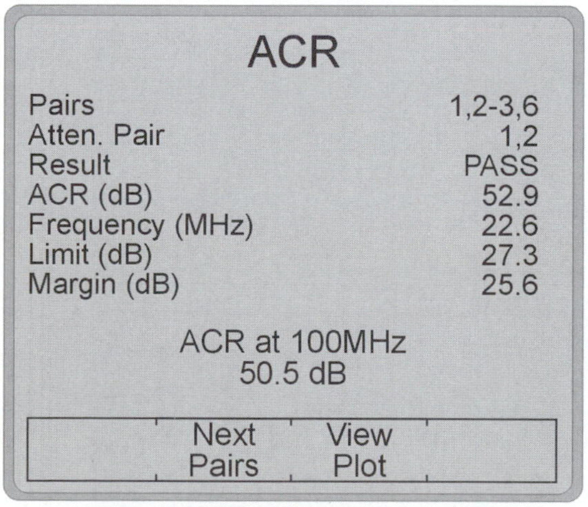

ATTENUATION

Pair	1,2
Result	PASS
Attenuation (dB)	1.1
Frequency (MHz)	100.0
Limit (dB)	24.0
Margin (dB)	22.9

Attenuation at 100MHz
1.1 dB

Next Pair	View Plot

Figure 4-41: Detailed Attenuation Test Results Screen

ACR

Pairs	Margin (dB)	
1,2-3,6	25.6	PASS
1,2-4,5	20.6	PASS
1,2-7,8	20.4	PASS
3,6-4,5	24.9	PASS
3,6-7,8	33.7	PASS
4,5-7,8	18.9	PASS

▲ ▼ to select pairs

View Result	View Plot

Figure 4-42: ACR Test Results Screen

ACR

Pairs	1,2-3,6
Atten. Pair	1,2
Result	PASS
ACR (dB)	52.9
Frequency (MHz)	22.6
Limit (dB)	27.3
Margin (dB)	25.6

ACR at 100MHz
50.5 dB

Next Pairs	View Plot

Figure 4-43: Detailed ACR Test Results Screen

TIP: The specific parameters reported for the selected pair include:

- **Pairs**–The specific wire pairs from which the crosstalk values used to calculate the ACR were taken.

- **Atten. Pair**–The wire pair from which the attenuation values used to calculate the ACR were taken.

- **Result**–The final determination as to whether the tested wire pairs meet the ACR requirements for the selected test standard. If the result shown is PASS, then the ACR for the tested pairs is higher than the standard specifications. A result that reports a FAIL indicates that the attenuation is lower than the standard.

- **ACR (dB)**–The ACR value considered to be the worst-case, or closest to exceeding the specifications. If the ACR does exceed specifications, the displayed value is the one that exceeds the specs by the greatest amount.

- **Frequency**–The frequency at which the worst-case ACR was calculated.

- **Limit**–For the worst-case frequency, the highest acceptable ACR value. The selected test standard defines the limit.

- **Margin**–The difference between the limit, and the worst-case ACR. If the worst-case ACR is higher than the limit, the number will be positive. If the worst-case ACR is lower than the limit, the number will be negative.

46. Record the ACR test results for this pair on your work sheet. If you detect a problem, be sure to clearly describe it, and mark the label on the cable accordingly.

47. View the detailed ACR results for the remaining pairs tested, record the results, and then return to the opening AUTOTEST screen.

48. Check with your instructor about saving the AUTOTEST results for storage in the Level II testers, or for printing out a report in a later activity. If you are instructed to save the results of this test, continue with the following steps. If not, disconnect the tested cable and go to step 52.

TIP: The control unit can store up to 500 AUTOTEST results. If the results are to be saved, this must be done prior to another AUTOTEST or SINGLE TEST being performed.

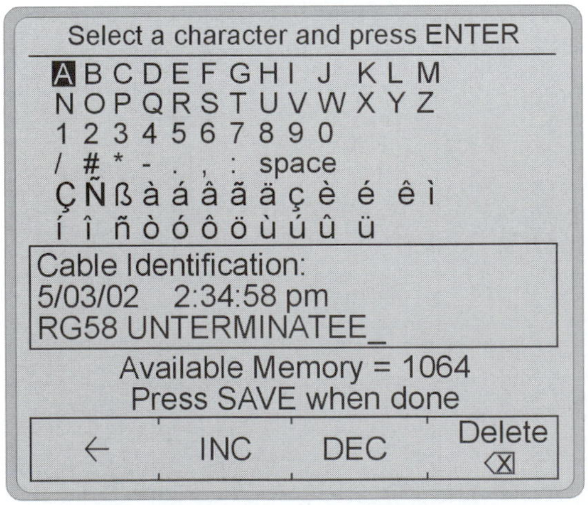

Figure 4-44: Saving AUTOTEST Results Screen

49. Once the AUTOTEST has been completed, press the SAVE button. The screen shown in Figure 4-44 appears.

50. Locate the Cable Identification block, and use the editing keys to place a name, for the cable you just tested, in the block. For example, CAT5 UTP JACKS1.

TIP: If necessary, read the Users Manual section entitled "Saving Autotest Results." Using a number for the ending character(s) for the cable identification name allows the Level II tester to automatically increment the name, provided that the auto increment function has been enabled. This is useful for testing a series of identical cables.

51. After completing the name for the cable you just tested, press the SAVE button to store the test results associated with the new cable name.

TIP: Two conditions may exist that will require some additional steps at this point. One of these is the situation where saving the AUTOTEST results fills the last remaining memory location in the Level II tester. This will cause a warning message to be displayed, indicating that the memory is full. The other condition is when the Level II tester's memory is already full. In this situation, any attempt to save additional test results will cause a message to be displayed indicating that additional data cannot be saved. At this point, previously saved test reports must first be deleted to make room for any new ones. Instructions for doing so can be found in the Users Manual.

Cable Test with RJ45/T568B Jack Terminations

52. Locate the CAT5 patch cable you created in Lab 5 (terminated with RJ45/T568B modular jacks).

53. Create an identification label for the CAT5 patch cable you created in Lab 5 (terminated with RJ45/T568B modular jacks), and attach it to the cable.

54. Repeat steps 4 through 47 for the CAT5 patch cable you created in Lab 5 (terminated with RJ45/T568B modular jacks).

TIP: The only difference here is that the cable under test will be wired for the TIA T568B configuration. The wire map should still indicate wire-to-wire continuity.

55. Record the test results in your Lab 18 work sheet for the CAT5 patch cable you created in Lab 5 (terminated with RJ45/T568B modular jacks).

56. Check with your instructor about saving the AUTOTEST results for storage in the Level II testers, or for printing out a report in a later activity. If you are instructed to save the results of this test, repeat steps 49 through 51 (using a unique file name where applicable). If not, disconnect the tested cable and go to step 57.

Cable Test with RJ45/T568A Plug Terminations

57. Locate the CAT5 patch cable you created in Lab 6 (terminated with RJ45/T568A modular plugs).

58. Create an identification label for the CAT5 patch cable you created in Lab 6 (terminated with RJ45/T568A modular plugs), and attach it to the cable.

59. Create one continuous cable by connecting a known good RJ45 patch cable (greater than or equal to 2 meters in length) to the far end of the CAT5 patch cable (with T568A termination jacks) you tested previously, from Lab 4.

60. Connect the CAT5 patch cable you created in Lab 6 (terminated with RJ45/T568A modular plugs) to the near end of the jacked CAT5 patch cable.

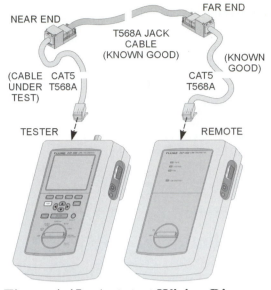

Figure 4-45: Autotest Wiring Diagram for T568A Plug Terminations

TIP: The idea here is to examine the wiring condition of the cable you created in Lab 6. Because the other two cables are known to be good, any problems that occur should be traced to the cable under test. See Figure 4-45, if necessary.

61. Repeat steps 5 through 47 for the CAT5 patch cable you created in Lab 6 (terminated with RJ45/T568A modular plugs).

62. Record the test results in your Lab 18 work sheet for the CAT5 patch cable you created in Lab 6 (terminated with RJ45/T568A modular plugs).

63. Check with your instructor about saving the AUTOTEST results for storage in the Level II testers, or for printing out a report in a later activity. If you are instructed to save the results of this test, repeat steps 49 through 51 (using a unique file name where applicable). If not, disconnect the tested cable and go to step 64.

Cable Test with RJ45/T568B Plug Terminations

64. Locate the CAT5 patch cable you created in Lab 7 (terminated with RJ45/T568B modular plugs).

65. Create an identification label for the CAT5 patch cable you created in Lab 7 (terminated with RJ45/T568B modular plugs), and attach it to the cable.

66. Create one continuous cable by connecting a known good RJ45 patch cable (greater than or equal to 2 meters in length) to the far end of the CAT5 patch cable (with T568B termination jacks) you tested previously, from Lab 5.

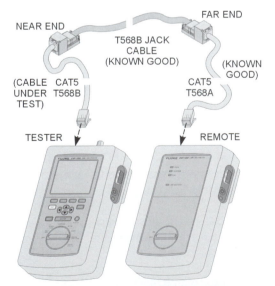

67. Connect the CAT5 patch cable you created in Lab 7 (terminated with RJ45/T568B modular plugs) to the near end of the jacked CAT5 patch cable.

TIP: The idea here is to examine the wiring condition of the cable you created in Lab 7. Because the other two cables are known to be good, any problems that occur should be traced to the cable under test. See Figure 4-46, if necessary, and take special note of the wire map test.

68. Repeat steps 5 through 47 for the CAT5 patch cable you created in Lab 7 (terminated with RJ45/T568B modular plugs).

Figure 4-46: Autotest Wiring Diagram for T568B Plug Terminations

69. Record the test results in your Lab 18 work sheet for the CAT5 patch cable you created in Lab 7 (terminated with RJ45/T568B modular plugs).

70. Check with your instructor about saving the AUTOTEST results for storage in the Level II testers, or for printing out a report in a later activity. If you are instructed to save the results of this test, repeat steps 49 through 51 (using a unique file name where applicable). If not, disconnect the tested cable and go to step 71.

Cable Test with Jack/Wallplate Combination

71. Locate the RJ45/T568A modular jack cable and wall plate combination you created in Lab 13 (terminated with an RJ45/T568A modular plug).

72. Create an identification label for the RJ45/T568A modular jack cable and wall plate combination you created in Lab 13 (terminated with an RJ45/T568A modular plug), and attach it to the cable.

73. Create one continuous cable by connecting a known good RJ45 patch cable (greater than or equal to 2 meters in length) into the jack end of the RJ45/T568A modular jack cable and wall plate combination.

TIP: The idea here is to examine the wiring condition of the cable you created in Lab 13. Because the cable you connected to its jack end is known to be good, any problems that occur should be traced to the cable under test. See Figure 4-47, if necessary.

74. Repeat steps 5 through 47 for the RJ45/T568A modular jack cable and wall plate combination you created in Lab 13 (terminated with an RJ45/T568A modular plug).

TIP: Consider the RJ45/T568A modular plug to be the far end of the cable.

75. Record the test results in your Lab 18 work sheet for the RJ45/T568A modular jack cable and wall plate combination you created in Lab 13 (terminated with an RJ45/T568A modular plug).

76. Check with your instructor about saving the AUTOTEST results for storage in the Level II testers, or for printing out a report in a later activity. If you are instructed to save the results of this test, repeat steps 49 through 51 (using a unique file name where applicable). If not, disconnect the tested cable and go to step 77.

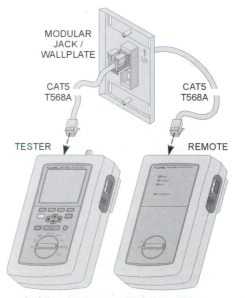

Figure 4-47: Autotest Wiring Diagram for Jack/Wallplate Terminations

Cable Test with Jack/Edge-Connector Combination

77. Locate the RJ45/T568A modular jack cable and edge-connector combination you created in Lab 14 (terminated with a T568A RJ45 unshielded modular plug).

78. Create an identification label for the RJ45/T568A modular jack cable and edge-connector combination you created in Lab 14 (terminated with an RJ45/T568A unshielded modular plug), and attach it to the cable.

79. Create one continuous cable by connecting a known good RJ45 patch cable (greater than or equal to 2 meters in length) into the jack end of the RJ45/T568A modular jack cable and edge-connector combination.

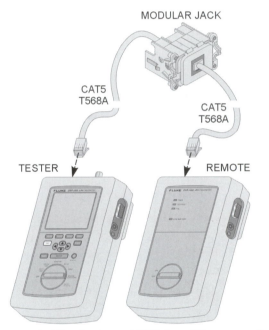

MODULAR JACK

CAT5
T568A

CAT5
T568A

TESTER

REMOTE

Figure 4-48: Autotest Wiring Diagram for Jack/Edge-Connector Terminations

TIP: The idea here is to examine the wiring condition of the cable you created in Lab 14. Because the cable you connected to its jack end is known to be good, any problems that occur should be traced to the cable under test. See Figure 4-48, if necessary.

80. Repeat steps 5 through 47 for the RJ45/T568A modular jack cable and edge-connector combination you created in Lab 14 (terminated with an RJ45/T568A modular plug).

TIP: Consider the RJ45/T568A modular plug to be the far end of the cable.

81. Record the test results in your Lab 18 work sheet for the RJ45/T568A modular jack cable and edge-connector combination you created in Lab 14 (terminated with an RJ45/T568A modular plug).

82. Check with your instructor about saving the AUTOTEST results for storage in the Level II testers, or for printing out a report in a later activity. If you are instructed to save the results of this test, repeat steps 49 through 51 (using a unique file name where applicable). If not, disconnect the tested cable and go to step 83.

Cable Test with RG58 Terminated with BNCs

83. Locate the 5-meter, RG58 coaxial cable (terminated with BNC connectors) that you created in Lab 9.

84. Create an identification label for the 5-meter, RG58 coaxial cable (terminated with BNC connectors) that you created in Lab 9, and attach it to the cable.

TIP: The idea here is to examine the wiring condition of the cable you created in Lab 9. Because this cable has already passed the preliminary continuity checks, any problems that occur will be related to other cable parameters.

85. Closely examine Figure 4-49.

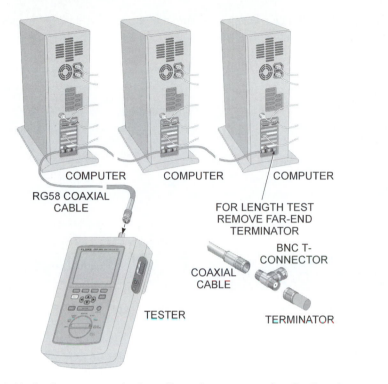

COMPUTER   COMPUTER   COMPUTER

RG58 COAXIAL
CABLE

FOR LENGTH TEST
REMOVE FAR-END
TERMINATOR

BNC T-
CONNECTOR

COAXIAL
CABLE

TESTER   TERMINATOR

**Figure 4-49:
Autotest Wiring
Diagram for RG58
Network**

TIP: Figure 4-49 displays a networked configuration connected to the Level II tester, consisting of several computers cabled together (with a terminator attached to the end node). Keep in mind that the cables could be routed through a patch panel, wall plate, or various wiring closets.

One thing to realize is that the Level II testers will not run an accurate impedance test on any coaxial cable, unless the cable itself is at least 16 feet in length (5 meters). If the cable is less than 16 feet long, a terminated (50 ohms) cable will always pass the impedance test, while an unterminated cable will always fail it. The RG58 coaxial cable (terminated with BNC connectors) that you created in Lab 9 should meet the 16-foot minimum requirement.

Remember that you don't need to have a network such as the one shown in Figure 4-49 to run some tests on the RG58 coaxial cable (terminated with BNC connectors) that you created in Lab 9.

As long as the cable is at least 16 feet long, it can be tested as a single section, as shown in Figure 4-50 (terminator attached). If you do run tests on an actual network, remember to turn OFF all of the computers connected to the cable(s) under test. This prevents any of the nodes from interfering with the readings being taken during the tests.

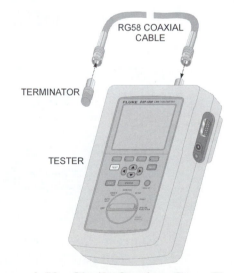

RG58 COAXIAL
CABLE

TERMINATOR

TESTER

86. Before beginning the test on the RG58 coaxial cable (terminated with BNC connectors) that you created in Lab 9, locate the Level II tester control unit's SETUP menu. Use the manual for the Level II tester for guidance if necessary.

**Figure 4-50: Single Section Coax Testing
Diagram**

87. Turn the Level II control unit on, and rotate its switch to the SETUP position.

88. From the SETUP menu, check to be sure that the test standard being used is the IEEE 10Base2, and that the cable type is RG58, as shown in Figure 4-51.

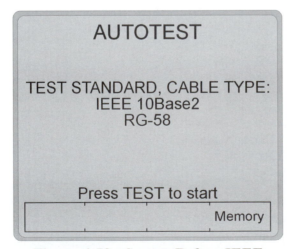

```
TEST STANDARD, CABLE TYPE:
     SELECT TEST STANDARD:
Aus/NZ Class C Channel
Aus/NZ Class C Basic Lnk
TIA Cat 3 Channel
TIA Cat 3 Basic Link
ISO11801 EN50173 Class B
ISO11801 EN50173 Class A
IEEE 10Base2
IEEE 10Base5
Coax Cables
                        Page 3 of 5
▲ ▼ and ENTER to select
               ┌──────┬──────┐
               │ Page │ Page │
               │ Up   │ Down │
```

```
TEST STANDARD, CABLE TYPE:
     SELECT CABLE TYPE:
10Base2
RG-58
RG-58 Foam
NVP = 66.0%




▲ ▼ and ENTER to select
```

Figure 4-51: IEEE 10Base2 Test Standard and RG58 Cable Type

89. Page down the screens to check that the correct date and time are reported by the Level II tester's control unit. Make any necessary adjustments, and press the SAVE button, if necessary.

TIP: Never change the settings on the tester to allow a test to pass that would fail at the correct settings.

90. Determine if the Level II tester should report cable length. If so, make sure that the cable to be tested does not have a terminator installed on its far end.

91. On the Level II tester control unit, rotate its switch to the AUTOTEST position.

```
        AUTOTEST



TEST STANDARD, CABLE TYPE:
      IEEE 10Base2
         RG-58




    Press TEST to start
                      ┌────────┐
                      │ Memory │
```

Figure 4-52: Screen Before IEEE 10Base2 RG58 Autotest

TIP: The screen on the Level II tester should display something similar to Figure 4-52.

92. Remove any cables that may be attached to the control unit's RJ45 connector.

93. Connect the near end of the RG58 coaxial cable (terminated with BNC connectors) that you created in Lab 15 to the BNC connector on the control unit.

94. Press the TEST button on the control unit to begin the testing.

95. Compare the results of your test with those shown in Figure 4-53.

TIP: Notice that the Resistance test shows a FAIL result, and is interpreted as an open circuit. This is because of the fact that the terminator was removed from the far end of the cable(s) under test.

96. Record the test results in your Lab 18 work sheet for the RG58 coaxial cable (terminated with BNC connectors) that you created in Lab 15, and be sure to stipulate that these results were for an unterminated cable.

97. With the Level II tester still displaying the previous information, use the BNC straight-through barrel to place the 50-ohm terminator onto the far end of the RG58 coaxial cable (terminated with BNC connectors) that you created in Lab 9.

```
TEST STANDARD, CABLE TYPE:
         IEEE 10Base2
            RG-58
AUTOTEST                      PASS
Resis.(Ω)       Limit(Ω)
  50.3          48.0-65.0     PASS
Length          Limit
  (ft)            (ft)
        No Reflection
Imped.(Ω)       Limit(Ω)
   51           42-58         PASS

┌─────────────────────────────┐
│           Memory            │
└─────────────────────────────┘
```

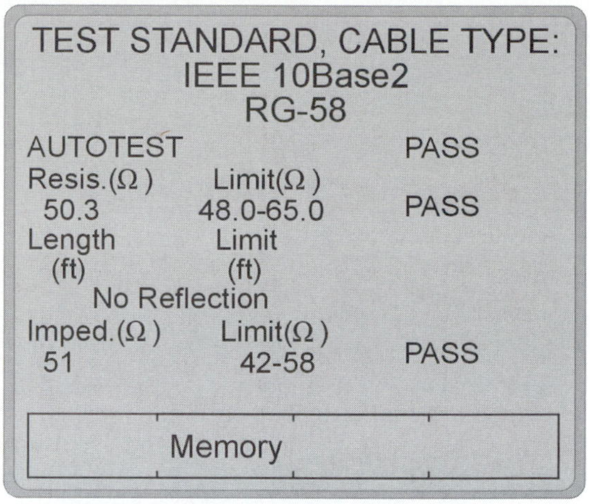

Figure 4-53: Autotest Results for IEEE 10Base2 RG58

98. Press the TEST button on the Level II tester.

99. Check with your instructor about saving the AUTOTEST results for storage in the Level II testers, or for printing out a report in a later activity. If you are instructed to save the results of this test, repeat steps 49 through 51 (using a unique file name where applicable). If not, go directly to step 100.

TIP: Keep in mind that when a cable installation is being made for a customer, it's an important consideration to make a second copy of the documentation to leave at the premises.

100. Press the TEST button on the Level II tester again.

101. Record the test results in your Lab 18 work sheet for the RG58 coaxial cable (terminated with BNC connectors) that you created in Lab 9, and be sure to stipulate that these results were for a terminated cable.

TIP: The NO REFLECTION report for the Length parameter indicates a correct terminating resistance. With no signal reflections, the Level II tester cannot measure the cable's length.

102. Check with your instructor about saving the AUTOTEST results for storage in the Level II testers, or for printing out a report in a later activity. If you are instructed to save the results of this test, repeat steps 49 through 51 (using a unique file name where applicable). If not, disconnect the tested cable and go directly to step 103.

Cable Test with RG8 Terminated with N Connectors

103. Locate the 5-meter, RG8 coaxial cable (terminated with N connectors) that you created in Lab 11.

104. Create an identification label for the 5-meter, RG8 coaxial cable (terminated with N connectors) that you created in Lab 11, and attach it to the cable.

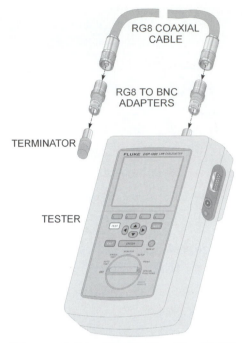

RG8 COAXIAL
CABLE

RG8 TO BNC
ADAPTERS

TERMINATOR

TESTER

**Figure 4-54: Single Section RG8
Coax Testing Diagram**

TIP: The idea here is to examine the wiring condition of the cable you created in Lab 11. Because this cable has already passed the preliminary continuity checks, any problems that occur will be related to other cable parameters.

105. Closely examine Figure 4-54.

TIP: Notice that this diagram is similar to Figure 4-46, except that the N connectors have been outfitted with the RG8 to BNC adapters for use with the control unit and the terminator.

106. Before beginning the test on the RG8 coaxial cable (terminated with N connectors) that you created in Lab 11, locate the Level II tester control unit's SETUP menu.

107. Turn the Level II control unit on, and rotate its switch to the SETUP position.

108. From the SETUP menu, check to be sure that the test standard being used is the IEEE 10Base5, and that the cable type is RG8, as shown in Figure 4-55.

```
TEST STANDARD, CABLE TYPE:
      SELECT TEST STANDARD:
Aus/NZ Class C Channel
Aus/NZ Class C Basic Lnk
TIA Cat 3 Channel
TIA Cat 3 Basic Link
ISO11801 EN50173 Class B
ISO11801 EN50173 Class A
IEEE 10Base2
IEEE 10Base5
Coax Cables
                      Page 3 of 5
▲ ▼ and ENTER to select
                  Page    Page
                   Up     Down
```

```
TEST STANDARD, CABLE TYPE:
      SELECT CABLE TYPE:
10Base5
RG-8
NVP = 84.0%




▲ ▼ and ENTER to select
```

Figure 4-55: IEEE 10Base5 Test Standard and RG8 Cable Type

109. Check for the correct date and time on the tester's control unit. Make any necessary adjustments, followed by pressing the SAVE button, if necessary.

110. Determine if the Level II tester should report cable length. If so, make sure that the cable to be tested does not have a terminator installed on its far end.

111. On the Level II tester control unit, rotate its switch to the AUTOTEST position.

TIP: The screen on the Level II tester should display something similar to Figure 4-56.

112. Remove any cables that may be attached to the control unit's RJ45 connector.

113. Connect the near end of the RG8 coaxial cable (terminated with N connectors) that you created in Lab 11 to the BNC connector on the control unit.

TIP: To do this, first connect the RG8 cable to the N side of one of the RG8 to BNC adapters. Then, connect the BNC side of the adapter to the Level II tester.

114. Press the TEST button on the control unit to begin the testing.

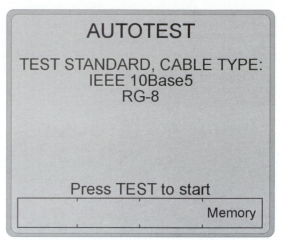

Figure 4-56: Screen Before IEEE 10Base5 RG8 Autotest

115. Compare the displayed results of your test with those shown in Figure 4-57.

TIP: Notice that the Resistance test shows a FAIL result, and is interpreted as an open circuit. This is because the terminator was removed from the far end of the cable(s) under test.

116. Record the test results in your Lab 18 work sheet for the RG8 coaxial cable (terminated with N connectors) that you created in Lab 11, and be sure to stipulate that these results were for an unterminated cable.

TEST STANDARD, CABLE TYPE:
IEEE 10Base5
RG-8

AUTOTEST		FAIL
Resis.(Ω)	Limit(Ω)	
OPEN	48.0-65.0	FAIL
Length (ft)	Limit (ft)	
16.2	1640	PASS
Imped.(Ω)	Limit(Ω)	
52	42-58	PASS

Memory

Figure 4-57: Autotest Results for IEEE 10Base5 RG8

117. With the Level II tester still displaying the previous information, place the 50-ohm terminator onto the remaining RG8 to BNC adapter. Then, connect the adapter to the far end of the RG8 coaxial cable (terminated with N connectors) that you created in Lab 11.

118. Press the TEST button on the Level II tester.

119. Check with your instructor about saving the AUTOTEST results for storage in the Level II testers, or for printing out a report in a later activity. If you are instructed to save the results of this test, repeat steps 49 through 51 (using a unique file name where applicable). If not, go directly to step 120.

TIP: Keep in mind that when a cable installation is being made for a customer, it's an important consideration to make a second copy of the documentation to leave at the premises.

120. Press the TEST button on the Level II tester again.

121. Record the test results in your Lab 18 work sheet for the RG8 coaxial cable (terminated with N connectors) that you created in Lab 11, and be sure to stipulate that these results were for a terminated cable.

TIP: The NO REFLECTION report for the Length parameter indicates a correct terminating resistance. With no signal reflections, the Level II tester cannot measure the cable's length.

122. Check with your instructor about saving the AUTOTEST results for storage in the Level II testers, or for printing out a report in a later activity. If you are instructed to save the results of this test, repeat steps 49 through 51 (using a unique file name where applicable). If not, disconnect the tested cable and go directly to step 123.

123. Check with your instructor about correcting any cables that failed their respective standards on the Level II testers, and make any required corrections.

124. Return the Level II testers (and charge units, if necessary) to their appropriate storage bags, along with any of the accessory cables you used.

125. Before leaving your lab work area, check to be sure that you are leaving it in a clean and orderly condition.

Lab 18 Questions

1. From cables you made during previous lab procedures, which, if any, failed their respective test standards on the Level II testers?

2. Of those tests performed during an AUTOTEST, which ones are specified by the TIA/EIA T568A or T568B standards?

3. Which tests performed on twisted pairs required the use of a smart remote unit?

4. On average, how long did the Level II testers take to provide the AUTOTEST results for one of the CAT5 twisted pair cables you tested?

5. What other cable or connector types besides CAT5 UTP RJ45 terminations are the Level II testers capable of testing?

6. What differences, if any, were evident during the various wire map tests using the CAT5 T568A and T568B cables? Explain.

Basic Link Test Configuration

When testing a basic link, it will verify that performance has been met on the permanent portion of the channel. During the testing, a maximum of 2 meters of patch cords are permitted at both ends of the link. This allows a total of 94 meters of cable. The testing procedure itself is simple to perform. One member of a two-person team goes to the wiring closet, unplugs a cable from the hub or patch panel, and attaches it to the testing device. The other connects the remote unit of the tester to the terminating end of the cable at the user's work area.

The device runs an autotest—a slate of predetermined tests—and the cable gets its grade. The results are stored in the testing unit, and the person at the terminating end moves to the next wall jack to repeat the process. But there's a little more to it than that. TSB 67 sets out the stipulations for testing cable. The spec defines the two test configurations, the Basic Link, as shown in Figure 4-58, and the Channel.

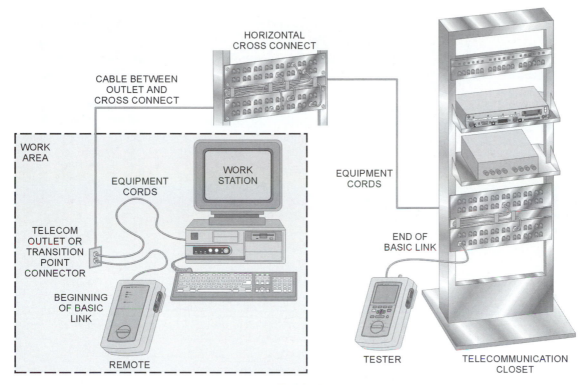

Figure 4-58: Basic Link Testing

As explained previously, Basic Link testing covers just the permanent portion of the cabling from the wall outlet to the first point of termination in the telecommunications closet. The cable contractor/installer is responsible for the testing of this portion, because it's usually installed before the hubs, PCs, and other hardware are in place.

Channel Test Configuration

Testing a channel verifies that the required performance has been met on the entire horizontal run. When testing a channel, including patch cords, a maximum of 100 meters is allowed. Because the Channel testing, shown in Figure 4-59, covers the entire cable run—cable, patch cords, and all connections in between—it's usually, but not always, the responsibility of the network manager to perform.

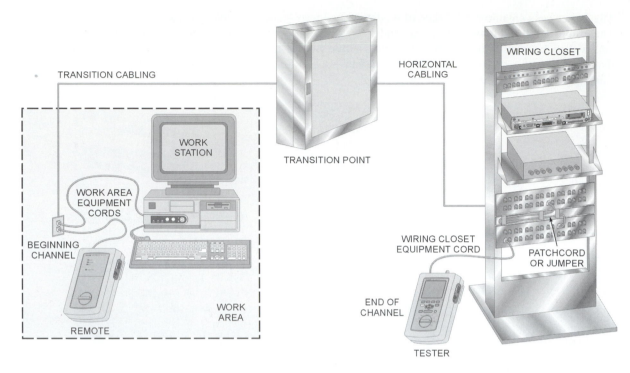

Figure 4-59: Channel Testing

CHAPTER SUMMARY

This chapter introduced you to the most important aspects involved in testing and trouble-shooting newly installed CAT5 cabling systems, and presented the basic criteria on which the certification of CAT5 installations are based. You first learned how the cable installer is protected from blame for later system failures by thoroughly testing the newly installed cabling according to the specifications contained in the TIA/EIA T568A standard and the TSB 67. Later certification of the installation will protect the reputation of the contractor responsible for the cabling work and the installer/s.

You learned the differences between a channel and a basic link, and the maximum allowable cable lengths for each. The importance of the use of quality workmanship and system components was stressed, while the various penalties to be paid in system performance were listed with respect to NEXT dB values. In addition, the subject of properly labeling all cable runs was revisited, with the ability to identify any individual wire run at any time during the life of the installation being the ultimate goal.

Much information was given about the adoption by the TIA/EIA of TSB 67, which defines the various required field tests to be performed prior to any CAT5 cabling installation being certified. You were given specific information regarding the four required tests (wire map, length, attenuation, and NEXT) for all CAT5 cabling installations, as well as several additional tests (ELFEXT, power sum ELFEXT, and return loss), which may be performed under certain circumstances.

The subject of delay skew was given special attention, and how it has become a source of trouble for cable manufacturers, especially when high-speed networks are installed for use on CAT5 cables using non-Teflon coatings. You were given some details about how CAT5 cable testing equipment provided erroneous information with regards to the length of individual pairs within the suspect cables. You learned how this situation brought about the need for cable testers that would measure delay skew, and why the TIA/EIA T568A standard will be changed in order to add this new testing parameter to the guidelines.

You were introduced to several types of test tools now used by cable installation crews to ensure that new wire plants will pass certification. While some types of checks can be done using simple continuity testers (low-end), you learned that more sophisticated tools are needed to perform most of the required field tests (high-end). You were given examples of low-end and high-end testers, as well as the names of the companies that manufacture the more sophisticated handheld cable testers, and several illustrations of their products. You also learned the difference between Level I and Level II testers.

Lab 16 provided an opportunity to get some hands-on experience with Level II testers. After performing the calibration procedure, you ran their internal self-testing routines and documented the results. Setting up the Level II testers was covered during Lab 17, dealing with such parameters as test standards, cable types, report identification, custom headers, remote end testing, auto increment, backlight time-out, power down time-out, audible tone, date, time, length units, numeric format, language, power line frequency, shield test, maximum frequency, and cable inside conduit.

Lab 18 was fairly lengthy, and dealt with making Level II tests on patch cables that you created during previous lab procedures. These included the CAT5 patch cable from Lab 4 (terminated with RJ45/T568A modular jacks), the CAT5 patch cable from Lab 5 (terminated with RJ45/T568B modular jacks), the CAT5 patch cable from Lab 6 (terminated with RJ45/T568A modular plugs), the CAT5 patch cable from Lab 7 (terminated with RJ45/T568B modular plugs), the RJ45/T568A modular jack cable and wallplate combination, the RJ45/T568A modular jack cable and edge connector combination, the RG58 coaxial cable from Lab 9 (terminated with BNC connectors), and the RG8 coaxial cable from Lab 11 (terminated with N connectors).

You then learned some more about the basic link test configuration, including how the testers are connected to perform the tests. This configuration was contrasted with the configuration for the channel tests, and the differences between the permanent and the changeable portions of a cable run were stressed.

REVIEW QUESTIONS

The following questions test your knowledge of the material presented in this chapter:

1. What is a channel, as it applies to a horizontal cable run?

2. How is a basic link different from a channel?

3. Why should network cabling be tested to a standard before the network equipment is connected to it?

4. What is the maximum horizontal cable length for a basic link?

5. How are Level I and Level II cable testers different?

6. State the maximum length of patch cords that can be used when testing a basic link.

7. If equipment is repositioned after a certification test has been performed, what action must be taken?

8. What is NEXT?

9. What action is required if a run passes a test with the minimum acceptable values?

10. How would wrapping a cable around a 3-inch pipe affect its NEXT performance rating?

CHAPTER
5

INDUSTRY STANDARDS

OBJECTIVES

Upon completion of this chapter and its related lab procedures, you should be able to perform these tasks:

1. Describe various ANSI/TIA/EIA standards for commercial buildings and residential settings.

2. Discuss the specific Technical Systems Bulletins that can affect telecommunications installations.

3. Recognize and use standard grounding and bonding requirements for telecommunications equipment.

4. Describe documentation methods that are used for cabling in a commercial building.

CHAPTER 5

Industry Standards

INTRODUCTION

It is very important for cable installation personnel to be familiar with the many different standards that may be encountered at one time or another while working in the industry. The following information represents extremely condensed versions of various standards that may be required as guidelines during a given installation. The installation technician should recognize the name and the content of each of these standards, and should be aware that these standards are constantly being revised to accommodate the changing technology.

STANDARDS AND BULLETINS

You will notice that many of the standards listed here include the ANSI acronym preceding the TIA/EIA with which you have already become familiar. Because the **American National Standards Institute (ANSI)** is also involved in the standardization aspect of telecommunications cabling, it is correctly displayed alongside the TIA/EIA when referring to the various standards. It wasn't shown previously in order to save space, but you should be aware that it belongs there.

American National Standards Institute (ANSI)

For 80 years, ANSI has served in its capacity as the administrator and coordinator of the private-sector voluntary standardization system of the United States. The Institute was founded in 1918 by five engineering societies and three government agencies, and remains a private, nonprofit membership organization supported by a diverse constituency of private- and public-sector organizations. ANSI facilitates the development of American National Standards (ANSs) by establishing a consensus among various qualified groups through its headquarters in New York City, and its satellite office in Washington, D.C. While remaining committed to supporting the development of national, and in many cases, international standards, ANSI-accredited developers are addressing the critical trends of technological innovation, marketplace globalization, and regulatory reform.

ANSI is the sole U.S. representative and dues-paying member of the two major non-treaty international standards organizations, the **International Organization for Standardization (ISO)**, and, via the U.S. National Committee (USNC), the **International Electrotechnical Commission (IEC)**. ANSI was a founding member of the ISO, and plays an active role in its governance. ANSI is one of five permanent members to the governing ISO Council, and one of four permanent members of ISO's Technical Management Board. Through ANSI, the United States has immediate access to the ISO and IEC standards development processes. In many instances, U.S. standards are taken forward, through ANSI or its USNC, to the ISO or IEC, where they are adopted in whole or in part as international standards.

International Organization for Standardization (ISO)

International Electrotechnical Commission (IEC)

The cabling standards that follow are in no way to be considered an all-inclusive listing. This is due to the fact that the standardization process is ongoing, with new documents being discussed and created, and older standards constantly being revised and updated.

The success of these efforts often is dependent on the willingness of U.S. industry and government to commit the resources required to ensure strong U.S. technical participation in the international standards process. The installation professional must keep informed of the latest additions and updates to the standards library, especially to those areas that directly impact his/her job performance.

ANSI/TIA/EIA 568

First released in 1991, **ANSI/TIA/EIA 568**, also known as the **Commercial Building Telecommunications Standard**, was developed as a response to the lack of universal Telco services, a situation that resulted in serious limitations with respect to system operation, adaptability, and maintenance.

The lack of a standard for building telecommunications cabling systems was a notable concern for companies representing the telecommunications and computer industries. At the beginning of 1985, the **Computer and Communications Industry Association (CCIA)** requested that the Electronic Industries Association (EIA) develop this necessary standard. It wasn't until July of 1991 that the first version of the standard was published as TIA/EIA 568. As such, TIA/EIA 568 was first to define a standard for generic telecommunications cabling systems that could accommodate varied applications. However, this standard only defined requirements for applications operating at frequencies up to 16 MHz.

ANSI/TIA/EIA T568A

In August of 1991 a Technical Systems Bulletin TSB 36 was published with specifications for higher grades (CAT4, CAT 5) of UTP. Then, in August of 1992, TSB 40 was published addressing higher grades of UTP-connecting hardware. In January of 1994 TSB 40 was revised to TSB 40A, to deal with UTP patch cords in more detail, and to clarify testing requirements for UTP modular jacks.

The existing 568 standard was revised, absorbing the previous TSB revisions to become TIA/EIA T568A. Just as TSB 36 and TSB 40A were absorbed into this revised standard, along with other revisions, additional revisions and TSBs can also be anticipated in the future.

First published in 1995, **ANSI/TIA/EIA T568A**, also known as the **Commercial Building Telecommunications Cabling Standard**, specifies a generic telecommunications cabling system for commercial buildings that will support a multi-product, multi-vendor environment, and provides information to be utilized for the design of telecommunications products for commercial enterprises. Obviously, since the publication of the ANSI/TIA/EIA T568A standard, ANSI/TIA/EIA 568 has been withdrawn and is no longer valid.

In order to fulfill the objectives set forth in ANSI/TIA/EIA T568A, the standard specifies the minimum requirements for telecommunications cabling within an office environment. It also includes recommended topologies, and maximum reliable cabling distances. Various performance guidelines are included, along with the required media parameters and proper connector/pin assignments to ensure reliable interconnectability. The expected useful life of telecommunications cabling systems that conform to ANSI/TIA/EIA T568A are reported as being in excess of ten years.

Outside the United States, the ISO has developed (also an ongoing process) a cabling standard called the ISO/IEC 11801 on an international basis, under the title **Generic Cabling for Customer Premises Cabling**, and there is an equivalent Canadian document that is called CSA T529. Some differences between TIA/EIA T568A and ISO/IEC 11801 exist, but nothing worth going to war over!

As you are already aware, this standard addresses the six major components of a structured wiring system. These components are:

- the Entrance Facilities—providing the point where outside cabling interfaces with the intra-building backbone cabling—the physical requirements of the network interface being defined in the ANSI/TIA/EIA 569 standard.

- the Main/Intermediate Crossconnect—located in the Equipment Room and serving as the interfacing between the Entrance Facilities and the Backbone Distribution System.

- the Backbone Distribution System—provides the interconnection between the Telecommunication Closets, Equipment Rooms, and Entrance Facilities, and consists of the backbone cables, Intermediate and Main Crossconnects, mechanical terminations, and the patch cords or jumpers used for backbone-to-backbone crossconnections.

- the Horizontal Crossconnect—provides the interconnection between the Work Areas and the Telecommunications Closets, typically on one floor of a building.

- the Horizontal Distribution System—extends from the Work Area telecommunications outlet to the Telecommunications Closet, and consists of the horizontal cabling, Telecommunications Outlet, cable terminations, and all related crossconnections.

- the Work Area(s)—components that extend from the Telecommunications Outlet to the station equipment, using wiring that is simple to interconnect, move, and change.

Once again, the purpose of this standard is to provide the means of planning and installing of a structured cabling system for commercial buildings, and this revision replaces the older ANSI/TIA/EIA 568 standard. See your instructor if you wish to have information regarding the purchase of the current ANSI/TIA/EIA T568A publication, or to find out if a copy is already available at your location, or education facility.

ANSI/TIA/EIA 569

ANSI/TIA/EIA 569, also known as the **Commercial Building Standard for Telecommunications Pathways and Spaces**, is a document that emphasizes the proper design of those building components applicable to the premise telecommunications infrastructure.

Therefore, this document seeks to standardize specific design and construction practices within, and between (primarily commercial), buildings that are in support of telecommunications media and equipment. Accordingly, there are standards given for rooms, areas, and pathways into and through which telecommunications equipment and media are installed.

The scope of ANSI/TIA/EIA 569 is limited to the telecommunications aspects of a commercial building's design and construction, specifically encompassing telecommunications considerations both within, and between, the associated buildings. Telecommunications aspects are generally the pathways into which the telecommunications media are placed, and the rooms/areas that are used to terminate the media, and install the accompanying telecommunications equipment.

Originating in 1990, this standard is a result of a joint effort by the **Canadian Standards Association (CSA)** and the Electronics Industries Association (EIA). Although it is published separately in the United States and Canada, the core sections of the document are both very similar. This standard addresses the following elements of building pathways and spaces:

- Backbone Pathways—consist of intra-building (within a building) and inter-building (between buildings) pathways, and may be either vertically or horizontally oriented.

   - Intra-building Pathways—provide the means for placing backbone cables from the entrance facilities or space, to telecommunications closets, and from equipment rooms, to either the entrance facilities or space, or telecommunications closets. They consist of conduit, sleeves or slots, and trays.

   - Inter-building Pathways—interconnect separate buildings, such as those in campus environments, consisting of underground, buried, aerial, or tunnel pathways.

- Entrance Facilities—consists of the telecommunications service entrance to the building, including the entrance point through the wall, and continuing to the entrance room or space, and may contain the backbone pathways that link to other buildings in campus environments. Separation of telecommunication circuits from power circuits and other electromagnetic energy sources is currently being studied by the standards committees of the ANSI/TIA/EIA.

- Equipment Room—centralized space for telecommunications equipment, which should house only equipment directly related to the telecommunications system and its environmental support systems.

   - Sizing—must meet the known requirements of specific equipment. If equipment requirements are unknown, plan for 0.75 square feet (0.07 square meters) of equipment room space for every 100 square feet (10 square meters) of workstation space. Regardless of equipment used, there must be a minimum of 150 square feet (14 square meters).

   - Special-use building sizing—must base ER size on the number of workstations.

- Horizontal Pathways—should provide the facilities for the installation of telecommunications cable from the telecommunications closet, to the work area telecommunications outlet. Design guidelines and procedures are specified for underfloor, access-floor, conduit, tray and wireway, ceiling, and perimeter facilities.

- Telecommunications Closet—must be dedicated to telecommunications functions and support facilities only, with a minimum of one closet per floor. Additional closets should be provided for each area up to 10,000 square feet (1,000 square meters) when either the floor area served exceeds 10,000 square feet (1,000 square meters), or the horizontal distance exceeds 300 feet (90 meters).

- Workstation—primarily concerned with the telecommunications outlets, because they are the usual connection point between the horizontal cable and the connecting devices in the work area. The reference is made to the overall box or face plate as opposed to individual outlets/connectors, with a minimum of one outlet face plate per workstation required. The typical workstation space allocation is one every 100 square feet (10 square meters).

Because specific standards are given for spaces and pathways into and through which telecommunications equipment and media are installed, it can be seen that part of the usefulness of this standard is its reference in documents such as bid requests, specifications, and contracts leading up to the construction of new facilities. Therefore, TIA/EIA 569 should also prove useful to the team that is responsible for delivering a well-designed facility to the customer—the architects, engineers, and the construction industry in general.

In review, ANSI/TIA/EIA 569 provides recommendations for the physical spaces and pathways used by modern telecommunications cabling systems and equipment, with recommendations made in the following areas:

- Telecommunications spaces, including work areas, telecommunications closets, equipment rooms, and entrance facilities. Specifications provided include:
  - The sizing of telecommunications spaces.
  - Floor loading requirements—dynamic and static.
  - Environmental considerations—floor coverings, lighting requirements, and so on.

- Telecommunications pathways, including horizontal and backbone cable distribution systems—conduits, sleeves, trays, and so on, and grounding pathways. Specifications provided include:
  - The number of cable pathways to install.
  - Acceptable types of pathways.
  - Installation procedures.

See your instructor if you wish to have information regarding the purchase of the current ANSI/TIA/EIA 569 publication, or to find out if a copy is already available at your location, or education facility.

ANSI/TIA/EIA 570

ANSI/TIA/EIA 570, also known as the **Residential and Light Commercial Telecommunications Wiring Standard**, is a document that provides recommendations for the selection and installation of cabling systems that are utilized specifically in residential, and smaller, commercial premises.

This document recognizes that residential properties are also legitimate environments for computing equipment components such as those found in commercial offices. Also recognizing the increasing need of individuals to work at home, and access remote networks, this document offers suggestions for designing the telecommunications cabling systems of residences, and smaller, commercial properties, for LANs, and for access to remote systems.

ANSI/TIA/EIA 570

Residential and Light Commercial Telecommunications Wiring Standard

While the overall reason for the existence of ANSI/TIA/EIA 570 is to address the wiring standards for residential and light commercial premises, its stated purpose is to provide the minimum requirements for the connection of up to four exchanges/access lines to various types of customer-premises equipment. It applies specifically to telecommunications premises wiring systems that are installed for end users within an individual building, residential housing unit (single family or multi-occupant), or light commercial establishment. See your instructor if you wish to have information regarding the purchase of the current ANSI/TIA/EIA 570 publication, or to find out if a copy is already available at your location or education facility.

ANSI/TIA/EIA 606

ANSI/TIA/EIA 606, also known as the **Administration Standard for the Telecommunications Infrastructure of Commercial Buildings**, provides recommendations for the documentation and administration of the premise telecommunications infrastructure.

Originally a bulletin, this standard has been incorporated into ANSI/TIA/EIA T568A, and was ratified in February, 1993. It promotes the use of an administration scheme that remains independent of the applications utilizing this infrastructure, and specifies the administrative requirements of the telecommunications infrastructure within a new, an existing, or a renovated (primarily commercial) building or campus.

This infrastructure can be thought of as the collection of those components (telecommunications spaces, cable pathways, grounding, wiring, and termination hardware) that provide the basic support for the distribution of all information within a building or campus.

The administration of the telecommunications infrastructure includes documentation (labels, records, drawings, reports, and work orders) of cables (media, terminations, splices), termination hardware, patching and crossconnect facilities, conduits, other cable pathways, telecommunications closets, other telecommunications spaces, grounding, and bonding.

The reason why the administration scheme should remain uniform, and independent of applications, is because most (if not all) applications will change several times throughout the life of a building, and its cabling system. The ANSI/TIA/EIA 606 standard establishes reliable guidelines for owners, end users, manufacturers, consultants, contractors, designers, installers, and facilities administrators involved in the administration of the telecommunications infrastructure, or related administration system.

This standard recommends documenting the cabling system in the following way:

- Assign a unique identifier to each element of the cabling system.

- Create a document for each of the identified elements.

- Link related records to each other.

Once all of the elements of the cabling system have been documented, the information gathered can be used to produce a number of reports. This is very important, because one of the most significant challenges for network managers is to maintain up-to-date records. If changes are made to the cabling system, and these changes are not registered, the situation becomes much the same as if the system was never documented in the first place.

Documenting the cabling system is a great benefit to the LAN administrator, because of the fact that LAN cabling is often blamed for any network problems. Being able to trace the individual components of the cabling system greatly simplifies any troubleshooting procedures that may be required.

See your instructor if you wish to have information regarding the purchase of the current ANSI/TIA/EIA 606 publication, or to find out if a copy is already available at your location, or education facility.

ANSI/TIA/EIA 607

ANSI/TIA/EIA 607, also known as the **Grounding and Bonding Requirements for Telecommunications in Commercial Buildings**, is a document that provides bonding and grounding recommendations for the premise telecommunications infrastructure.

This standard specifies the requirements for a uniform telecommunications grounding and bonding infrastructure, that should be followed within commercial buildings where telecommunications equipment is intended to be installed. Its purpose is to enable the planning, design, and installation of telecommunications grounding systems within a building, with or without prior knowledge of the telecommunication systems that will subsequently be installed. Obviously then, the grounding and bonding requirements must be satisfied prior to the start of any cabling installation.

This telecommunications grounding and bonding infrastructure should support a multivendor, multi-product environment, as well as the grounding practices for the various systems that may later be installed on customer premises. With this in mind, ANSI/TIA/EIA 607 specifies the requirements for:

- A ground reference for telecommunications systems within the telecommunications entrance facility, the telecommunications closet, and equipment room.

- Bonding and connecting pathways, cable shields, conductors, and hardware at telecommunications closets, equipment rooms, and entrance facilities.

- Interconnectivity to other building grounding systems.

Make no mistake! Telecommunications systems must be grounded in accordance with NEC Article 250 (Grounding), and ANSI/TIA/EIA 607. Entire systems, including all telecommunication racks within the building, must be connected to a dedicated ground via No. 6 (or larger) insulated green ground wire. It is no longer an acceptable practice to use water pipes for grounding purposes, and this standard is very specific regarding what must be done in order to comply. That is not to say, however, that unique problems cannot arise from time to time concerning the proper grounding technique to be used under certain circumstances.

For example, suppose that you are performing some telecommunications upgrading on one floor of a high-rise office building, and you discover that there is no existing telecommunications ground wire installed. You now have a decision to make. Should you take the time, and increase the cost of this installation to the customer, to run an entirely new ground wire? Or would it be permissible to ground one end of the metallic sheath of a new riser cable to the electrical ground at the penetration point, and the other end to a power panel box (connected by a licensed electrician, of course) where the riser terminates on the floor?

Actually, the action you have been contemplating has in fact been the practice of the telephone operating companies for years. While the ANSI/TIA/EIA 607 was intended primarily to provide direction for the design of new commercial buildings, it is strongly suggested that you use it as a guide for the upgrade or retrofit of existing buildings as well. It is the opinion of some experts in the field that ANSI/TIA/EIA 607 attempts to do for network integrity, what the National Electrical Code (NEC) does for the safety of life and limb.

In point of fact, if the riser cable was an intra-building backbone cable, the 1996 NEC does not require you to ground its cable shield. However, for an inter-building backbone cable, Article 800-33 of the 1996 NEC insists that an insulating joint, or equivalent device, be used to ground the metallic sheath of any communications cables entering a building. It further emphasizes that the cable must be grounded as close as is practical to its point of entrance, which is considered to be the point of emergence through an exterior wall, a concrete floor slab, or from a grounded rigid metal conduit, or a grounded intermediate metal conduit.

It must be noted that these recommendations do not replace the bonding and grounding requirements of national and local electrical codes, which can be obtained from each city, or other current safety regulations. This document makes bonding and grounding recommendations for the purpose of meeting standard telecommunications performance requirements.

As such, the ANSI/TIA/EIA 607 document should be useful to anyone engaged in the telecommunications design, maintenance, renovation, or retrofit of new or existing buildings. This standard should also be useful to manufacturers of telecommunications equipment and to those responsible for purchasing, installing, or operating such equipment and devices. See your instructor if you wish to have information regarding the purchase of the current ANSI/TIA/EIA 607 publication, or to find out if a copy is already available at your location, or education facility.

TSB 36, 40, & 40A

TSB 36

Additional Cable Specifications for Unshielded Twisted Pair Cables

TSB 40

TSB 40A

Additional Transmission Specifications for Unshielded Twisted Pair Connecting Hardware

The information about category cable ratings was released in two Technical Systems Bulletins, TSB 36 and TSB 40. These bulletins recommended several revisions and additions to ANSI/TIA/EIA 568, and added the Category rating system, consequently replacing the old Level system.

TSB 36, also known as the **Additional Cable Specifications for Unshielded Twisted Pair Cables**, provided more stringent requirements for the transmission characteristics of high-performance unshielded twisted pair cables. These cables were not current for the 1991 ANSI/EIA/TIA 568 specification standard, because they were still under development, and their performance levels were not firmly established when the standard was published. As previously mentioned, this bulletin was incorporated into ANSI/EIA/TIA T568A.

TSB 40, while also covering some additional specifications for UTP cables, added specifications for connecting devices, such as jacks, crossconnect blocks, and patch panels. As already mentioned, this TSB has also been integrated into the main body of the ANSI/TIA/EIA T568A standard document.

TSB 40A, also known as the **Additional Transmission Specifications for Unshielded Twisted Pair Connecting Hardware**, specified the transmission performance requirements for UTP-connecting hardware, that were consistent with the three categories of UTP cable specified in ANSI/TIA/EIA TSB 36, as well as some additional requirements for crossconnect jumpers, and for cables used for UTP patch cords.

This document contains the minimum set of transmission parameters, and their associated limits, necessary to assure that properly installed connectors will have minimal effects on cable performance. These requirements apply only to individual UTP connectors, including telecommunications outlets, patch panels, transition connectors, and crossconnect blocks. As with its older siblings, TSB 40A was also incorporated into ANSI/TIA/EIA T568A. See your instructor if you wish to have information regarding the purchase of the current ANSI/TIA/EIA T568A publication, or to find out if a copy is already available at your location or education facility.

TSB 53

TSB 53, also known as the **Extended Specifications for 150-Ohm STP Cables and Data Connectors**, was a document of 1992 vintage that extended the frequency operating range of 150-ohm STP cabling systems (cables and compatible connectors) from 20 MHz (Type 1) to 300 MHz (Type 1A). Such cabling was to be known as STP-A.

Obviously, because this technical bulletin covered cable and connector specifications extending up to 300 MHz, the previously existing specifications were no longer supported, except that all of the normally valid component and installation practices were still to be met, or exceeded, including the applicable building and safety codes.

Specifications for STP-A telecommunications cable include:

- Color-coding pair 1 as Red/Green, and pair 2 as Orange/Black.

- Shielded, 150-ohm 2-pair cable using 22 AWG (0.63 mm) solid conductors.

- Should be marked as "150 ohm STP-A", in addition to any markings required by local or national safety codes.

- Same mechanical and transmission requirements to be applied to backbone and horizontal cables, with additional requirements provided for outdoor runs.

Specifications for STP-A connectors include:

- Standard outlet interface and pair assignments are similar to the ISO 8802-5 Token-Ring connector (IEC 807-8), except that performance requirements are more severe.

- Marked as "STP-A", with other markings required by local or national safety codes.

Specifications for STP-A patch cables include:

- Stranded conductors at 26 AWG (0.4 mm).

- Allows for an overall shield, rather than individually shielded pairs.

- Less severe attenuation and NEXT loss requirements than horizontal cable.

Specifications for STP-A installation practices include:

- Shields should be grounded at the Telecommunication Grounding Busbar.

- Shield voltage to ground should not exceed 1 Vrms at the work area.

- Shield resistance to ground should not exceed 3.5 ohms at the work area.

- For runs between buildings at different ground potentials, fiber is recommended.

As published, the second edition of the T568 cabling standard (ANSI/TIA/EIA T568A) takes precedence over this technical bulletin. See your instructor if you wish to have information regarding the purchase of the current ANSI/TIA/EIA T568A publication, or to find out if a copy is already available at your location or education facility.

TSB 67

TSB 67

Transmission Performance Specifications for Field Testing of UTP Cabling Systems

TSB 67, also known as the **Transmission Performance Specifications for Field Testing of UTP Cabling Systems**, is a document that specifies the electrical characteristics of field testers, test methods, and minimum transmission requirements for UTP cabling. The same committee that originated the ANSI/TIA/EIA T568A standard and the TSB 67 requirements for field testing of CAT5 cabling links, is now working to add several new guidelines that will meet the needs of 1000baseT (Gigabit Ethernet) systems.

Take a good look at Table 5-1 to see where the successors to TSB 67 will be taking us.

Table 5-1: TIA/EIA and ISO/IEC Standards Specifications

Proposed Standard	Transmission Parameter	Channel Performance	Improvement over Previous Standard	Highest Frequency
New CAT5	NEXT @ 100 MHz	27.1 dB		100 MHz
	PSNEXT @ 100 MHz	N.A.		
	Attenuation @ 100 MHz	24.0 dB		
	ELFEXT @ 100 MHz	17.0 dB		
	PSELFEXT @ 100 MHz	14.4 dB	Recommendations for 1000BaseT	
	Return Loss @ 100 MHz	8.0 dB		
	Propagation Delay @ 10 MHz	555 ns		
	Delay Skew	50 ns		
	Attenuation-to-Crosstalk Ratio	3.1 dB		
Enhanced CAT5 (proposed)	NEXT @ 100 MHz	30.1 dB	+ 3.0 dB	100 MHz
	PSNEXT @ 100 MHz	37.1 dB	new	
	Attenuation @ 100 MHz	24.0 dB	same	
	ELFEXT @ 100 MHz	17.4 dB	+ 0.4 dB	
	PSELFEXT @ 100 MHz	14.4 dB	same	
	Return Loss @ 100 MHz	10.0 dB	+ 2.0 dB	
	Propagation Delay @ 10 MHz	555 ns	same	
	Delay Skew	50 ns	same	
	Attenuation-to-Crosstalk Ratio	6.1 dB		
	PS Attenuation-to-Crosstalk Ratio	3.1 dB		
CAT6 (proposed)	NEXT @ 100 MHz	39.9 dB	+ 9.8 dB	250 MHz
	PSNEXT @ 100 MHz	37.1 dB	+ 10.0 dB	
	Attenuation @ 100 MHz	21.2 dB	− 2.8 dB	
	ELFEXT @ 100 MHz	23.2 dB	+ 5.8 dB	
	PSELFEXT @ 100 MHz	20.2 dB	+ 5.8 dB	
	Return Loss @ 100 MHz	12.0 dB	+ 2.0 dB	
	Propagation Delay @ 10 MHz	555 ns	same	
	Delay Skew	50 ns	same	

Table 5-1: TIA/EIA and ISO/IEC Standards Specifications (cont.)

Proposed Standard	Transmission Parameter	Channel Performance	Improvement over Previous Standard	Highest Frequency
New Class D	NEXT @ 100 MHz	27.1 dB		
	PSNEXT @ 100 MHz	24.1 dB		
	Insertion Loss	24.0 dB		
	ELFEXT @ 100 MHz	17.0 dB		
	PSELFEXT @ 100 MHz	14.4 dB		
	Return Loss @ 100 MHz	10.0 dB		
	Propagation Delay @ 10 MHz	555 ns		
	Delay Skew	50 ns		
	Attenuation-to-Crosstalk Ratio	3.1 dB		
	PS Attenuation-to-Crosstalk Ratio	0.1 dB		
Class E (proposed)	NEXT @ 100 MHz	39.9 dB		
	PSNEXT @ 100 MHz	37.1 dB		
	Insertion Loss	21.7 dB		
	ELFEXT @ 100 MHz	23.2 dB		
	PSELFEXT @ 100 MHz	20.2 dB		
	Return Loss @ 100 MHz	12.0 dB		
	Propagation Delay @ 10 MHz	555 ns		
	Delay Skew	50 ns		
	Attenuation-to-Crosstalk Ratio	18.2 dB		
	PS Attenuation-to-Crosstalk Ratio	15.4 dB		

Check over the new proposed guidelines for CAT5, CAT5E, and CAT6. The newer test parameters for CAT5 and Class D+, such as ELFEXT, return loss, propagation delay, and delay skew, should help to determine the ability of installed CAT5 cabling to support the Gigabit Ethernet LAN standard.

CAT5E includes several tightened specifications that take into account the additional Far-End Crosstalk (FEXT) contributed by the cross connections, while CAT6 will be specified for transmission speeds of up to 250 MHz.

Another critical option to consider is fiber. Technology is paying more and more attention to fiber optic systems, as the push to address the demand for gigaspeeds continues to drive the market. With this type of momentum building up, fiber stands a good chance of making the big jump all the way to a direct computer connection. The new buzz words now circulating are **Fiber To The Desktop (FTTD)**, and FTTD is the new buzz acronym. Although this course is specifically oriented towards CAT5 cabling certification objectives, the cabling installer may wish to augment his/her certification training by completing a Fiber Cabling Installation Certification course as well. Consult the **Evolving Technologies Association International (ETAI)** for details concerning any specifically recommended certification curriculum.

One of the questions cable installers may be asking themselves is whether it is worth buying a new tester for testing to CAT5E (or ClassD+ for ISO/IEC) certification? This is a valid question, considering the price of Level IIE, handheld testers. These standards only require a bandwidth of 100 MHz, and it is likely that your current CAT5 tester comes close to verifying the performance of both the new CAT5 and CAT5E/ClassD+ standards.

Fiber To The Desktop (FTTD)

Evolving Technologies Association International (ETAI)

The reality of your situation might not dictate that CAT5 and CAT5E/ Class D+, of and by themselves, would warrant the purchase of new tester. However, if your tester is to be relied on to test a CAT6/ClassE installation, it will need to have the ability to interface with the various CAT6/ClassE connectors, while accurately testing channel links. Additionally, it should be capable of testing fiber (assuming fiber certification), provide excellent diagnostics, and do it all fairly quickly. Testing all the runs of a major installation, while using tediously slow verification techniques, may test the sanity of the installing technician as well.

Information leading to a practical solution to the issues related to field-testing is presented in ANSI/TIA/EIA TSB 67, and much of the detail concerning this standard was presented in the previous chapter. See your instructor if you wish to have information regarding the purchase of the current ANSI/TIA/EIA TSB 67 publication, or to find out if a copy is already available at your location or education facility.

TSB 75

TSB 75, also known as the **Additional Horizontal Cabling Practices of Open Offices**, is a document that specifies the optional practices of open office environments, for any horizontal telecommunications cabling recognized in ANSI/TIA/EIA T568A. It specifies optional connection schemes and topologies that are easily modified, when portions of the horizontal cabling and pathways supported by office furniture, or movable partitions, are frequently reconfigured.

Because part of the horizontal cabling in a typical open office normally runs directly through furniture pathways, or connects to the furniture itself, moving the furniture becomes difficult. It usually involves replacing the entire horizontal cabling run.

Because of the fact that user organizations began demanding offices that were much easier to reconfigure, these types of issues were addressed by the ANSI/TIA/EIA TSB 75 bulletin. The standard was intended to make open-office reconfiguration less difficult, and was expected to harmonize with ISO/IEC 11801, the international generic cabling standard.

Churn and Flexibility

The escalating calls for higher productivity, lower cost, and efficient customer service have led to increased levels of **churn** (high rates of reorganization and rearrangement in offices). This situation is compounded by the assignment of increasing numbers of special projects that require quick completion. As substantial numbers of temporary employees are retained, they require that networking connections be made readily available. If critical specialists are shared (moving from project to project quickly, as needed), most of their furniture and computers will be portable in nature, requiring that network and power connections be ready to fire up. When the special project is completed, these connections must remain unobtrusive until/unless they are needed again.

If a conference room is not available on short notice, an ad hoc work team meeting could necessitate the rearrangement of its own space, and the reconfiguration of individual work areas may involve moving walls or panels. In this age of increasing manipulation of knowledge, the requirements of individual privacy for concentrated work conflict with the open-spaces concept that facilitates instant, face-to-face communications. Shifting walls or panels can once again solve the problem.

The negative aspects of this continuous office reconfiguration can be reduced significantly by reducing, or eliminating, the need to alter the cabling runs through building pathways. This would obviously save time, and if the necessary changes can be made more quickly, downtime can be greatly reduced. The cost of downtime and lost productivity has often been the deciding factor when a needed reconfiguration is delayed, or rejected.

Workspace Strategy

Suppose the work areas were outfitted with remateable connections at the points where cabling attaches to the outlets. The furniture could be moved, or replaced, without disturbing the cabling runs within building spaces. Such a scenario suggests that users could make the necessary connection changes for themselves, rather than waiting for the facilities staff to do it for them. This would reduce the time wasted between work-area reconfigurations.

This idea is not without controversy, however. The administration of the network itself could suffer problems, due to the possibility of unskilled users making connection errors, or the outright loss of control by network managers. A compromise between network management and workspace efficiency will undoubtedly have to be worked out on a case-by-case basis.

If new developments allowed for the easy managing of networks and infrastructures, everyone could soon be a network manager. After all, the idea that everyone could have a telephone was once a ridiculous notion, before the dial telephone was invented. Before that time, placing a call was totally dependent on a live telephone operator making the necessary trunk connections. The dial telephone now makes everyone using it, in effect, an operator.

Distribution efforts in both electric and telephone service have proven the fact that by placing the distribution equipment closer to the end user, costs can be reduced and system flexibility can be increased. While this trend has not been greatly reflected in TSB 75 as yet, one should remember that cabling standards remain in a constant state of flux, and are subject to change at regular intervals.

TSB 75 Provisions

Prior to the development of modular devices, telecommunications outlets typically required connection points on the mating cable in the back, and another at the workstation side. With reconfigurations becoming more and more of a natural business requirement, what may appear to be a simple matter of moving the furniture and equipment can create havoc on a formally structured cabling system. The existing cables can get switched or damaged, and the required slack parameters may be difficult to maintain. If distances are substantially increased, the horizontal runs from the TC may have to be replaced and reinstalled. The expense and disruption inherent in these kinds of activities can make the very word "reconfiguration" symbolic of foreboding disaster.

Figure 5-3 depicts a simple work area reconfiguration that is greatly simplified by the use of a MUTOA. In this particular example, each MUTOA is designed to service a maximum of four work areas. Notice how the horizontal cabling running to the TC remains intact when the configuration changes from A to B. The only wiring changes involved the patch cords running between the MUTOAs and the work areas.

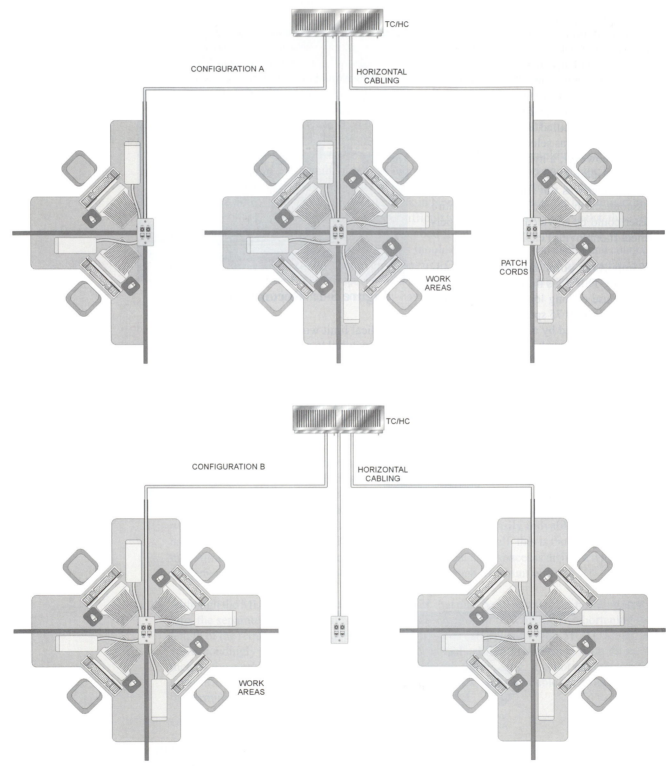

Figure 5-3: Using a MUTOA to Reconfigure a Work Area

CP

When work-area cables connect the telecommunications outlets directly to the terminal equipment, the interconnection between horizontal cables in building pathways and horizontal cables passing through work-area pathways (furniture) is provided by the Consolidation Point (CP). This is different from the MUTOA in that the interconnection point takes place completely within the horizontal cabling, as shown in Figure 5-4.

Whereas the MUTOA actually serves as the termination point for the horizontal cables, the CP actually extends the horizontal cabling right into telecommunications outlets within individual office locations. Though the CP is located very near the furniture clusters, similar to MUTOA, in order to reach the individually dedicated work-area outlets, short horizontal cables are run from the CP. This concept allows most of the horizontal cabling to remain intact during the reconfiguration process, but not all.

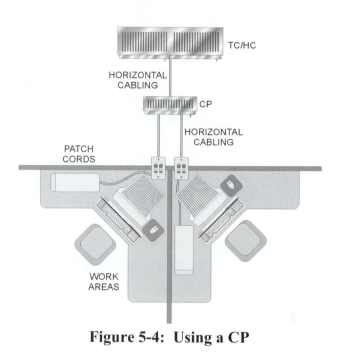

Figure 5-4: Using a CP

As you might guess, the CP is designed for use within office space that is occasionally reconfigured, but not often. While the placement of the CP is such that disruption of the horizontal cables located in the building pathways does not occur, the short horizontal cables may be removed or reinstalled as required. Figure 5-5 depicts the process for a CP that goes from configuration A to configuration B.

The required work can be performed quickly and inexpensively, while no disruption need occur within any other cluster. The outlets in the work area can be configured with any of the equipment interfaces that may be required.

The CP is housed in a suitable enclosure, as is the MUTOA, and consists of connecting hardware that serves as an interconnect, rather than a crossconnect. This is because when using a CP, no patch cords are permitted. The CP makes sense where there are individual workspaces formed from movable partitions or panels. Some important requirements of CPs are:

- Consolidation points shall not be located in any obstructed area.

- Consolidation points shall be located in fully accessible, permanent locations such as building columns and permanent walls.

- Consolidation points shall not be installed in furniture unless that unit of furniture is permanently secured to the building structure.

- Consolidation-point accessibility and marking should follow the ANSI/TIA/EIA T568A recommendation that a consolidation point should be readily accessible and its location visibly marked, for ease of routine maintenance and reconfiguration.

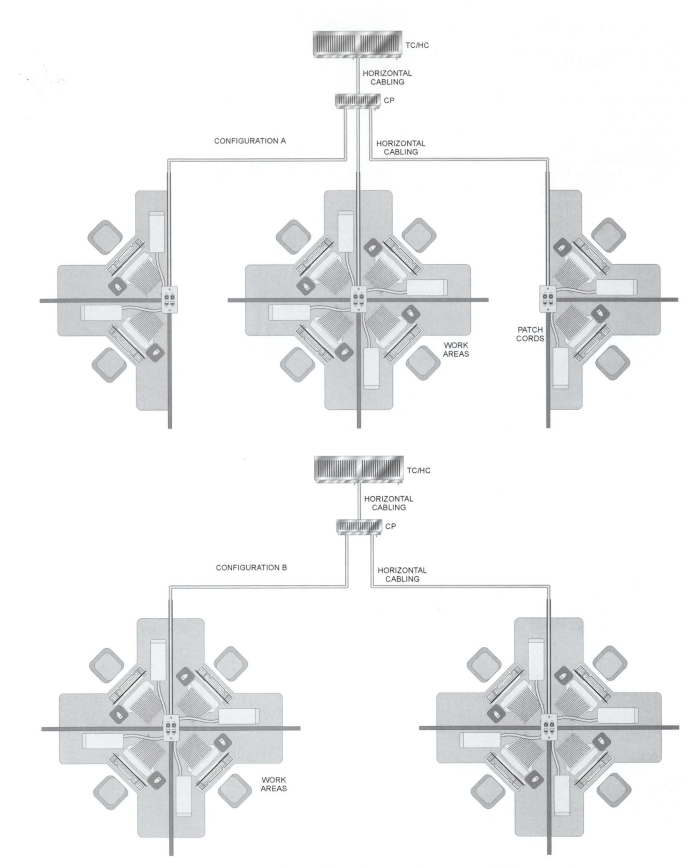

Figure 5-5: Using a CP to Reconfigure a Work Area

MUTOA and CP Performance

The performance implications of the CP should be negligible when compared with the standard TIA channel-performance model in ANSI/TIA/EIA T568A, Annex E, which includes a **Transition Point** (**TP**) between round cable and flat, under-carpet cable. The TP and CP have the same performance impact, because there is only one generic specification for connecting hardware. The CP or TP reduces headroom—Attenuation-to-Crosstalk Ratio (ACR)—at 100 MHz by about 2.5 decibels. Assuming worst-case components, combined in the least favorable way, this still leaves 3.5 dB of positive headroom.

Transition Point (TP)

As already mentioned, TSB 75 allows the use of work-area cables longer than the customary 3-meter limit. Nevertheless, it should be remembered that the attenuation of stranded work-area cables is about 20% greater than that of the solid strand used in horizontal cabling. Accordingly, longer work-area cables are permitted only where horizontal links are correspondingly shorter, so that maximum channel attenuation does not exceed specified values.

Practical Issues

One of the problems that cable installers have been concerned about is the risk of losing track of a CP that has been located in a ceiling, having faced similar situations with the use of **zone cabling** (open-office, reconfigurable cabling). The practice of adding a CP and some additional horizontal cable within a building pathway, when the need for a horizontal extension arises, presents the risk that the process might be repeated later, leading to a run of multiple CPs, and the situation of declining performance.

zone cabling

This is why TSB 75 recommends that downstream CP cabling be restricted to pathways within the work area (where people work), and not in building spaces such as floors or ceilings. If the spirit of the document is followed, then the problem of lost CPs would no longer exist, although this recommendation is not a definitive requirement, because there seems to be less of a problem with CPs in access floors than in ceilings.

Users of a MUTOA need to make sure that each piece of equipment is plugged into its proper service receptacle, and that the jumper and horizontal cable length combinations do not exceed the permitted maximums. TSB 75 labeling conventions will effectively deal with these concerns. Consider the following as legitimate open office locations for a CP or a MUTOA:

- Utility columns attached to furniture panels.

- Overhead storage cabinets.

- Enclosures at floor level.

- Freestanding utility columns.

Mobile furniture is probably best served by a MUTOA secured to a building structure. Outlets on mobile furniture would require it to be tethered, destroying its mobility. Floor locations can support mobility where really needed, but they create aesthetic problems and present a hazard if located where people can trip on cords. Cables downstream of the MUTOA are typically in pathways on desks, tables, or storage areas.

Next Steps

It is probable that the next thing people will want to move is the utility column itself. And, rather than wait for facilities staff, they'll undoubtedly want to do it themselves. However, using the current standardized 4-pair connectors, it would be difficult to create a quick-disconnect for the MUTOA. Twenty-five-pair connectors have not yet been standardized, and have only recently attained CAT5 performance. If the cabling industry fails to recognize the need to support this, users may just go wireless.

Some organizations with many competent users and heavy usage patterns have already begun to install full-service crossconnects, and active equipment in the work area. This trend will doubtless become more prevalent as the equipment becomes easier to use. See your instructor if you wish to have information regarding the purchase of the current ANSI/TIA/EIA TSB 75 publication, or to find out if a copy is already available at your location or education facility.

FCC Part 68

coupler

Before 1972, in order to connect anything to the public telephone network that was not directly supplied by the phone company, you had to arrange with the company to lease a **coupler**, which was simply a "black box" that served as an interface between the public network, and the non-telephone company product that you wanted to connect. In 1972, the Federal Communications Commission (FCC) examined the technical considerations involved in direct electrical connection of **Customer Provided Equipment** (**CPE**) to the public network. As a result of their examination, Docket 19528 was established, which asserted that any equipment meeting the standard specifications of electrical performance could be connected directly to the public telephone network, without the use of a coupler.

Customer Provided Equipment (CPE)

FCC Part 68

FCC Part 68 of the docket specified/assigned registration numbers to this type of equipment, verifying that the equipment was safe for connection to the telephone network. Following a series of delays brought about by the public telephone companies, a court stay was lifted on June 16, 1976, supposedly permitting the registration program to go into effect for toll restrictors, answering machines, and data modems. However, the phone companies objected to this docket, and kept the matter tied up legally until, finally, on October 17, 1977, the Supreme Court denied the phone companies' argument, and docket 19528 went into effect. By the end of 1977, the consumer's right to freely connect privately owned and manufactured terminal equipment to the phone lines, anywhere in the U.S., was well established.

NEC Article 800

National Electric Code (NEC) Article 800

Communications Circuits

The **National Electric Code (NEC) Article 800**, called **Communications Circuits**, covers the installation of the following communications equipment and circuits:

- Telephone.
- Telegraph (except radio telegraph).

- Outside wiring for fire alarms, burglar alarms, and similar central station systems.

- Telephone systems not connected to a central station system, but using similar types of equipment.

For example, the NEC Substitution Chart shown in Table 5-2 delineates various fire resistance categories for cables used for communications circuits.

Fire Resistance Level	Test Requirement	Article 800 Rating
Plenum Cables (highest)	UL910 (Steiner Tunnel) CSA-FT6 (Steiner Tunnel)	MPR/CMP
Riser Cables Multiple Floors	UL-1666 (Vertical Shaft) CSA-FT4 (Vertical Tray)	MPR/CMR
General Purpose Cables	UL-1581 (Vertical Tray)	MP/CM
Residential Cables (lowest)	CSA-FT4 (Vertical Tray) UL-1581/VW-1	CMX

Table 5-2: NEC Substitution Chart

Cables with a higher fire resistance level may always be substituted for those with a lower fire resistance level, but not vice versa. In addition, non-fire rated, outside plant, telephone cables may not be run outside of a rigid, metal conduit more than 50 feet from the point of entrance into a building. Those cables that are rated CMG or CM may be used in runs penetrating one floor (NEC 800-53). Table 5-3 describes some of the wire coding.

Type	Description
CM	Communications Wires and Cables
CMG	Communications and General Purpose Cables
CL2/CL3	Class 2 and Class 3 Remote-Control, Signaling, and Power-Limited Cables
FPL	Power-Limited Fire Alarm Cables
MP	Multipurpose Cables
PLTC	Power-Limited Tray Cables

Table 5-3: NEC Wire Coding

Figure 5-6 depicts the way in which cables with higher levels of fire resistance may be substituted for those requiring a lower rating.

Updated every three years, the NEC is considered to be the bible of the electrical industry, containing sections on emerging technologies, and other important changes that you, as a cable installer, should be familiar with. The 1996 NEC provides the most current and complete safety criteria for all electrical installations, and is the most widely used electrical code in the US, and around the world. There are many jurisdictions in which the NEC has been adopted as law.

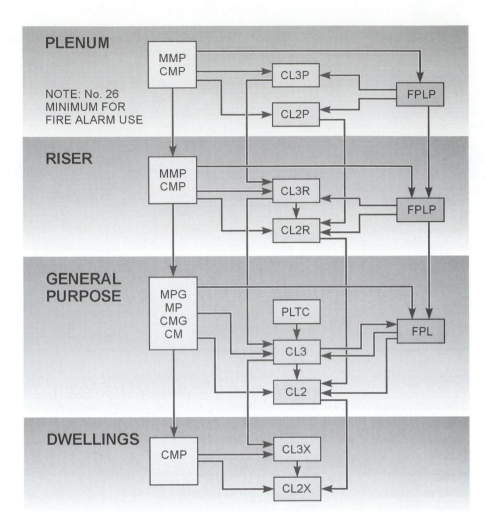

Figure 5-6: Cable Substitution Hierarchy

See your instructor if you wish to have information regarding the purchase of the current NEC, or to find out if a copy is already available at your location or education facility.

CHAPTER SUMMARY

This chapter presented a review of the various standards now observed by the cabling industry. You learned that as a cabling installation technician, you will be responsible for being aware of the standards that are applicable to the installation on which you are working. You also now realize that these standards are subject to change as time goes on.

You read about the American National Standards Institute (ANSI), and the role it plays in the national and international standardization process. You learned how and why ANSI/TIA/EIA 568 came about, and some of its initial shortcomings. You discovered what Technical Systems/Service Bulletins (TSBs) are, and how TSB 36, TSB 40, and TSM 40A were absorbed into ANSI/TIA/EIA T568A. Along with details about ANSI/TIA/EIA T568A, you learned how other revisions and TSBs are continually being added to it as time goes on.

You familiarized yourself with the ANSI/TIA/EIA 569 standard, and how it emphasizes design and construction practices conducive to premise telecommunications infrastructure. You learned about ANSI/TIA/EIA 570, and how it standardizes cabling installations for residential and light commercial premises. You were made aware of the importance of accurate documentation and administration of premise telecommunications infrastructure, and how ANSI/TIA/EIA 606 (later incorporated into ANSI/TIA/EIA T568A) promotes an administration scheme that is independent of any operating applications. You were introduced to the ANSI/TIA/EIA 607 standard, along with the importance of employing proper grounding and bonding techniques on any cabling installation.

Other TSBs you learned about included TSB 53, which extended the frequency range of 150-ohm STP cabling systems, TSB 67, which specifies field testing parameters and minimum transmission requirements for UTP cabling, and TSB 75, which specifies the optional cabling practices for open offices. With regards to TSB 75, you learned about churn, and the importance of maintaining flexibility within office environments that undergo frequent reconfiguration. You read how MUTOAs and CPs are used to deal with churn, and how the choice of office furniture can affect the reconfiguration process. Finally, you reviewed the history of FCC Part 68, and how users were finally allowed to connect personal terminating equipment to the public telephone system. The importance of NEC Article 800 was also stressed, along with various cable ratings and requirements.

REVIEW QUESTIONS

The following questions test your knowledge of the material presented in this chapter:

1. What is the purpose of the ANSI/TIA/EIA T568A standard?

2. Name the six items covered by the ANSI/TIA/EIA standard.

3. State the purpose of the ANSI/TIA/EIA Telecommunications Pathways and Spaces certification.

4. What is the ANSI/TIA/EIA 570 standard and how is it different from the T568A standard?

5. Name the three items that the ANSI/TIA/EIA 606 Administration Standards for the Telecommunications Infrastructure of Commercial Buildings standard recommends for documentation of a telecommunication cabling system in a commercial building.

6. List five types of telecommunications pathways.

7. List four types of telecommunications spaces.

8. How did the FCC Part 68 decision affect the telecommunications network in the United States?

9. Which bonding and grounding guidelines take precedence in a given application, ANSI/TIA/EIA standards or National/Local Electrical Code standards?

CHAPTER

6

PULLING CABLE

Upon completion of this chapter and its related lab procedures, you should be able to perform these tasks:

1. Explain why the ability to pull longer lengths of cable has decreased installation times and labor costs.

2. Differentiate between the installation techniques of pulling cable versus blowing cable.

3. Define the terms "tensile strength" and "jam ratio."

4. Describe the importance of the coefficient of friction with regards to cable installations.

5. Explain the difference between straight conduit pulls and pulls through bent conduit, from the standpoint of COF usage.

6. Demonstrate how to calculate the jam ratio for a three-cable pull.

7. Given the inner diameter of the conduit and the outer diameter of the cable, determine the likelihood as to which configuration a three-cable pull will assume.

8. Contrast between water-based and silicone-based pulling lubricants.

9. Describe the negative aspects of using mini-roller lubricants for cable pulls through multiple-bend conduit.

Pulling Cable

Perhaps the most critical aspect of a cable installation is the actual pulling of the cable through the various sections of conduit already embedded within the walls of the building. The major concerns that confront the installer revolve around the diametric size of the conduit, along with the number and angle of bends through which the cable(s) must pass.

During the past 20 years, the ability to pull longer and longer lengths of cable through preinstalled sections of conduit has greatly contributed to the decrease in both installation times and labor costs. Consider the greater amounts of time and effort that must be exerted for an installation where cables must be pulled over shorter lengths of conduit, reinserted into the next section, repulled, and so on. The improvements that installers are realizing today can be attributed as much to a better understanding of the practical physical limitations of modern communications cable, as to the development of several classes of pulling lubricants, and the equipment that has been devised not only to pull the cable, but to lubricate it as it is being moved through the conduit.

Cable manufacturers have implemented techniques designed to increase the strength of the outer jackets of their products, and their resistance to the destructive effects of pull tension. In addition, the recent lessons learned about the spread of fire through cabling systems (the World Trade Center, for example) has led to the manufacture of fire-retardant cables using specially formulated materials designed to resist ignition and the spread of flame.

Nevertheless, these innovations alone do not explain why today's cable installers are currently able to pull longer lengths of cable through layouts of conduit, many of which have various bend angles. Stronger cable has been combined with modern lubricant technology to permit cable pulls covering greater and greater distances. Friction coefficients have been greatly reduced, resulting in corresponding reductions in the amount of tension suffered by the cable during the pulling process.

But again, the manufacturers of cable-pulling lubricant must take into account the risk of fire spread. A combustible wax residue could negate the effectiveness of a flame-suppressing cable jacket, and much of the lubricant, or dried residue, will remain dispersed throughout the conduit long after the cable is installed.

Recent studies have shown that wax-based lubricants can easily melt and ignite, spreading fire throughout the conduit in which they are present. Even certain polymer-based lubricants were found to ignite and spread flame. Given these findings, the ability of a lubricant to self-extinguish should be as important as it is with regard to a fire-retardant cable.

Cable installers need to be aware of any performance specifications that include the fire suppression capabilities of both the cable and the lubricants used to install it.

CABLE IN CONDUIT

The realities of modern cable installations dictate that communications cable be placed in conduit or duct pipe, whether located outside of the plant, or inside of the building(s). The primary job of the conduit, of course, is to provide protection for the cable from either environmental damage or outright physical abuse.

The major threats to underground cable installations are from unsupervised surface digging, subterranean animal activity, or shifting earth (mudslides, earthquakes, shifting rock, and so on). Conduit can protect underground cable installations from each of these perils. In addition, underground cable that populates conduit can be replaced or upgraded fairly easily. Additional digging or excavation becomes unnecessary when the old cable can simply be pulled out of the conduit, and the new cable pulled in immediately afterwards.

Groups of conduits are frequently deployed in large metropolitan areas, called "duct banks." Once these banks are placed into position, cable can be placed into an empty duct whenever the city's cable infrastructure grows or changes significantly. This inevitable growth is then accommodated without the major traffic disruptions inherent whenever large-scale excavations occur in the street.

A comparison between 100mm diameter twisted pair copper cable and 10-20mm fiber optic cable, shows that because fiber has a lower breaking strength, and is more easily damaged than twisted pair, a large percent of underground fiber optic cable is installed in conduit, providing it with much needed protection.

Pulling

During the past fifty years, there have been literally millions of miles of electrical and communications cable installed using a simple and basic method of cable pulling. In fact, cable pulling is the most common method of installing cable into conduit, and is well known to cable installers throughout the world.

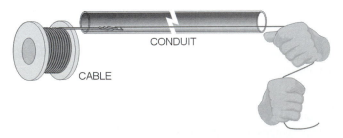

Figure 6-1: Pulling Cable in Conduit

The installer first threads a line through the conduit. Then, the line is attached to the cable and is used to drag the cable back through the conduit, as shown in Figure 6-1. Hand-over-hand pulling of the rope supplies the force necessary to pull the cable. Alternately, some type of equipment is used to supply the pulling force needed to overcome the cable's frictional resistance to movement through the conduit. The maximum force allowed on the cable limits the length that can be pulled at any one time.

As simple as the preceding description sounds, cable pulling has become quite a technological exercise! The main concern here is the determination of the maximum distance that a cable can be pulled without it being damaged in some way. Longer pulls will serve to minimize the need for splices, and, in turn, reduce the overall expense of an installation. Having a good idea about the maximum stress a newly installed cable can withstand will enable the cable installer to provide more efficient cable installations, producing less damaged cable and longer useful cable life.

Blowing

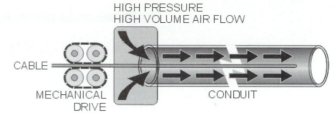

For lightweight fiber optic cable, there is an alternate method of installing cable in conduit. It is called "cable blowing," and uses a smooth, uninterrupted flow of air to provide high-speed air installation, as shown in Figure 6-2. Often, a piston is placed within the conduit and attached to the end of the cable being blown.

Figure 6-2: Blowing Cable through Conduit

While most outside plant personnel are very familiar with the pulling method of installing cable, they are not quite as knowledgeable about cable blowing methods, although air-assisted cable installations must overcome the same frictional force to move the cable as traditional pulling methods do. The way in which this is accomplished is quite different, however.

Two different methods are combined to guide the cable through the conduit. The first method involves the use of a mechanical device to push the cable into the source end of the conduit. Secondly, air is forced into the conduit and across the cable jacket towards the destination end of the conduit. The combination of these two forces provide the thrust to move the cable through. This is known as **High Air Speed Blowing** (**HASB**), and the purpose of the airflow through the wide-open conduit is to push the cable forward at the speed provided by the mechanical front-end pusher. The dragging force of the air is equally distributed along the length of the cable, and its volume can reach levels of between 300 to 600 cubic feet per minute (cfm).

In the alternate **push/pull** method, as shown in Figure 6-3, the force of the air on a piston, missile, or carrier at the front end of the cable provides the secondary impetus in combination with the mechanical pushing device at the front end. Because the conduit is not wide-open, due to the air-blocking properties of the attached piston, airflow does not need to be as vigorous as in the HASB method. Typical air flows of 200 to 300 cfm are common, and as might be expected, the cable itself does not propagate through the conduit as quickly as it does using HASB.

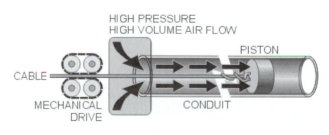

While cable blowing is typically reserved for underground duct, it will work for above-ground conduit installations as well. It contrasts greatly with the pulling method, because the cable itself is not required to go

Figure 6-3: Push/Pull Cable Blowing

through a series of conduit bends. The exponential frictional force, created when the cable is pulled into a bend, is many times greater than what would occur if only the weight of the cable opposed the pulling force. The long lengths associated with cross-country fiber optic pulls are not possible through conduit that hosts an accumulation of bend angles.

Because of the lightweight and flexible properties of fiber cable, long and uninterrupted lengths of installed cable are possible, and desired. In conduit lacking the hard bends inherent in premises cabling, it is easy to understand why cable blowing methods are preferred for outside plant fiber cable installations.

TENSION AND JAMMING

One of the main concerns for a cable installer when pulling any type of cable is the question of how hard the cable can be pulled without damaging it. How much pulling force can be applied to it before it is rendered useless? In this regard, the size and type of a particular cable should dictate the maximum allowable tensions that it can be expected to tolerate.

The ability of a material to resist a force that tends to tear it apart is known as its **tensile strength**. Typical tensile strengths for fiber optic and coaxial cables range between 200 to 2,500 Newtons. A **Newton** is described as the unit of force required to accelerate a mass of one kilogram (kg) one meter per second per second. Tensile strengths for large copper cables can typically range as high as 15 kN. While the manufacturer can provide the maximum installation tensions for any of its particular cables, the installer is responsible for properly pulling the cable. Therefore, the installer must know and respect these maximum allowable pulling tensions, and be prepared to conduct an installation that does not exceed them.

When a cable pulling session involves multiple cables, the number of parameters with which an installer must be concerned are multiplied as well. Damage to the cables becomes even more difficult to avoid, and the importance of understanding how to use a **Jam Ratio** in predicting the possibility of trouble cannot be overestimated.

Frictional Force

Because of the need to keep the pulling force below the cable's maximum allowable tension, a natural limit exists in the length of cable that can be pulled through a given conduit. The cable's frictional resistance to movement increases the force that is required to pull the cable through the conduit. As the length of cable entering the conduit during a pull increases, the tension along the cable, especially at the bends, goes up as well.

Coefficient of Friction

The **Coefficient of Friction (COF)** is a mathematical tool that is used to measure frictional resistance. A simple physics class example might help to explain what a coefficient of friction is. If a wooden block weighing 5 kg was seated on a horizontal steel plate, and it required 2 kg force (19.6 N) to pull the block across the plate, what would the coefficient of friction be for this action?

The coefficient of friction (wood on steel) is defined as the ratio of the dragging force (2 kg) to the normal force (the 5 kg weight of the wooden block). Therefore, for this example, the friction coefficient would be 0.4. Because the COF does not describe width, height, or length, it is considered to be a dimensionless number.

Suppose the wooden block from the example was replaced with a 5 kg rubber block. Common sense would dictate that it will take a greater force to drag the rubber block. If it required 6 kg of force to drag the rubber block across the steel plate, the measured coefficient of friction (rubber on steel) would be 1.2, or three times as large as with the wooden block.

The main observation to be made from these two examples is that there is no one coefficient of friction. As the surfaces that rub together change, the COF will vary as well. If the block is replaced with cable, and the plate is replaced with conduit, the mechanics involved with cable pulling become clearer, more or less.

In reality, there are a few more complications involved, such as:

- Neither the cable nor the conduit is flat.

- There may be more than one cable being pulled, complicating the dynamics between the rubbing surfaces.

- Pulls are not conducted through totally straight conduit, adding forces other than gravitational weight into the equation, especially at the conduit bends.

Equations for Pulling Success

As we have already learned, excess tension during a cable pull can damage cable and reduce its life. Planning a pull ahead of time can avoid damage, as well as time and money lost through aborted installations. Cable-pulling theory allows installers to estimate pulling tension, based on the physical law that the force required to move an object across a surface is equal to the force between the surfaces times their coefficient of friction.

In cable pulling, the cable is the object, and the conduit is the surface. The required pulling force depends on both the gravitational weight of the cable itself, and the resistance of friction when the cable is being pulled around bends in the conduit. The end result is a series of equations that require specific input on cable weight, conduit length and direction, location of conduit bends, and the coefficient of friction between the cable jacket, the conduit, and any pulling lubricant that may be used.

Straight Conduit Equation

Assuming that a valid COF has been determined, the expected cable pulling tension can be determined through mathematical calculation. Depending on whether or not the conduit through which the cable being pulled is straight, or contains bends, specific equations can be used. For example, if the conduit is straight, with no bends, the equation used is:

$$T_{out} = T_{in} + LW\mu$$

where T_{out} is equal to the tension out, T_{in} is equal to the tension in, L is equal to the length of the straight run, W is equal to the weight of the cable (per length), and μ is equal to the COF.

Bent Conduit Equation

If the conduit contains bends, and consideration is taken for the non-gravitational forces in the conduit bends, the equation used is:

$$T_{out} = T_{in}\, e^{\mu\theta}$$

where T_{out} is equal to the tension out, T_{in} is equal to the tension in, e equals the natural log base (approximately 2.71828), μ equals the COF, and θ equals the angle of bend.

The significance of the COF residing in the exponent should not be underestimated. Tension calculations will be grossly inaccurate if the COF is unrealistic. When pulls are being negotiated over several bends, the larger values of tension and sidewall pressure will ultimately result in disaster in cases where the COF is erroneous.

Tension Limits

All the calculations in the world will be meaningless unless the cable installers know the maximum force they can apply to a cable without risking damage to it. Tension limits apply to all types of cables, and in addition, some types have sidewall-pressure limits as well.

These tension limits vary with the type, construction, and size of the cable, and they are normally listed on a cable's spec sheet. Allowable tensions as low as 30 to 50 pounds exist for some types of lightweight coaxial cable, while large fiber optic cables, outfitted with strength members, can usually withstand from 400 to 800 pounds of tension without suffering damage. For multi-pair copper cable, the maximum allowable tension will be determined from the number, and gauge, of the conductors.

Once you know the tension limits of the cable being installed, you can start calculations using full pulling equations. These can become very complex when angles, bends, and direction are factored in for the different segments of a pull.

Some companies offer software programs that can easily calculate these complex equations, and analyze even the most difficult cable pulls, including parameters such as pulling tension, conduit fill, jam ratio, and sidewall pressure. Quick changes can be input for conduit system design, tension, lubricant quantity requirements, and cable parameters. These programs are also capable of measuring tension during an actual pull, including the coefficient of friction calculation.

Multi-Cable Jamming

There are other problems for cable installers to consider when cable pulls are performed on several cables simultaneously. Problems occur when the combined diameters (three for example) of the cables being pulled approach the interior diameter of the chosen conduit. As they are pulled around a bend, they may line up in a straight line (linearly). Forced towards the inside of the bend, the cables risk being wedged against the wall of the conduit. If this happens, the cables become stuck in the wedged position. This undesired phenomenon is known as **jamming**.

jamming

The main problem in dealing with multi-cable jamming is the possibility of ruining one or more of the cables when and if enough pulling tension is mustered to move them through the bend. Cable jackets may be ripped off, or various segments of insulation destroyed. In such cases, the cable would have to be removed, discarded, and replaced. Obviously, this situation would result in additional expense and time being wasted on the installation.

Calculating Jam Ratios

In order to avoid the situation described above, an equation has been developed that compares the **Outer Diameter (OD)** of the cable to the **Inner Diameter (ID)** of the conduit. The resulting calculation is called the **Jam Ratio**, and it gives the cable installer a reliable indication as to whether a planned pull is favorable or unfavorable. Sometimes an unfavorable jam ratio cannot be avoided, and an undesired risk will have to be taken. Hopefully, these occasions will be few and far between for an experienced cable installation team.

Three-Cable Equation

The formula for the calculation of the jam ratio for a three-cable pull is:

$$J = 1.05 \text{ ID/OD}$$

where J is equal to the jam ratio, ID equals the interior diameter of the conduit, and OD equals the outer diameter of the cable. The factor of 1.05 takes into consideration the fact that the conduit assumes a more oval shape at the bends.

Table 6-1 indicates the probability for jamming to occur for various ranges of three-cable jam ratios. Notice that certain ranges provide extremely favorable jam ratios.

Table 6-1: Three-Cable Jamming Probability

Jam Ratio	Probability of Jamming
2.4 to 2.5	Low
2.5 to 2.6	Medium
2.6 to 2.9	High
2.9 to 3.0	Medium
3.0 to 3.2	Low

Other COF Issues

Recent studies in cable pulling have revealed that when compared with single-cable pulls, three-cable pulls have demonstrated higher than expected pulling tensions than would be expected using theory alone to predict what the COFs would be. Therefore, whenever three cables are being pulled at sidewall pressures less than 150 pounds per foot, the recommended three-cable COF should be somewhere between 10 to 100 percent larger than that for a single-cable pull.

There is a device called a **trifurcater** that helps to maintain a triangular orientation between three cables as they are fed through the conduit. The idea is to prevent the change of positions between the cables where they roll over each other and jam. However, questions remain.

- What are the exact ranges of jam ratios that indicate pulling problems?

- What constitutes an accurate COF for three-cable pulls?

- What is the optimal conduit feed position for three-cable pulling?

- What if the conduit already contains cable that is functioning properly?

Pulling Experimentation

Unless some specific testing procedure is undertaken, questions such as those above cannot be adequately answered. Fortunately, companies that manufacture conduit, cable, and cable-pulling lubricants have conducted enough testing to provide reliable data about multi-cable pulls.

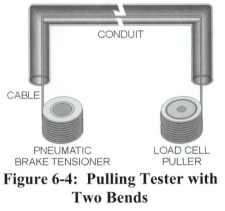

Figure 6-4: Pulling Tester with Two Bends

Using the apparatus shown in Figure 6-4, a section of conduit was oriented so as to permit the pulling of cable around two conduit bends of 90 degrees. Using a pneumatic breaking device, the incoming cable tension could be varied. On the opposite end of the cable, a pulling device applied the pulling tension under measurable conditions. The data provided from this testing setup allowed for the calculation of the COF, and the measurement of jamming effects. The variable parameters included cable positioning, jacket type, conduit format, and the lubrication formula being used, if any.

The pulling device was outfitted with an electronic load measuring circuit that transmitted information to a PC via an RS232 interface. During a pulling session, the PC was instructed to sample the data once per second. Once the data was averaged (20 tension readings per data point), the COF could be calculated.

Testing Cable

Variables such as jacket, conduit, and lubricant types were held constant in order to focus on the three-cable criteria. Although various XHHW cable sizes were tested, the same manufacturer was chosen for all of them. XHHW is copper or aluminum cable capable of handling high-heat environments under wet or dry conditions. The conduit through which the pulls were conducted was 2-inch (50mm) Schedule 40 **PolyVinyl Chloride (PVC)**. Schedule 40 PVC (2-inch diameter) is manufactured with an outer diameter (OD) of 2.375 inches, and an inner diameter (ID) of 2.067 inches. The thickness of the conduit's wall is 0.154 of an inch.

Various sizes of cable were pulled, ranging from 2/0 **American Wire Gauge (AWG)** to 500 **Multiple Circular Mils (MCM)**. Using the pulling tester, and the selected cable sizes, calculations were made for three-cable pulls through the 2-inch conduit. Table 6-2 shows the jam ratios, percentage of conduit fill, and the remaining clearance through the conduit. All of the calculations were based on the OD of the cables and the ID of the conduit.

PolyVinyl Chloride (PVC)

American Wire Gauge (AWG)

Multiple Circular Mils (MCM)

Table 6-2: Cable Calculations

Cable Size	OD inches (mm)	Jam Ratio	Fill Percentage	Clearance inches (mm)
2/0	0.50 (12.7)	4.3	18.4	1.0 (26.4)
3/0	0.57 (14.5)	3.8	23.6	0.90 (22.9)
4/0	0.59 (15.0)	3.7	25.8	0.85 (21.6)
250	0.67 (17.0)	3.2	33.0	0.68 (17.3)
300	0.70 (17.8)	3.1	36.0	0.61 (15.5)
350	0.77 (19.6)	2.8	43.6	0.45 (11.5)
400	0.82 (20.8)	2.6	49.4	0.33 (8.4)
500	0.89 (22.6)	2.4	58.9	0.13 (3.3)

Configurations

There are two basic configurations that three-cable pulls will assume as they ride through the conduit. Figure 6-5 depicts these two configurations, known as (a) **triangular** and (b) **cradled**.

When the diameters of the cables are large enough, the cables will adopt the triangular configuration. Experiments have shown that the triangular configuration is preferred when the OD of the pulled cable is greater than 40% of the conduit's ID (OD/ID = > 40%).

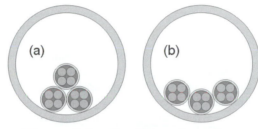

Figure 6-5: Configurations for a Three-Cable Pull

By controlling the **feed position** of the cables as they entered the conduit, an attempt can be made to force a specific configuration. For example, in order to force a triangular configuration, two cables are fed to the inside half of the conduit while one cable is fed to the outside half. To achieve a cradled configuration, only one cable is fed to the inside half of the conduit. The remaining two cables are fed to the outside half.

Cable Weight

The weight factor for a three-cable pull is somewhat more complicated than when dealing with only one cable. A single cable will make contact with the bottom of the conduit due simply to the pull of gravity. The situation for a multi-cable pull results in increased force being applied to the cable due to weight. This additional force, called the **Weight Correction Factor**, is not attributable to gravity alone. The additional parameters associated with the weight correction factor are the cable to conduit size ratio and the particular configuration (triangular or cradled) of the cable within the conduit.

Table 6-3 displays the weight correction factor for both cradled and triangular cable configurations using the various cable sizes selected for testing.

Table 6-3: Weight Correction Factor

Size	Cradled	Triangular
2/0	1.14	1.06
3/0	1.21	1.09
4/0	1.23	1.10
250	1.33	1.15
300	1.37	1.18
350	1.51	1.27
400	1.62	1.37
500	1.84	1.65

Notice that the weight correction factor for the cradled configuration is always higher than that for the triangular configuration at a given wire size. If all of the other factors remain the same, it's obvious that the cradled configuration will produce higher levels of tension than the triangular configuration. Not so obvious is the fact that COF calculations will suffer for accuracy if the wrong configuration is assumed.

Table 6-3 indicates that with an increase of cable size, there is a corresponding increase in the weight correction factor for both configuration types. This, of course, translates into higher tension values as well. By using the weight correction factor in determining tension, and then the COF, the resulting data on friction should be useful when comparing different sizes of cable. This would not be the case if only direct tension values were known and compared.

Lab 19 – Pulling CAT5 Cable

Lab Objective

To properly pull CAT5 cable through multiple sections of conduit, and test the resulting run

Materials needed

- Lab 19 work sheet
- Simulator grid panel*
- Conduit, PVC, ½-inch x 4-inch section (2)
- Conduit, PVC, ½-inch x 9-inch section, with 90-degree bend (7)

- Conduit, PVC, ½-inch x 2-inch coupling (2)

- Bracket, conduit, ½-inch (14)

- Pliers, diagonal cutters

- Cable, CAT5 UTP 100-ohm, non-plenum (spool)

- Termination plug, RJ45, unshielded (2)

- Crimping tool, RJ45

- Fish tape, steel, 50-foot

- Electrical tape

- Electricians' scissors

- Tension scale/Tape measure

- Twine, roll

- Pulling lubricant (if necessary)

- Screwdriver, Phillips, small

- Machine screw, Phillips, pan head, ½-inch, 10-24x (28)

- Wing nut, 10-24x (28)

- Level II testers, w/case & cables (control unit and remote)**

- Users manual for Level II testers

*The simulator grid panel should already be mounted on a large desk with its two feet bolted first to the panel itself, and then to the desk top. The panel needs to be at eye level, with its backside accessible for ease of mounting the various cabling components.

**Prepare ahead of time for any lab procedures that involve the use of the Level II testers by charging the Nickel-Cadmium battery packs in both the control unit and the smart remote. In an emergency situation, the units will operate during a charge operation, but they will have to remain near an ac outlet during this time.

Instructions

Before beginning this lab, read through all of the following steps at least once, thoroughly. You will actually perform two separate pulls during this procedure to avoid building up too much tension over the entire layout of the conduit.

1. Locate the spool of 4-pair CAT5 UTP cable, and use the diagonal cable cutters to separate a 25-foot section from the spool. Set this section of cable aside, temporarily.

TIP: This may seem to be a rather lengthy piece of cable, but keep in mind that a length of at least 16 feet is required for several of the tests that are conducted on CAT5 cable by the Level II testers. This gives you enough leeway for recovering from any wiring errors that may occur and still have enough length left to take meaningful Level II measurements.

2. Begin this procedure by closely examining Figure 6-6. Then, locate and gather the seven, 9-inch sections (with 90-degree bends) of ½-inch PVC cable conduit.

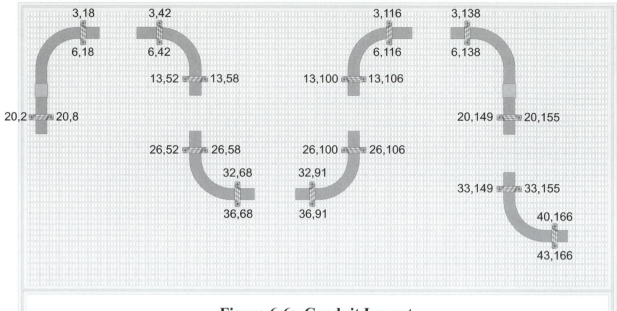

Figure 6-6: Conduit Layout

TIP: Figure 6-6 displays the conduit layout for the simulator grid panel. You will use this layout for lab procedures that will follow, so it is important to locate the PVC sections exactly as shown in Figure 6-6. The diagram is accurate regarding the rows (50) and columns (180) of openings in the grid panel. If necessary, you may use it to ensure an accurate layout of the PVC conduit sections.

3. Locate and gather the two, 4-inch sections of ½-inch PVC conduit, and the two, 2-inch conduit couplings.

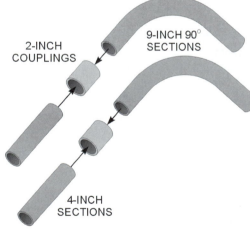

Figure 6-7: Fitting the Conduit Sections Together

4. Select two of the 9-inch sections (with 90-degree bends) of conduit, and insert one of the 2-inch couplings on one end of each section, as shown in Figure 6-7.

5. Next, insert one of the 4-inch pieces of straight conduit into the remaining end of each coupling, and ensure that these 3-piece combinations of conduit fit snugly together.

6. On Figure 6-6, observe the small numbers located near where the mounting brackets are holding conduit section 1 into place.

TIP: These numbers represent the row (from top to bottom) and column (from left to right) coordinates for the ½-inch, 10-24x pan head machine screws (Phillips). These screws thread through the ½-inch mounting brackets, the grid panel, and then into the wing nuts. Obviously, the wing nuts tighten the brackets down to hold the PVC sections into place. Conduit 1 is one of the 3-piece sections you just assembled.

7. Locate two mounting brackets, four pan head machine screws, and matching wing nuts. Then, place them within easy reach.

8. Observing the coordinates shown for conduit 1 in Figure 6-6, orient a section of PVC conduit as shown, and position one of the brackets over the proper coordinates.

9. Insert a pan head machine screw through a bracket and a matching coordinate in the simulator grid panel.

10. Thread a wing nut onto the protruding machine screw on the back of the grid panel, and tighten the wing nut slightly.

TIP: Leave enough play in the conduit bracket assembly to make small adjustments. If you tighten the first wing nut too far, the second machine screw may not reach through the grid panel far enough to start the second wing nut.

11. Insert a pan head machine screw through the remaining hole in the mounting bracket and its matching coordinate in the simulator grid panel.

12. Thread a wing nut onto this protruding machine screw on the back of the grid panel, and tighten the wing nut slightly.

TIP: Try to leave some play in the assembly at this point, if possible. Once the remaining bracket is mounted, you may tighten the section of conduit down.

13. Use the coordinates shown in Figure 6-6 to install the remaining bracket over conduit 1 and into the grid panel, using the remaining machine screws and wing nuts.

14. Once the conduit is mounted at the proper coordinates, tighten the wing nuts as required. The grid panel should now appear as shown in Figure 6-8.

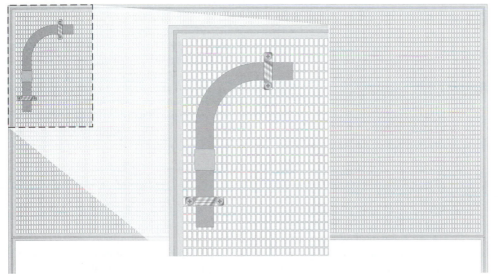

Figure 6-8: Conduit 1 Mounted

15. Using the same techniques you used for conduit 1, mount conduits 2 through 7 on the simulator grid panel. Continue to use the grid coordinates shown in Figure 6-6 for the machine screws and mounting brackets.

TIP: Conduit section 6 will be the remaining 3-piece combination you assembled earlier.

16. When you have completed mounting the sections of conduit, check your simulator grid panel against the layout shown in Figure 6-6 once again, for accuracy.

TIP: If you discover any deviations between Figure 6-6 and your assembly, take the time to make the necessary corrections before moving on.

17. Observe Figure 6-9 carefully. Then, locate the fish tape.

18. Starting from the bottom of conduit 1, route the end of the fish tape through the sections of conduit 1 through 3, as shown by the directional arrows in Figure 6-9.

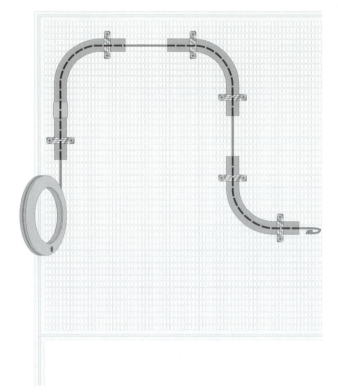

Figure 6-9: Routing the Fish Tape

TIP: As the fish tape protrudes through each section of conduit, guide it to the opening in the next section. The layout of conduit you have installed represents an installation requiring two separate cable pulls. Real-world installations may involve multiple pulls within the same lengths of conduit. In an effort to reduce the buildup of friction, your layout involves two pulls using the same cable. In the event that the friction becomes a critical factor at your site (due to variations in conduit type, cable jacket type, temperature, and so on), apply lubricant to the cable as directed by your product's manufacturer. The following guidelines should be kept in mind whenever you are using fish tape:

- Protect the fish tape from contacting live circuits or from inadvertent bending by keeping it in its reel when not in use.

- Periodically check the fish tape for damage such as cracks, rust, sharp bends, nicks, or gouges. These types of flaws can cause the tape to break under the stress of a pull.

- Never open a reel of fish tape without first extending the tape completely. The tension built up across coiled steel is formidable and dangerous.

- Because a large amount of tension can build up during a cable pull, always maintain an adequate recovery posture in case this tension is suddenly released while pulling. Keep away from trip or landing hazards while controlling a cable under tension.

- If pulling cable through conduit containing existing cable, check to be sure that none of this wiring is under power during the pull.

- Never try to create a bend at the end of cold fish tape for pulling purposes. The steel will fracture unless it is torch heated (red hot) before a bend is made.

- Avoid creating sharp bends in fish tape being pulled. If such a bend appears in fish tape you are using, cut the end off below the bend, and reform the end using a torch.

- When heating fish tape with a torch, allow the tape to cool completely before using it.

- Fish tape can be damaged or broken if used with slip joint or locking pliers.

19. Locate the 25-foot section of CAT5 UTP cable you set aside earlier, as well as the twine and the electrical tape.

20. Coil the 25-foot section of CAT5 UTP in the front of the grid panel with one of its ends facing conduit 3, as shown in Figure 6-10.

21. Locate the twine and the electrical tape, and place them nearby.

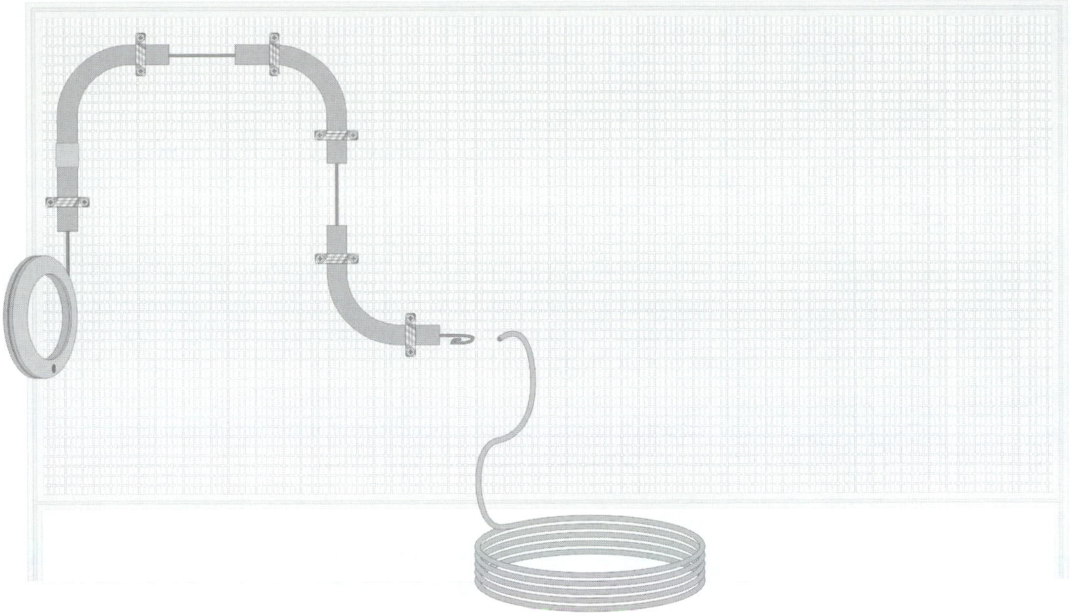

Figure 6-10:
Positioning the
CAT5 Cable

22. Observe Figure 6-11 and prepare the end of the CAT5 cable that faces conduit 3 around the hook at the end of the fish tape, as shown.

23. Roll out enough twine from the spool to thread through the conduit sections already occupied by the fish tape.

24. Next, insert the twine through the hook at the end of the fish tape and twist it, as shown in Figure 6-11.

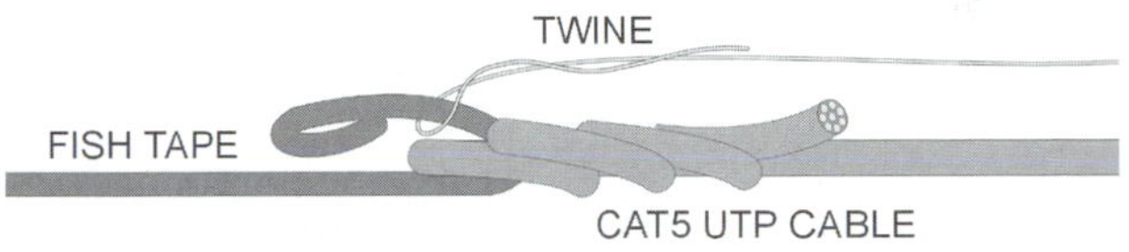

Figure 6-11: Preparing a Pull

TIP: The practice of pulling twine through the conduit, along with the cable being installed, is a universal courtesy extended to other cable installation workers who may do follow-up work once the initial installation has been completed. It allows the pulling of another cable through the conduit at some later date. It would be expected that more twine would be pulled through at that time for the next job, and so on.

25. Use the electrical tape to secure the cable and the twine to the end of the fish tape, as shown in Figure 6-12.

TIP: Once the cable has been secured to the fish tape, it's ready to be pulled. At this point, the pull rating of the cable itself must be considered. If this value is exceeded during a pull, the cable itself may be damaged and rendered useless for its intended purpose.

CAT5, 4-pair UTP cable normally bears a rating of 25 pounds. This means that if more than 25 pounds of force are needed to pull this cable through the intended conduit path, damage to the cable may result. The individual on the pulling end needs to be able to identify the point at which the pulling tension exceeds this value. Check with your instructor to see if any pulling lubricant is recommended.

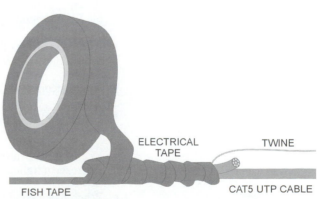

Figure 6-12: Securing a Cable Pull

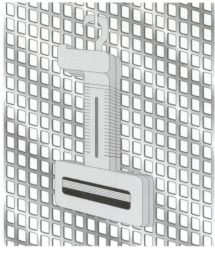

Figure 6-13: Securing the Tension Scale

26. Locate the Tension scale/Tape measure and secure its pulling hook around one of the openings in the grid panel, as shown in Figure 6-13.

TIP: The simulator grid panel should be securely mounted at this point, and any lubricant that may be required should have been applied as directed by your instructor.

27. Place one hand against the grid panel to brace it (or have your lab partner do it), and pull the Tension scale/Tape measure toward you until its pointer reads 25 pounds (almost 11.5 kgs) of force being applied.

28. Repeat this procedure several times until you think you know what 25 pounds of tension feels like.

29. Without looking at the meter, pull on the Tension scale/Tape measure until you feel the tension is equal to 25 pounds of force. Then, have your lab partner read the tension value shown on the scale.

30. Allow your lab partner to repeat steps 26 through 29 while you read the scale.

TIP: The following tips should be kept in mind during a cable pull:

- If trouble occurs on an installed cable run, pull a new wire rather than splicing the defective one.

- If a spliced cable must be used, never locate it behind a wall or some other area where it cannot be reached.

- General pulling limits for 4-pair UTP cable is 25 pounds. For STP cable the limit should be kept below 55 pounds.

- On long cable runs through conduit, it is not recommended to pull cable through more than two 90-degree bends at once. This suggestion should not, however, prevent you from successfully pulling CAT5 UTP cable through the conduit layout in this lab procedure.

- Avoid pulling cable through more than 100 feet (30 meters) of conduit at one time.

- Provide support for every 5 feet of cable in order to prevent the strain on hanging cable from supporting its excess weight.

- Avoid conditions that result in sharp bends in the cable. For 4-pair UTP cable, the bend radius limit is 4 times the cable's diameter.

- Avoid stepping on the cable at all times while you work around it.

- Protect the outer jacket of the cable from stress such as staple fasteners, or overtightened cable ties.

- Keep the cable run away from sources of electrical noise, such as electrical motors, or fluorescent lights.

- When using conduit, leave a pull cord in the conduit with the pulled cable, so that more cable can be pulled at a later time.

- Keep cable pulls away from heat sources such as heat ducts and hot water pipes.

- Try to avoid fire barriers or thick layers of insulation when installing cables by locating a run through the inner walls.

- Maintain the Tip and Ring polarity designations for all connections.

31. With your lab partner bracing the simulator grid panel, pull the fish tape, with the cable/twine attached, through the conduit sections.

TIP: Try to estimate how much force you are using during the pull, so as not to exceed the 25-pound tension limit.

32. When the fish tape exits from the bottom of conduit 1 (with the cable/twine still attached), pull approximately 2 feet of cable/twine through the conduit before unwrapping the electrical tape from the end of the fish tape, and untwisting the cable/twine.

TIP: You are now ready to prepare the second cable pull.

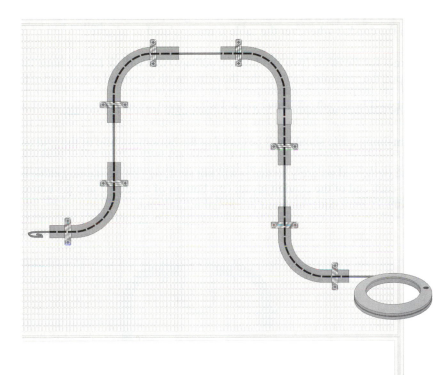

Figure 6-14:
Routing the Fish
Tape for Pull
Number Two

33. Starting from the bottom-right of conduit 7, route the end of the fish tape through the sections of conduit 7 through 4, as shown by the directional arrows in Figure 6-14.

34. Locate the remaining free end of the 25-foot section of CAT5 UTP cable from the front of the simulator grid panel, and face it towards the bottom-left of conduit 4.

35. Roll out enough twine from the spool to thread through the conduit sections now occupied by the fish tape (plus two feet), and cut the twine from the end of its spool.

TIP: You'll want enough twine to provide two feet beyond the bottom-right end of conduit 7.

7. Place a wing nut on the machine screw and lightly tighten it by hand, or by bracing it with a finger while using the screwdriver.

8. If necessary, adjust the block to make sure it is completely level, then tighten both screws snugly, but not too tight.

9. Compare your simulator grid panel with that shown in Figure 6-23, and make any necessary adjustments.

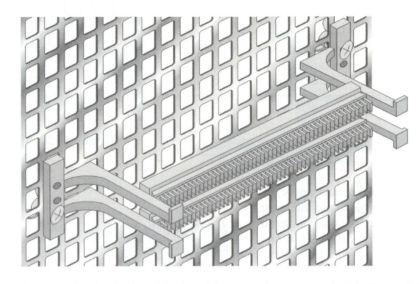

Figure 6-23: 110-Type Termination Block Mounted

TIP: It's time to make the first break in the cable run, at the point marked in Figure 6-19 with the dotted line. Later, you will compare the first test results on the cable run with the ones you get following the wiring of the 110-type termination block.

10. Locate the diagonal cable cutters, and observing the proper safety precautions, make a clean cut through the 4-pair, CAT5 UTP cable about halfway between conduit 2 and conduit 3 (dotted line in Figure 6-19).

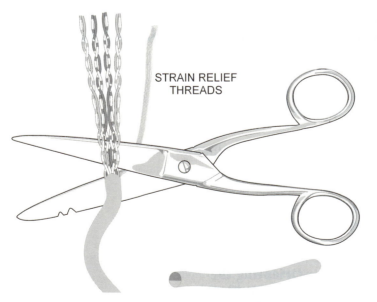

STRAIN RELIEF THREADS

Figure 6-24: Stripping the Sheath

TIP: This is a good time for you to decide how good your cable preparation skills are. If you are confident, you may choose to terminate the CAT5 cable without leaving much slack between conduit 2 and conduit 3. If, on the other hand, you do not want to risk leaving too short a piece of cable to work with (in case you make a mistake and have to clip some bad cable off an end), you may choose to leave the existing slack (from the loop) in the run as a precaution. Remember, the only solution to a cable run that is too short is to replace it with a longer one.

11. Without cutting or nicking any of the conductors, strip approximately one and a half (1.5) inches of cable sheath (not to exceed 2 inches) from the section of CAT5 UTP cable running out of the bottom of conduit 2. Snip the strain relief threads with electricians' scissors, as shown in Figure 6-24.

12. Insert a 110-type blade into the punch-down tool, making sure that the cutting edge faces outward, as shown in Figure 6-25.

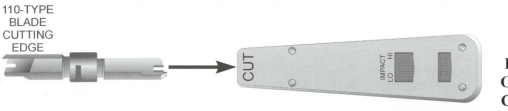

110-TYPE
BLADE
CUTTING
EDGE

**Figure 6-25:
Orienting the
Cutting Edge**

TIP: The cutting top of the blade should be oriented so that it is located beside the letters "CUT" on the punch-down tool.

13. Place the setting of the punch-down tool in the "LO" impact position.

TIP: The impact tool has two settings on it. They are Hi and Lo. The Hi setting is used to terminate wire on a 66 block. The Lo setting is for the 110 block. Be sure to check the settings before you terminate any wire on any block. Improper settings can damage the block you are working with.

14. Spread the four pairs without untwisting any of them. Then, orient the CAT5 UTP cable for entry into the 110-type block.

TIP: If you decided to keep the run as short as possible, you will probably want to bring the cable from under the top-left of the block. Longer lengths of cable can enter the block through the slot under the left side.

15. Next, lace the tip wire of pair one "T1" (mostly white with blue stripe) into the first, or leftmost, slot on the 110-type block.

TIP: Be sure that the cable sheath is located as close as possible to the termination point, in order to prevent excess exposure of the insulated pairs. Just because you stripped 1 ½ inches of sheath off the cable doesn't mean that you want that much distance between the termination point and the remaining sheath.

16. Lace the ring wire of pair one "R1" (mostly blue with white stripe) into the next slot to the right of T1, taking care to maintain the twists as close as possible to the termination point, as shown in Figure 6-26.

TIP: The teeth on the 110-type block are designed to separate the pairs without interfering with the integrity of their twist.

17. Next, lace the tip wire of pair two "T2" (mostly white with orange stripe) into the next slot to the right of R1 on the 110-type block.

18. Lace the ring wire of pair two "R2" (mostly orange with white stripe) into the next slot to the right of T2, taking care to maintain the twists as close as possible to the termination point.

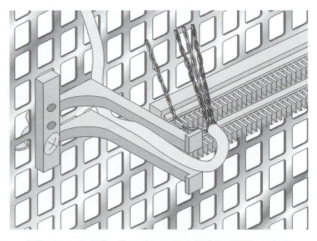

Figure 6-26: Lacing the Ring Wire R1

19. Next, lace the tip wire of pair three "T3" (mostly white with green stripe) into the next slot to the right of R2 on the 110-type block.

20. Lace the ring wire of pair three "R3" (mostly green with white stripe) into the next slot to the right of T3, taking care to maintain the twists as close as possible to the termination point.

21. Next, lace the tip wire of pair four "T4" (mostly white with brown stripe) into the next slot to the right of R3 on the 110-type block.

TIP: Remember that you do not have to untwist the pairs, because the pins on the 110-type block are designed to individualize the wires for you.

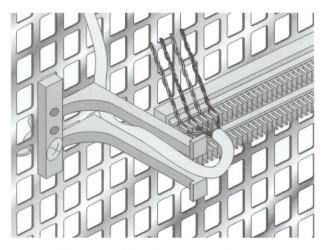

Figure 6-27: Four Pairs Laced

22. Lace the ring wire of pair four "R4" (mostly brown with white stripe) into the next slot to the right of T4, taking care to maintain the twists as close as possible to the termination point. With all four pairs laced, your wiring should now appear similar to that shown in Figure 6-27.

23. Locate the punch-down tool, and observe that the letters "CUT" have been etched into its side.

TIP: These letters should face the side of the block containing the conductors to be cut off. A rule of thumb suggests that these letters should face up when cutting the wires on the top row, and face down to cut the wires on the bottom row.

24. Utilizing the punch-down tool, cut and terminate the individual conductors of the CAT5 UTP cable that you just laced into the 110-type termination block, as shown in Figure 6-28.

Figure 6-28: Cutting and Terminating Conductors

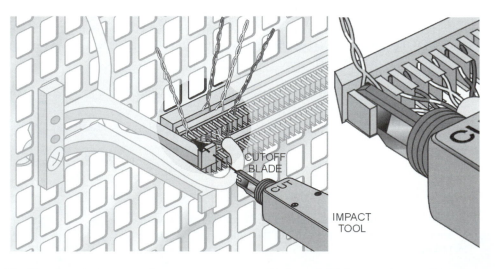

25. If necessary, you may choose to repeat steps 11 through 24 for practice, until you have gained the desired level of proficiency for this lab procedure.

TIP: Check with your instructor about this, because each time you repeat these steps the cable will grow 1.5 inches shorter. However, if practice is needed, do it now while you still have two feet of cable protruding from the bottom of conduit 1.

26. Without cutting or nicking any of the conductors, strip approximately one and a half (1.5) inches of cable sheath (not to exceed 2 inches) from the section of CAT5 UTP cable running out of the top of conduit 3. Snip the strain relief threads with electricians' scissors.

27. Spread the four pairs without untwisting any of them. Then, orient the CAT5 UTP cable for entry into or onto the 110-type block.

TIP: If you decided to keep the run as short as possible, you will probably want to bring the cable under the bottom-left of the block. Longer lengths of cable can enter the block through the slot on the left side.

28. Locate the 110-type connector block, and orient it directly over the connections you just terminated (from the CAT5 cable coming from conduit 2) on the 110-type termination block.

TIP: Unlike the metal connector clip of the 66-type block, the IDC design of the 110-type connector block is surrounded by plastic to decrease signal coupling.

TIP: The blue housing marker should be on the left (leaving the brown housing marker on the right), as shown in Figure 6-29. These markers make it easier to locate the termination points for the four colors. What you are getting ready to do is to terminate the bottom section of pulled cable in such a way as to simultaneously connect to the top section you just terminated. The 110-type connector block makes this job easy.

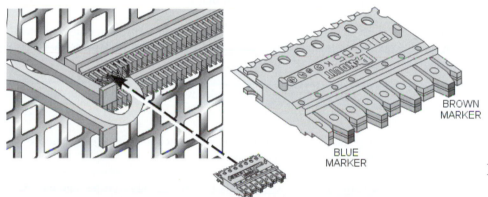

BROWN MARKER

BLUE MARKER

Figure 6-29: Blue Housing Marker

Alternately, you could choose to terminate the bottom section of cable to some other area on the 110 termination block. In fact, this is the way most terminations are accomplished in a real wiring closet. Such terminations would represent an open circuit for each of the four wire pairs. They would remain open until such time as they were bridged, or patched, to other terminations on this or another block, in order to complete a circuit. For a 110-type block, this would require a patch cord (110-to-110) to make the required connection, and would represent another opportunity for the signals being carried over these cables to be degraded.

Manufacturers have recently introduced products that provide combination terminal blocks that sport RJ45 jacks for convenient patching between the various terminated circuits. A cable installer should be ready to use any of the newer technologies that are shown to improve the quality and versatility of network cabling installations.

29. Reverse the blade, and use the 110-type punch-down tool to install the 110-type connector block right on top of the area where the first four wire pairs were terminated.

30. Next, lace the tip wire of pair one "T1" (mostly white with blue stripe) into the first, or leftmost, slot on the 110-type connection block.

TIP: Again, be sure that the cable sheath is located as close as possible to the termination point, in order to prevent excess exposure of the insulated pairs. You want as little distance as possible between the sheath and the termination point.

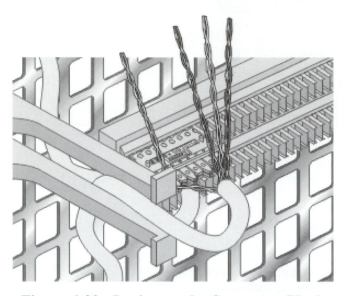

Figure 6-30: Lacing on the Connector Block

31. Lace the ring wire of pair one "R1" (mostly blue with white stripe) into the next slot on the 110-type connector block, to the right of T1, taking care to maintain the twists as close as possible to the termination point, as shown in Figure 6-30.

32. Next, lace the tip wire of pair two "T2" (mostly white with orange stripe) into the next slot to the right of R1 on the 110-type connector block.

33. Lace the ring wire of pair two "R2" (mostly orange with white stripe) into the next slot on the connector block, to the right of T2, taking care to maintain the twists as close as possible to the termination point.

34. Next, lace the tip wire of pair three "T3" (mostly white with green stripe) into the next slot to the right of R2 on the connector block.

35. Lace the ring wire of pair three "R3" (mostly green with white stripe) into the next slot to the right of T3, taking care to maintain the twists as close as possible to the termination point.

36. Next, lace the tip wire of pair four "T4" (mostly white with brown stripe) into the next slot to the right of R3 on the 110-type connector block.

TIP: Don't untwist any pairs, because the pins on the connector block are designed to individualize the wires for you.

37. Lace the ring wire of pair four "R4" (mostly brown with white stripe) into the next slot to the right of T4. With all four pairs on both cables laced, your wiring should now appear similar to that shown in Figure 6-31.

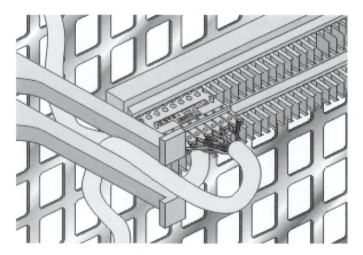

Figure 6-31: All Pairs Laced

TIP: If your wiring has been done correctly, your pulled cable should once again be a continuous run, even with the 110-type block installed between conduits 2 and 3.

38. Locate the punch-down tool and reverse the blade once again so that the cutting blade is active.

39. Utilizing the punch-down tool, cut and terminate the individual conductors of the CAT5 UTP cable that you just laced into the 110-type connector block.

40. Once both sections of cable have been terminated and connected at the 110-type block, use what you have learned from previous activities to perform the required Level II tests on the 100-ohm CAT5 UTP cable pull.

41. Record the results of the Level II tests in your Lab 20 work sheet.

42. Compare these results with those you recorded following the initial cable pull.

43. Check with your instructor to see if you are required to print a report for this cable test. If so, follow the instructions in the Users manual for the Level II testers regarding the printing of a report on this test, and print the required report.

44. According to the information presented by the Level II testers, determine with your instructor if the installation can still be certified, or if it must be corrected or qualified in some way.

45. After completing the testing, return the Level II testers (and charge units, if necessary) to their storage bags, along with any of the accessory cables you used.

46. Check to be sure that you are leaving your lab area in a clean and orderly condition.

Lab 20 Questions

1. What is the only solution to a cable run that is too short?

2. Explain the use of each end of a 110-type blade.

3. Why is the punch-down tool set to the LO impact setting?

4. Why are the pins of the 110-type block designed to individualize the wire pairs?

5. Which parameters, if any, displayed significant changes during the Level II testing from the values recorded following the initial cable pull?

Lab 21 – Terminating CAT5 to a 66 M1-50 Block

Lab Objective

To properly terminate CAT5 cable to a 66 M1-50 block, and test the resulting run

Materials Needed

- Lab 21 work sheet

- Simulator grid panel*

- Termination block, 66

- Connector bridge clip, 66 (8)

- Punch-down tool, steel (with a 66-type blade)

- Bracket, 89-series (1)

- Electricians scissors

- Pliers, diagonal cutters

- Machine screw, Phillips, pan head, ½-inch, 10-24x (2)

- Wing nut, 10-24x (2)

- Level II testers, w/case & cables (control unit and remote)**

- Users manual for Level II testers

- Screwdriver, Phillips, small

*The simulator grid panel still has 7 pieces of ½-inch PVC conduit and one 110-type termination block mounted, along with the 25-foot section of 4-pair, CAT5 UTP cable pulled from the previous lab procedure.

**Prepare ahead of time for any lab procedures that involve the use of the Level II testers by charging the Nickel-Cadmium battery packs in both the control unit and the smart remote. In an emergency situation, the units will operate during a charge operation, but they will have to remain near an ac outlet during this time.

Before beginning this lab, read through all of the following steps at least once, thoroughly.

1. Examine Figure 6-32 carefully.

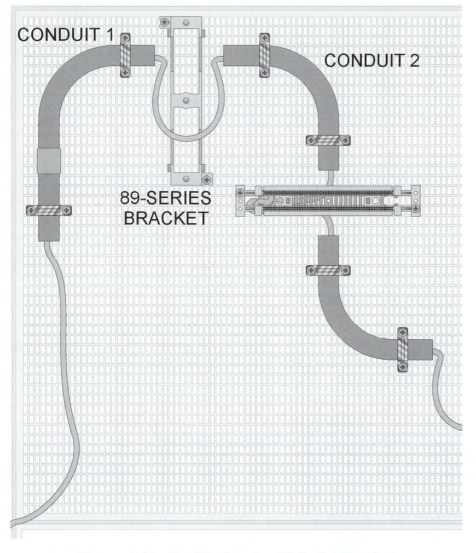

Figure 6-32: Positioning the 89-Series Bracket

2. Locate the 89-series bracket against the grid panel of the simulator in a vertical orientation, as indicated in Figure 6-32, between conduit 1 and conduit 2.

TIP: The horizontal screw tab should be located at the top-left, and the vertical screw tab should be located at the bottom-right.

3. While holding the 89-series bracket in place, thread a Phillips, pan head, ½-inch 10-24x machine screw through the vertical screw tab at the top left side of the bracket, and into the hole in the grid of the simulator panel, as shown in Figure 6-33.

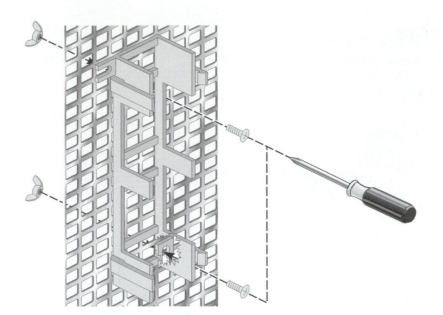

Figure 6-33: Placing an 89-Series Mounting Screw

4. Use a wing nut to tighten the block to the panel.

TIP: If there isn't enough space behind the grid panel to turn the wing nut freely, brace the wing nut with one finger while using the screwdriver to tighten the machine screw. The machine screw/wing nut combination should only be tight enough to keep the bracket from shifting.

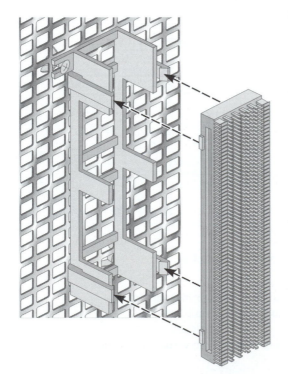

Figure 6-34: Positioning in the 66 M1-50 Block

5. With the 89-series bracket oriented vertically, place the second machine screw through the horizontal screw tab and into a hole in the simulator's grid panel, also shown in Figure 6-33.

6. With the bracket oriented vertically, place a wing nut on the machine screw and lightly tighten it.

7. Before tightening the screws or wing nuts snugly, be sure the 89-series bracket is properly oriented in a vertical direction. Do not over-tighten!

8. Snap the 66 M1-50 block onto the 89-series bracket with the horizontal screw tab on the block to the top-left of the assembly, just as with the 89-series bracket, as shown in Figure 6-34.

9. Compare your simulator grid panel with that shown in Figure 6-35, and make any necessary adjustments.

TIP: It's time to make the second break in the cable run, at the point marked in Figure 6-35 with the dotted line. Later, you will compare the previous test results on the cable run with the ones you get following the wiring of the 66-type termination block.

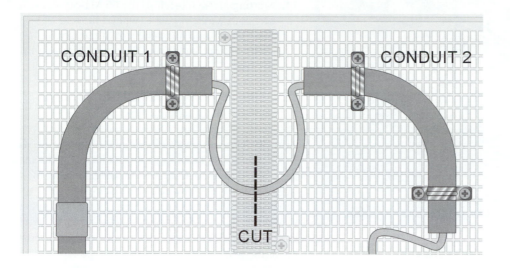

Figure 6-35: Type-66 M1-50 Termination Block Mounted

10. Locate the diagonal cable cutters, and observing the proper safety precautions, make a clean cut through the 4-pair, CAT5 UTP cable about halfway between conduit 1 and conduit 2 (on the dotted line in Figure 6-35).

TIP: Care should be taken at this point with your wiring. The margin for error is narrow for the cable section that runs through conduit 2. You still have some slack, however, with the cable section running through conduit 1. By this time, your cable preparation skills should be good enough to terminate this block properly with a minimum of problems.

11. Without cutting or nicking any of the conductors, strip approximately one and a half (1.5) inches of cable sheath (not to exceed 2 inches) from the section of CAT5 UTP cable running from the top-right of conduit 1. Snip the strain relief threads with electricians' scissors.

12. Insert a 66-type blade into the punch-down tool, making sure that the cutting edge of the blade faces outward, as shown in Figure 6-36.

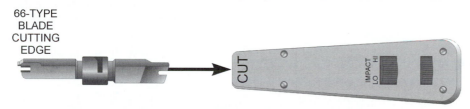

Figure 6-36: Inserting a 66-Type Blade

13. Place the setting of the punch-down tool in the "HI" impact position.

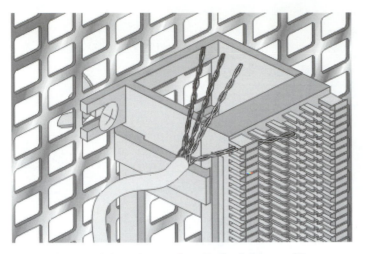

Figure 6-37: Inserting Pair 1 into a Slot

14. Starting at the top-left of the 66 M1-50 block, insert both wires of pair 1 (white/blue and blue/white) into the same slot of the fanning strip, without untwisting the pairs, as shown in Figure 6-37.

15. At this point, divide the pair into the correct quick-clips for termination.

TIP: The pair can be divided in one of two ways. The first method is by orienting both conductors in the same direction. The second method involves placing the tip wire (white/blue) upward, and placing the ring wire (blue/white) downward. When using the second method, be sure to turn the tool to achieve the proper cutting angle. Locate the cable sheath as close as possible to the slots on the side of the 66-type block.

16. Once the pair 1 conductors are being held in the proper quick-clips, use the punch-down tool by inserting the blade, from right to left, into the corresponding channel in the block for the top conductor (white/blue).

17. Extend the blade of the tool, firmly, to the very end of the channel, until the wire is cut off and terminated, as shown in Figure 6-38.

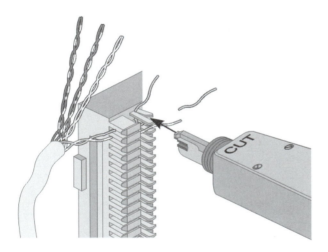

Figure 6-38: Terminating Pair 1

18. Insert the blade, from right to left, into the corresponding channel in the block for the bottom conductor (blue/white), and extend the blade of the tool, firmly, to the very end of the channel, cutting and terminating the wire.

19. Moving down to the appropriate slots in the fanning strip, repeat steps 14 through 18 for pair 2 (white/orange and orange/white), pair 3 (white/green and green/white), and pair 4 (white/brown and brown/white).

20. When you have completed your work on the left side of the 66-type block, compare it to the properly terminated 66-type block appearing in Figure 6-39. Your work should appear similar.

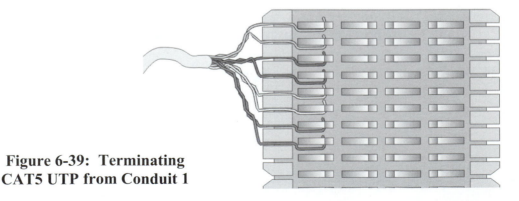

Figure 6-39: Terminating CAT5 UTP from Conduit 1

21. Without cutting or nicking any of the conductors, strip approximately one and a half (1.5) inches of cable sheath (not to exceed 2 inches) from the section of CAT5 UTP cable running from the top-left of conduit 2. Snip the strain relief threads, with electricians' scissors.

22. Using what you have already learned, start at the top-right of the 66 M1-50 block and terminate the 4-pair, CAT5 UTP cable running from conduit 2.

23. When you have completed your work on the right side of the 66-type block, compare it to the properly terminated 66-type block appearing in Figure 6-40. Your work should appear similar.

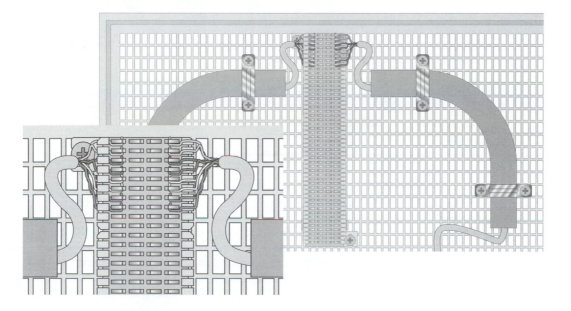

Figure 6-40: Terminating CAT5 UTP from Conduit 2

TIP: The question now becomes, how will the terminations that you've wired to the 66 M1-50 block be connected? As things now stand, an open circuit exists for each of the four pairs that lie on either side of conduit 1. The solution involves the 66-type connector clips.

24. Locate the eight 66-type connector clips.

25. Beginning with the two, top-middle quick-clips on the 66 M1-50 block, place a 66-type connector clip across and over them.

26. Push the connector clip down until it seats completely over the top-middle quick-clips of the 66 M1-50 block.

27. Install the second connector clip over the pair of quick-clips immediately below the middle pair you just clipped on the 66 M1-50 block.

TIP: Each of the twisted pairs will require two connector clips (one for each wire).

28. Using the remaining six 66-type connector clips, populate the next three pairs of middle quick-clips, pushing them completely down for a firm seat, as indicated in Figure 6-41.

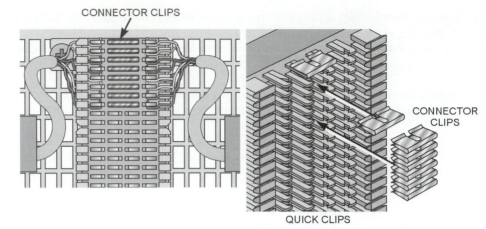

CONNECTOR CLIPS

CONNECTOR CLIPS

QUICK CLIPS

Figure 6-41:
Connecting the
Quick-Clips

TIP: If your wiring has been done correctly, with the connector clips in place, you should once again have a complete circuit between both ends of the CAT5 UTP cable run.

29. Once both sections of cable has been terminated and connected at the 66-type M1-50 block, use what you have learned from previous activities to perform the required Level II tests on the 100-ohm CAT5 UTP cable pull.

30. Record the results of the Level II tests in your Lab 21 work sheet.

31. Compare these results with those you recorded following the initial cable pull and following the terminations performed on the 110-type block.

32. Check with your instructor to see if you are required to print a report for this cable test. If so, follow the instructions in the Users manual for the Level II testers regarding the printing of a report on this test, and print the required report.

33. According to the information presented by the Level II testers, determine with your instructor if the installation can still be certified, or if it must be corrected or quali-fied in some way.

34. After testing has been completed, return the Level II testers (and charge units, if nec-essary) to their storage bags, along with any of the accessory cables you used.

35. Before leaving your lab work area, check to be sure that you are leaving it in a clean and orderly condition.

Lab 21 Questions

1. By splitting the wire pairs into the correct quick-clips for termination, what is it that is maintained for each pair?

2. List the tip and ring colors for each of the four wire pairs of a CAT5 cable.

3. Why was the punch-down tool set to the HI impact setting?

4. Which parameters, if any, displayed significant changes during the Level II testing from the values recorded following the initial cable pull?

Lab 22 – Testing CAT5 Terminations Between Blocks

<u>Lab Objectives</u>

To properly terminate CAT5 cable onto 66-type M1-50 blocks, and 110-type blocks

To properly test cable runs, and document the test results

<u>Materials Needed</u>

- Lab 22 work sheet
- Simulator grid panel*
- Termination block, 110 (2)
- Connector block, 110 (2)
- Termination block, 66 (2)
- Connector bridge clip, 66 (16)
- Bracket, 89-series (2)
- Machine screw, Phillips, pan head, ½-inch, 10-24x (8)
- Wing nut, 10-24x (8)
- Punch-down tool, steel (with 110-type and 66-type blades)
- Electricians' scissors
- Pliers, diagonal cutters
- Level II testers, w/case & cables (control unit and remote)**
- Manuals for Level II testers
- Screwdriver, Phillips, small

*The simulator grid panel still has 7 pieces of ½-inch PVC conduit, one 110-type termination block, and one 66-type M1-50 termination block mounted, along with the 25-foot section of 4-pair, CAT5 UTP cable run. The blocks were terminated during previous lab procedures.

**Prepare ahead of time for any lab procedures that involve the use of the Level II testers by charging the Nickel-Cadmium battery packs in both the control unit and the smart remote. In an emergency situation, the units will operate during a charge operation, but they will have to remain near an ac outlet during this time.

<u>Instructions</u>

Before beginning this lab, read all of the following steps thoroughly, at least once.

1. Using Figure 6-42 as a guide, mount two additional type-66 and two additional type-110 termination blocks on the simulator grid panel.

TIP: You may decide at this point to wire both 66-type or 110-type blocks as a group. By doing them as groups, you will not have to change the punch-down blade as often and will be doing the same procedure twice instead of changing back and forth.

TIP: You should have enough slack remaining in the large section of uncut CAT5 UTP cable to allow the additional blocks to be mounted without damaging any part of the run. Follow the layout shown in Figure 6-42.

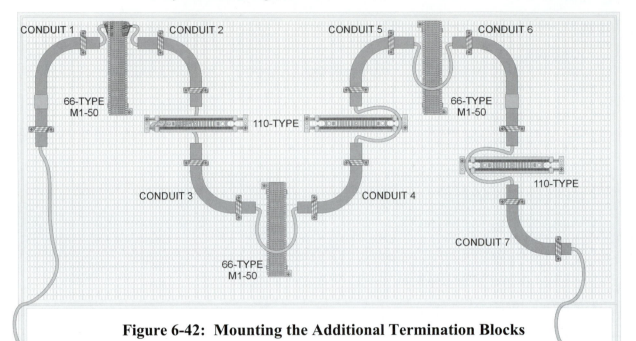

Figure 6-42: Mounting the Additional Termination Blocks

2. Beginning at the right side of conduit 3, use the techniques you have learned to cut, prepare, and terminate that section of the cable run to the left side of the 66-type M1-50 block that you just mounted there.

3. On the right side of the 66-type M1-50 block, prepare and terminate the respective end of the CAT5 UTP cable coming from the left side of conduit 4.

4. Locate eight 66-type connector clips and place one across and over the two, top-middle quick-clips on the 66 M1-50 block.

TIP: The clips on the 66-type blocks used for Category 5 are designed to reduce reactive coupling.

5. Push the connector clip down until it seats completely over the block's top-middle quick-clips. Then, repeat this process with the remaining connector clips for the remaining seven middle quick-clips.

TIP: At this point, your layout should look similar to that shown in Figure 6-43.

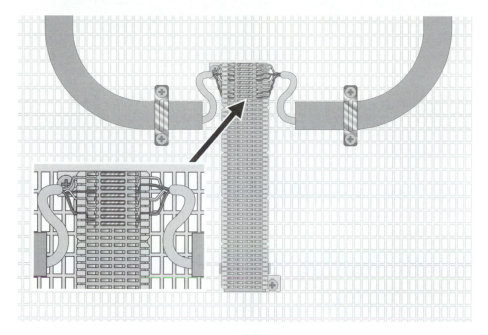

Figure 6-43: Second 66 M1-50 Block Terminated and Connected

6. Once both sections of cable have been terminated and connected at the second 66-type M1-50 block, use what you have learned from previous activities to perform the required Level II tests on the entire 100-ohm CAT5 UTP cable pull.

7. Record the results of the Level II tests in your Lab 22 work sheet, and compare these results with those you recorded following both the initial cable pull and the other block terminations.

8. Check with your instructor to see if you are required to print a report for this cable test. If so, follow the instructions in the Users manual for the Level II testers regarding the printing of a report on this test, and print the required report.

9. Once again determine with your instructor if the installation can still be certified, or if it must be corrected or qualified in some way.

10. From the top of conduit 4, use the techniques you have learned to cut, prepare, and terminate the cable run to the top-left of the second 110-type termination block.

11. At the bottom-left of the block, prepare, terminate and connect the end of the CAT5 UTP cable running from the top of conduit 5, using a 110-type connector block.

TIP: At this point, your layout should look similar to that shown in Figure 6-44.

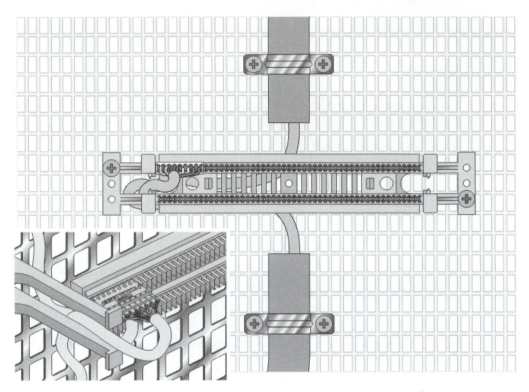

Figure 6-44: Second 110-Type Block Terminated and Connected

12. Once again, use what you have learned from previous activities to perform the required Level II tests on the entire 100-ohm CAT5 UTP cable pull, and record the results of the Level II tests in your Lab 22 work sheet.

13. Compare these results with those you recorded following both the initial cable pull and the other block terminations, and print any required Level II test reports as directed by your instructor.

14. Once again determine with your instructor if the installation has maintained its CAT5 UTP certification, or if it must be corrected or qualified in some way.

15. Utilizing the same techniques as with the previously installed blocks, prepare, terminate, connect, and test the remaining 66 M1-50 and 110-type blocks.

TIP: Once all of the blocks have been connected to the run (on the simulator grid panel), the completely wired system should appear similar to Figure 6-45.

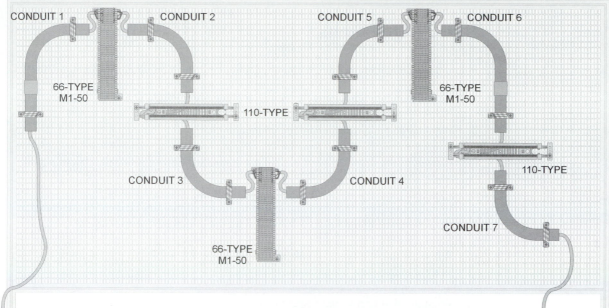

CONDUIT 1   CONDUIT 2   CONDUIT 5   CONDUIT 6

66-TYPE
M1-50

66-TYPE
M1-50

110-TYPE

110-TYPE

CONDUIT 3   CONDUIT 4

CONDUIT 7

66-TYPE
M1-50

Figure 6-45: Six Blocks Terminated and Connected

16. Compare the test results for the completed run (with all of the blocks connected), with those you recorded following the initial cable pull. Make a mental note of the most significant areas of parameter degradation.

17. Determine with your instructor if the completed installation has maintained a CAT5 UTP certification, or if it must be corrected or qualified in some way.

18. Make a final visual inspection of your CAT5 cable terminations at each of the blocks, from left to right, for neatness and accuracy. Then, make any required corrections or adjustments.

TIP: Neatness counts. Get in the habit of labeling all runs, and keeping them neatly organized. The first and last blocks in the run will not be connected together.

19. Once all of the required corrections and/or adjustments have been completed, use the Level II testers to perform the indicated EIA/TIA T568A tests, one at a time. Document these final test results in your Lab 22 work sheet. Include such parameters as:

    • Wire pairs containing the heaviest crosstalk.

    • Distance (in feet) to the crosstalk.

    • Impedance anomalies (location and percentage of each).

TIP: Always document the results of any tests you perform, using the EIA/TIA standards.

20. Once testing has been completed, return the Level II testers (and charge units, if necessary) to their storage bags, along with any of the accessory cables you used.

21. Before leaving your lab work area, check to be sure that you are leaving it in a clean and orderly condition.

Lab 22 Questions

1. What is the advantage of wiring all of the 110-type or 66-type blocks as a group?

2. Which tests are conducted by the Level II testers for EIA/TIA CAT5 cables?

3. What is the purpose of a 66-type connector clip?

4. How does the installation of a 110-type connector block reduce the risk of signal degradation?

5. Which Level II parameters underwent the most significant amount of degradation when comparing the initial cable pull with the completed terminations?

6. Which of the two types of termination blocks used in this lab procedure would you, as a cable installer, prefer to use? Why?

7. What good does a visual inspection of the run do, once all of the Level II tests have been successfully completed?

Lab 23 – Altering and Testing Block Terminations

<u>Lab Objectives</u>

To alter the number of cable terminations from Lab 22, and to compare the test results with documentation from Lab 22

<u>Materials Needed</u>

- Lab 23 work sheet

- Simulator grid panel*

- Punch-down tool, steel (with 110-type and 66-type blades)

- Cable, CAT5 UTP, 100-ohm, non-plenum (spool)

- Electricians' scissors

- Pliers, diagonal cutters

- Level II testers, w/case and cables (control unit and remote)**

- Manuals for Level II testers

- Screwdriver, Phillips, small

*The simulator grid panel still has 7 pieces of ½-inch PVC conduit, three 110-type termination blocks, and three 66-type M1-50 termination blocks mounted, terminated, and connected with the 25-foot section of 4-pair, CAT5 UTP cable run that was completed and tested during lab procedure 22.

**Prepare ahead of time for any lab procedures that involve the use of the Level II testers by charging the Nickel-Cadmium battery packs in both the control unit and the smart remote. In an emergency situation, the units will operate during a charge operation, but they will have to remain near an ac outlet during this time.

Instructions

Before beginning this lab, read all of the following steps thoroughly, at least once.

1. Begin by locating the second 66-type termination block, as you count from left to right along the run on the simulator grid panel, shown in Figure 6-46.

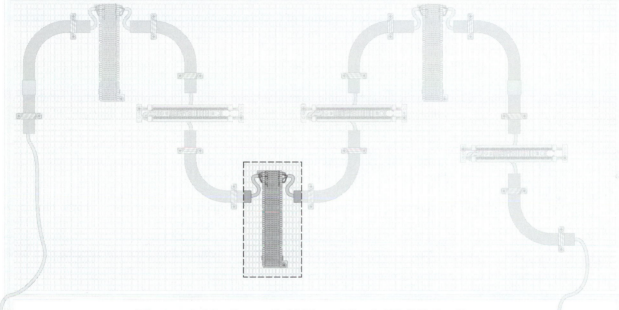

Figure 6-46: Second 66-Type Block Highlighted

2. Using the diagonal cable cutters, snip the CAT5 cables (on both the left and right sides) connecting the selected 66-type termination block to the run. Make the cuts as close as possible to the block, as shown in Figure 6-47.

3. Unsnap the selected 66-type termination block from its 89-series bracket, with the CAT5 cable pieces still attached, and set it aside.

4. Using the Phillips screwdriver, loosen the screws and remove the wing nuts holding the 89-series bracket to the simulator grid panel.

5. Remove the 89-series bracket from the simulator grid panel and set it aside.

6. Locate the two 110-type termination blocks that were formerly wired to the 66-type block you just removed (located to the upper left and upper right).

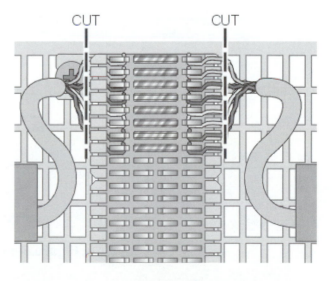

Figure 6-47: Cutting Cables to the 66-type Block

7. As close to these two 110-type blocks as possible, use the diagonal cable cutters to snip the CAT5 cables connecting them to the run.

8. Once these two 110-type termination blocks are severed from the run, remove them from the simulator grid panel, and set them aside.

9. Remove the leftover sections of CAT5 cable from conduit 3 and conduit 4, and place them in the cable scrap heap.

10. Using the diagonal cable cutters, snip the CAT5 cable that runs through conduit 2, connecting to the right side of the 66-type termination block on the left. Make the cuts as close as possible to the block.

11. Remove the leftover section of CAT5 cable from conduit 2, and remove the twisted pair fragments from the 66-type block terminals as well.

TIP: The 66-type blade can pry open the terminals to help in removing the wire fragments.

12. Using the diagonal cable cutters, snip the CAT5 cable that runs through conduit 5, connecting to the left side of the 66-type termination block on the right. Make the cuts as close as possible to the block.

13. Remove the leftover section of CAT5 cable from conduit 5, and remove the twisted pair fragments from the 66-type block terminals as well.

TIP: You should now have two, 66-type termination blocks and one, 110-type termination block remaining on the simulation grid panel, along with all of the conduit sections, as shown in Figure 6-48. Use a new piece of CAT5 cable to connect between the remaining 66-type termination blocks, if the originally cut CAT5 sections won't reach across the grid panel.

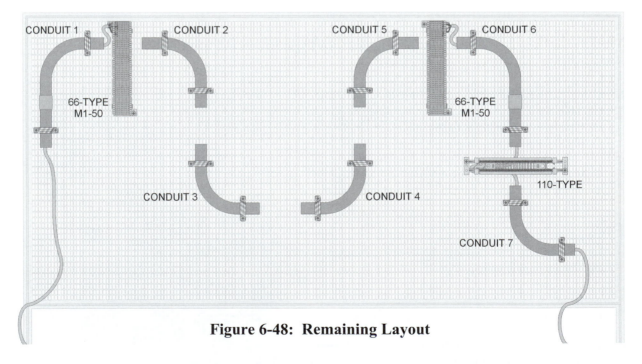

Figure 6-48: Remaining Layout

14. Locate the CAT5 cable spool directly below the 66-type block between conduit 5 and conduit 6.

15. Begin to pull cable from the spool and manually (no fish tape) thread it through conduit 5 from right to left.

16. When the CAT5 cable emerges from the top of conduit 5, pull approximately 5 feet of cable through.

17. Now thread the 5 feet of CAT5 cable completely through conduit 4, from the top to the bottom, leaving a slight amount of slack between it and conduit 5.

TIP: No more CAT5 cable should need to be taken from the spool.

18. Thread the remaining free cable through conduit 3, from right to left (and out the top), and take up any slack.

19. Then, thread the remaining length of CAT5 cable through conduit 2, from the bottom to the top (and out the left), removing any slack.

20. Use the diagonal cable cutters to cut the CAT5 cable running from the spool at about 5 inches from the right end of conduit 5, and set the spool of remaining cable aside.

21. Using the techniques that you have already learned, terminate the remaining blocks together in a single run, as shown in Figure 6-49.

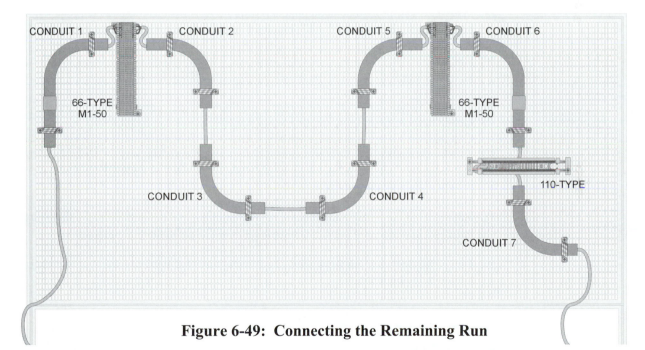

Figure 6-49: Connecting the Remaining Run

TIP: The required connector clips should still be in place on the remaining 66 M1-50 blocks.

22. Once again, use what you have learned from previous activities to perform the required Level II tests on this rewired 100-ohm CAT5 UTP cable pull, and record the results of the Level II tests in your Lab 23 work sheet.

23. Compare these results with those you recorded following both the initial cable pull, and the other block terminations (particularly step 7 in Lab procedure 22), and print any required Level II test reports as directed by your instructor.

24. Determine with your instructor if the installation has maintained its CAT5 UTP certification, or if it must be corrected or qualified in some way.

25. Following a final visual inspection of your CAT5 cable terminations for neatness and accuracy, make any required corrections or adjustments.

TIP: Neatness counts. Get in the habit of labeling all runs and keeping them neatly organized. The first and last blocks in the run will not be connected together.

26. Once all of the required corrections and/or adjustments have been completed, use the Level II testers to perform the indicated EIA/TIA T568A tests, one at a time.

27. Document these final test results in your Lab 23 work sheet. Be sure to include information on such parameters as:

   • Wire pairs containing the heaviest crosstalk.

   • Distance (in feet) to the crosstalk.

   • Impedance anomalies (location and percentage of each).

TIP: Always document the results of any tests you perform using the EIA/TIA standards.

28. When the testing has been completed, return the Level II testers (and charge units, if necessary) to their storage bags, along with any of the accessory cables you used.

29. Remove the fragments of UTP cable from the termination blocks you took off of the grid panel, as well as any 66-type connector clips still mounted.

30. On the 110-type termination blocks, pry off the 110-type connector blocks still attached to them first. Then, the first set of punched wires can be removed.

TIP: This can be tricky for 110-type connector blocks because of the way they grip the termination block. Don't tear your fingers up trying to get them off. Use a small flat-blade screwdriver to pry one side up at a time.

31. Return all of the termination blocks, connector clips, and connector blocks that were removed from the grid panel's cable run back to the designated storage area.

32. Before leaving your lab work area, check to be sure that you are leaving it in a clean and orderly condition.

Lab 23 Questions

1. What were the key differences in the test results between Lab 22 and Lab 23?

2. How can these test results differences be explained?

3. Which section(s) in the grid panel setup run would you designate as a "basic link"?

4. How did the Level II test results for this procedure compare with those of Lab procedure 22 (step 7)?

Lab 24 – Differentiating Between T568A and T568B

<u>Lab Objectives</u>

To understand the differences between the wiring configurations of T568A and T568B, and the importance of utilizing the same wiring configuration on both ends of a cable run

<u>Materials Needed</u>

- Lab 24 work sheet

- Simulator grid panel*

- Punch-down tool, steel (with 110-type and 66-type blades)

- Electricians' scissors

- Pliers, diagonal cutters

- Patch panel, 16-port

- Level II testers, w/case &cables (control unit and remote)**

- Manuals for Level II testers

- Screwdriver, Phillips, small

- Termination plugs, RJ45, unshielded (2)

- Crimping tool, RJ45

- Cable, CAT5 UTP RJ45 jack/wall plate combination (from previous lab)

- CAT5 UTP cable, 12-inch scrap piece

*The grid panel still has 7 pieces of ½-inch PVC conduit, one 110-type termination block, and two 66-type M1-50 termination blocks mounted, terminated, and connected with the 25-foot section of 4-pair, CAT5 UTP cable run that was tested during Lab procedure 23.

**Prepare ahead of time for using the Level II testers by charging the Nickel-Cadmium battery packs in both the control unit and the smart remote. In an emergency, the units will operate during a charge operation, but they will have to remain near an ac outlet during this time.

<u>Instructions</u>

Before beginning this lab, read all of the following steps thoroughly, at least once.

1. Using the diagonal cable cutters, snip the CAT5 cable protruding from the top-right of conduit 1, making the cut as close as possible to the block. Then, pull the severed CAT5 cable out of the bottom of conduit 1, at the left side of the run.

2. Using the diagonal cable cutters, snip the CAT5 cable protruding from the top-left of conduit 7, making the cut as close as possible to the block. Then, pull the severed CAT5 cable out of the right of conduit 7, at the right side of the run.

3. Using the diagonal cable cutters, snip the CAT5 cable sections running between the two 66-type termination blocks and the remaining 110-type termination block.

4. Unsnap the two 66-type termination blocks from their 89-series brackets, with the CAT5 cable pieces still attached, and set them aside.

5. Using the Phillips screwdriver, loosen the screws, and remove the wing nuts holding the 89-series brackets to the simulator grid panel.

6. Remove the 89-series brackets from the simulator grid panel, and set them aside.

7. Locate the remaining 110-type termination block that was formerly wired to one of the 66-type blocks you just removed (located to the lower-right).

8. Using the Phillips screwdriver, loosen the screws, remove the wing nuts holding this remaining 110-type block to the grid panel, and remove it.

9. Remove the shorter sections of CAT5 cable from all of the conduits and place them in the cable scrap heap.

TIP: The two large sections of CAT5 UTP cable, each with a RJ45 plug installed (T568A configuration), should remain, while all of the conduits are now empty.

10. Remove the twisted pair cable fragments, and the connector clips, from the terminals on the two 66-type blocks. Then, return the blocks and clips to the parts storage area.

TIP: The 66-type blade can pry open the terminals to help in removing the wire fragments.

11. Remove the fragments of UTP cable from the connector block attached to the 110-type termination block you just took off of the grid panel.

12. On the 110-type termination block itself, pry off the 110-type connector block still attached. Then, the first set of punched wires can be removed.

TIP: This can be tricky for a 110-type connector block because of the way it grips the termination block. Don't tear your fingers up trying to get it off. Use a small flat-blade screwdriver to pry one side up at a time.

13. Return this termination block and its associated connector block to the storage area.

TIP: The simulator grid panel should now appear as shown in Figure 6-50.

14. Remove the brackets holding the sections of conduit from the simulator grid panel, and return the screws, wing nuts, brackets, and conduit sections to the storage area.

15. At this time, return the simulator grid panel to the designated storage area as well.

16. Being careful not to nick or cut any of the individual conductors, strip approximately 1.5 inches of cable sheath (not to exceed 2 inches) from the free end of one of the remaining sections of CAT5 UTP cable.

17. Cut off the strain-relief threads from this free end as well, using the electricians' scissors or cable sheath removers.

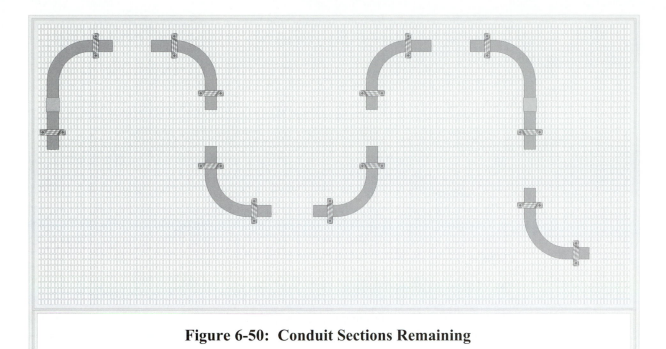

Figure 6-50: Conduit Sections Remaining

18. Untwist each of the cable pairs for a length of approximately 1 inch.

TIP: Keep the individual pairs grouped together for ease of identification, because some tip wires may not have any visible trace of color and may appear to be solid white.

19. Arrange the cable colors according to the TIA/EIA T568A pin configuration.

20. Using the electricians' scissors, or cable sheath remover, trim the CAT5 conductors to approximately 5/8 of an inch, including the strain-relief threads as well.

21. Insert these CAT5 conductors into one of the RJ45 modular plugs.

22. Make sure that the conductors are inserted as far as possible into the plug.

23. Remove the conductors momentarily to examine their ends for evenness.

24. If the wires in the center of the bunch appear to be longer than the wires at the ends, use the cable cutters to snip enough to make all of the conductors even, as shown in Figure 6-51. Then, reinsert the trimmed wires into the RJ45 modular plug.

25. Now, place the wired RJ45 (T568A configuration) modular plug into the crimp tool.

26. Before crimping, ensure that the CAT5 conductors are fully inserted into the plug, and that the cable sheath is resting above the crimp edge inside the plug.

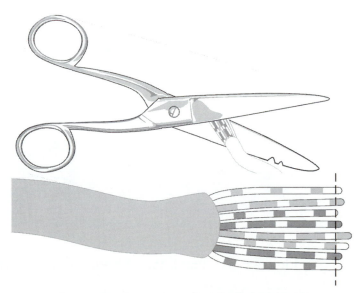

Figure 6-51: Trimming Twisted Pairs

27. Apply pressure to the cable to make sure that the conductors remain inserted to the ends of the plug, and use firm, even pressure with the crimping tool to attach the RJ45 modular plug.

28. Remove the crimped end of the cable from the tool, with the RJ45 plug attached.

29. Temporarily set this cable aside, and locate the other CAT5 UTP cable section. Then, repeat steps 16 through 28 (except steps 19 and 25) using this section.

30. For steps 19 and 25, arrange the cable colors according to the TIA/EIA T568B pin configuration instead of the T568A configuration.

TIP: You should now have two cable sections terminated with RJ45 plugs. One cable section is wired straight through, with the T568A configuration plugs at both ends. The other cable is cross wired with an RJ45 plug wired for the T568A configuration at one end, and an RJ45 plug wired for the T568B configuration at the other.

31. Locate the 16-port patch panel.

TIP: In order to make meaningful patches from the front of the patch panel, certain wiring must be accomplished at the rear terminals. This is wiring that would normally occur in the telecommunications closet and terminate at the patch panel. Patch cords would be used to connect various circuits together from the front panel jacks. However, for the purposes of this procedure, a pair of jacks (5 and 8 suggested) will be hard-wired together, straight through, at the rear of the patch panel. This will allow the testing of the cable sections you have created using the Level II meters.

32. Locate the 12-inch piece of scrap CAT5 UTP cable.

33. Being careful not to nick any of the conductors, strip 1.5 inches of cable sheath (not to exceed 2 inches) from each end.

34. Use the electricians' scissors or cable sheath removers to cut off the strain-relief threads from both ends.

35. Separate each of the cable pairs for a length of approximately 1 inch.

TIP: Keep the individual pairs grouped together for ease of identification, because some tip wires may not have any visible trace of color and may appear to be solid white.

36. On each cable pair untwist the wires for approximately 1/4 of an inch, but no further.

37. Locate the 16-port patch panel and observe the two, 8-port punchdown blocks mounted on the back.

38. Pay particular attention to the block mounted on the right, as shown in Figure 6-52.

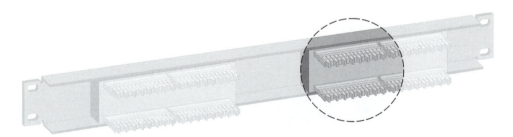

Figure 6-52: Patch Panel 8-Port Punchdown Blocks

TIP: Each 8-port punchdown block is numbered 1 through 8, with the right-most block corresponding to ports 1 through 8 on the front panel, and the left-most block corresponding to ports 9 through 16.

39. Locate the marker indicating the internal wiring scheme of the 8-port block.

TIP: This marker may indicate either T568A or T568B. Depending on the equipment that is terminated in the telecommunications closet, the patch panels being used would be wired specifically to interface with that equipment. If it wasn't, the risk of connecting mismatched wiring schemes would be high. Because your cables have been wired as straight-through connections, they can be tested normally without being concerned with the scheme of the patch panel's internal wiring.

40. Insert a 110-type blade into the punch-down tool, making sure that the cutting edge faces outward.

TIP: The cutting top of the blade should be oriented so that it is located beside the letters "CUT" on the punch-down tool.

41. Place the setting of the punch-down tool in the "LO" impact position.

42. Spread the four wire pairs on the 12-inch piece of scrap CAT5 without untwisting them any further.

43. Then, orient the piece of CAT5 UTP cable for entry into the right-most block at ports 5 and 8, as shown in Figure 6-53.

44. Lace the tip wire of pair one "T1" (mostly white with blue stripe) into the first, or left-most slot of port 5.

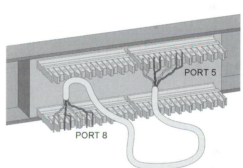

TIP: Be sure that the cable sheath is located as close as possible to the termination point, in order to prevent excess exposure of the insulated pairs. Just because you stripped 1.5 inches of sheath from off the cable doesn't mean that you want that much distance between the termination point and the remaining sheath.

Figure 6-53: Wiring the 16-Port Hub

45. Lace the ring wire of pair one "R1" (mostly blue with white stripe) into the next slot to the right of T1, taking care to maintain the twists as close as possible to the termination point.

TIP: The teeth on the block are designed to separate the pairs without interfering with the integrity of their twist.

46. Next, lace the tip wire of pair two "T2" (mostly white with orange stripe) into the next slot to the right of R1 on the block.

47. Lace the ring wire of pair two "R2" (mostly orange with white stripe) into the next slot to the right of T2, taking care to maintain the twists as close as possible to the termination point.

48. Next, lace the tip wire of pair three "T3" (mostly white with green stripe) into the next slot to the right of R2 on the block.

49. Lace the ring wire of pair three "R3" (mostly green with white stripe) into the next slot to the right of T3, taking care to maintain the twists as close as possible to the termination point.

50. Next, lace the tip wire of pair four "T4" (mostly white with brown stripe) into the next slot to the right of R3 on the 110-type block.

TIP: Remember that you do not have to untwist the pairs, because the pins on the 110-type block are designed to individualize the wires for you.

51. Lace the ring wire of pair four "R4" (mostly brown with white stripe) into the next slot to the right of T4, taking care to maintain the twists as close as possible to the termination point.

TIP: With all four pairs laced into port 5, your wiring should now appear similar to that shown in Figure 6-54.

52. Locate the punch-down tool, and observe that the letters "CUT" have been etched into its side.

TIP: These letters should face the side of the block containing the conductors to be cut off. A rule of thumb suggests that these letters should face up when cutting the wires on the top row, and face down to cut the wires on the bottom row.

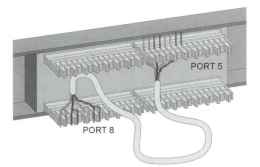

Figure 6-54: Wiring for Port 5 Completed

53. Utilizing the punch-down tool, cut and terminate the individual conductors of the CAT5 UTP cable scrap that you just laced into the port 5 portion of the block.

54. Substituting port 8 for port 5, repeat steps 40 through 54 to wire the remaining end of the CAT5 scrap cable to the port 8 portion of the block.

55. Compare your work with Figure 6-55 and make any necessary adjustments.

TIP: The 16-port hub should now be wired to allow its use for testing the Level II wiring parameters of the CAT5 UTP cables you have previously created.

56. Locate the Users manual for the Level II meters, and review the list of activities that can be performed from the control unit's SINGLE TEST menu.

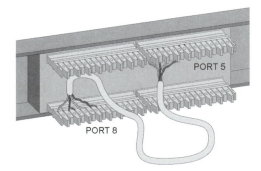

Figure 6-55: Wiring for Hub Completed

TIP: Notice that the single test versions of the wire map, resistance, TDR, and TDX analyzation operations allow you to repeatedly run a test, and update its results on the display each time, using the scanning function. This is an ideal testing mode to use when problem conditions on a cable are intermittent.

57. Locate the RJ45/T568A modular jack cable and wall-plate combination you created in Lab 13 (terminated with an RJ45/T568A modular plug).

58. Connect a known good RJ45 patch cable (greater than or equal to 2 meters in length) into the jack end of the RJ45/T568A modular jack cable and wall plate combination.

59. Connect the plug end of the RJ45/T568A modular jack cable and wall plate combination into the appropriate jack (either port 5 or port 8) on the 16-port patch panel.

60. Locate the CAT UTP cable segment on which you installed the RJ45/T568A termination plug earlier in this procedure.

61. Connect one end of this cable into the appropriate jack on the 16-port patch panel.

TIP: The appropriate jack would be the one that is directly wired to the patch panel jack into which the RJ45/T568A modular jack cable and wall plate combination is connected.

62. Turn the rotary function selector on the smart remote unit to the ON position.

TIP: The remote unit will emit several audible tones.

63. Connect the end of the combination cable that represents the far end of the network to the smart remote unit (using the RJ45 jack located in the unit's top).

TIP: This is the free end of the cable now plugged into the wall plate jack (step 59).

64. Turn the rotary function selector on the control unit clockwise to the SINGLE TEST menu position.

TIP: The control unit will emit several audible tones.

65. Check the settings displayed on the control unit's screen for accuracy, to be sure that the test being conducted is for the cable type under test. Then, if necessary, make any required adjustments from the control unit's SETUP menu.

TIP: Make sure that the meter is set to measure the CAT5 channel, rather than the basic link.

66. Connect the end of the combination cable that represents the near end of the network to the control unit (using the RJ45 jack located in the unit's top).

TIP: This is the free end of the cable on which you installed the RJ45/T568A termination plug earlier, now plugged into the 16-port patch panel (step 62). This is the cable being tested at this point! Your setup should look similar to that shown in Figure 6-56.

67. Meanwhile, the display on the control unit should appear similar to that shown in Figure 6-57, showing a listing of the individual tests that are available.

TIP: Notice that most of the tests listed are identical to the ones that are performed by the AUTOTEST series. The exceptions to this are the **Time Domain Crosstalk (TDX) Analyzer** and the **Time Domain Reflectometry (TDR)** tests. The TDX test can actually indicate where, along the cable run being tested, the crosstalk is occurring. Depending upon which display format is being used, you can view the largest value of crosstalk detected (list) or all of the crosstalk values (plot).

Time Domain
Crosstalk (TDX)
Analyzer

Time Domain
Reflectometry (TDR)

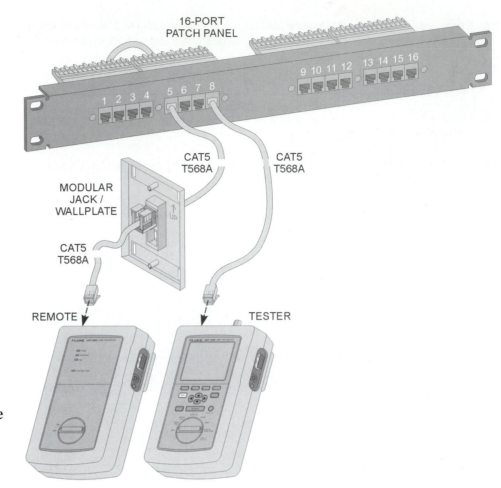

16-PORT
PATCH PANEL

CAT5
T568A

CAT5
T568A

MODULAR
JACK /
WALLPLATE

CAT5
T568A

REMOTE

TESTER

**Figure 6-56:
Testing First
CAT5 UTP Cable
Segment**

Although the TDX test reports the approximate crosstalk levels, its main usefulness is in locating the various sources of crosstalk along the run under test. For a definitive test on crosstalk levels, the NEXT test would be used.

The TDR test is concerned with locating areas along a cable run where impedance mismatches, called anomalies, are being detected. These **anomalies** cause signal reflections and can be caused by short or open circuits, poor connections, or mismatched cable types. This is one reason that cable installers always use the same cable type (preferably from the same manufactured spool allotment) along the complete run. Reflections that show up on this test are those exceeding the selected test standard limits.

anomalies

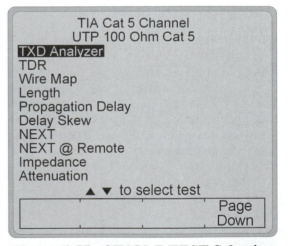

TIA Cat 5 Channel
UTP 100 Ohm Cat 5
TXD Analyzer
TDR
Wire Map
Length
Propagation Delay
Delay Skew
NEXT
NEXT @ Remote
Impedance
Attenuation
▲ ▼ to select test

Page
Down

**Figure 6-57: SINGLE TEST Selection
Screen**

68. Highlight the TDX Analyzer test in the control unit's display menu. Then, run the test on the first CAT5 UTP cable segment by pressing the TEST button on the control unit.

TIP: Depending on the wiring setup and the battery condition of the Level II testers, several things may happen. If for some reason the connection to the remote unit is not detected, or if the remote's batteries are low, you may receive a continuous Scanning for Remote message, rather than a very short one. In this case the selected Single Test will not be run until a properly charged remote unit is detected. Or, if a calibration message appears on the control unit's screen, perform the calibration procedure presented in Lab 16.

If the data that was generated from a previous test procedure was not saved, pressing the TEST button will give the control unit an opportunity to warn the user about this before going on. The user can choose to save the previous results, or discard the previous results by pressing the TEST button again.

69. Observe the first TDX Analyzer screen to be displayed. It is presented in the list format, shown in Figure 6-58.

TIP: The specific parameters reported for the selected pairs include:

- **Pairs**–The wire pairs for which the displayed data is relevant.

- **Peak**–The highest value of crosstalk measured between the two selected pairs. Although the reported levels of crosstalk are adjusted in order to compensate for the normal attenuation occurring across the cable, a result that reports a peak value greater than 50 would exceed the selected test standard.

- **Distance**–The measured distance along the cable from the control unit to the peak value of crosstalk.

- **View Plot**–Indicates the button to press to view a plot of all the locations along the tested cable where crosstalk is detected.

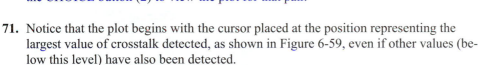

Figure 6-58: TDX Analyzer List Format Screen

70. Highlight one of the wire pairs and press the ENTER button or the CHOICE button (2) to view the plot for that pair.

71. Notice that the plot begins with the cursor placed at the position representing the largest value of crosstalk detected, as shown in Figure 6-59, even if other values (below this level) have also been detected.

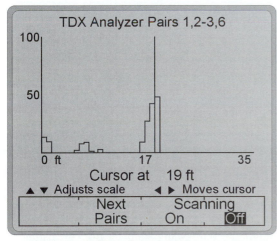

Figure 6-59: TDX Analyzer Plot Format Screen

On the horizontal line representing the cable run, the 0 ft marker indicates the location of the control unit, while the right or left arrow buttons can be used to move the cursor along the run. The up or down arrows can be used to adjust the scale along the X-axis of the display.

72. Record the TDX Analyzer test results in your Lab 24 work sheet for each wire pair of the first CAT5 UTP cable segment.

73. Press the EXIT button appropriately to return to the SINGLE TEST menu.

74. From the SINGLE TEST menu, highlight the TDR parameter from the listing, and press the TEST button to conduct the test and view the information.

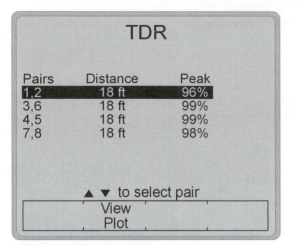

Figure 6-60: TDR Analyzer List Format Screen

75. Observe the first TDR test screen to be displayed. It is presented in the list format, shown in Figure 6-60.

TIP: The specific parameters reported for the selected pair include:

- **Pair**–The specific wire pair for which the displayed data is relevant.

- **Distance**–The first distance reported is the length of the cable as measured from the test tool to the cable's end. If a second distance is displayed, this is the length of the cable that lies between the test tool and the greatest reflection anomaly that exceeds the limit set by the selected test standard.

- **Peak**–The greatest percentage of reflected test signal measured at the anomaly.

- **View Plot**–Indicates the button to press to view a plot of all the locations and percentages of reflection for all of the detected anomalies.

76. Highlight a wire pair and press the ENTER button or the CHOICE button (2) to view the plot for that pair.

TIP: Notice that the plot begins with the cursor placed at the position representing the largest reflection percentage detected, as shown in Figure 6-61, even if other values (below this level) have also been detected. Any values above the 0% line indicate impedances greater than the cable's characteristic impedance, while values below the 0% line indicate impedances lower than the cable's characteristic impedance.

The position of the cursor along the run marks the location corresponding to the reflection information provided. The right or left arrow buttons can be used to move the cursor along the cable run, while the up or down arrows can be used to adjust the scale of the display's X-axis.

77. Record the TDR test results in your Lab 24 work sheet for each wire pair of the first CAT5 UTP cable segment.

78. Press the EXIT button appropriately to return to the SINGLE TEST menu.

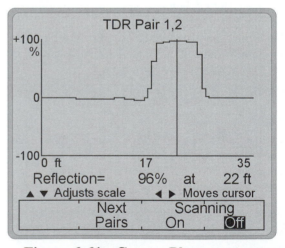

Figure 6-61: Cursor Placement on Reflection Plot

79. From the SINGLE TEST menu, highlight the Wire Map parameter from the listing, and press the TEST button to conduct the test and view the information.

TIP: You should already be familiar with this test from previous AUTOTEST activities.

80. Record the Wire Map test results in your Lab 24 work sheet for each pair of the first CAT5 UTP cable segment, and then return to the SINGLE TEST menu.

81. Using the ANSI/TIA/EIA T568A standard, conduct the remaining tests from the SINGLE TEST menu for each pair of the first CAT5 UTP cable segment.

TIP: These tests should include Length, Propagation Delay, Delay Skew, NEXT, NEXT @ Remote, Impedance, Attenuation, and Resistance.

82. Record all these test results in your Lab 24 work sheet for each pair of the first CAT5 UTP cable segment.

TIP: Now it's time to perform the SINGLE TEST series on the second CAT5 UTP cable segment you created earlier in this lab procedure.

83. Disconnect the first CAT5 UTP cable segment from the cable run you've been testing, and set it aside.

TIP: This would be the cable running from the Level II test control unit to the patch panel. All of the other parts of the cable run should remain intact.

84. Locate the CAT UTP cable segment on which you configured the RJ45/T568B termination plug earlier in this procedure.

85. Connect one end of this cable into the appropriate jack on the 24-port patch panel, and the other end to the Level II control unit.

TIP: Recall that this cable is wired for T568B on one end and T568A on the other.

86. Press the TEST button on the control unit and observe the report from the display.

TIP: The Level II meters detect the fact that this cable segment has a wiring problem. The display indicates a condition similar to that shown in Figure 6-62. Although the tests can be continued at this point (by pressing CHOICE button 4), a miswired cable would normally be rewired or corrected (if possible) before continuing with any further testing.

87. Check with your instructor about any further testing, and record the results in the Lab 24 work sheet.

88. Once all testing has been completed, turn the Level II testers OFF, disconnect the various cables from between the Level II testers and from the patch panel, and remove the scrap piece of CAT5 that you hardwired to the panel's rear block.

89. Return the cables you tested, and the 16-port patch panel, back to their designated storage areas and return the Level II testers (and charge units, if necessary) to their appropriate storage bags, along with any of the accessory cables you used.

90. Before leaving your lab work area, check to be sure that you are leaving it in a clean and orderly condition.

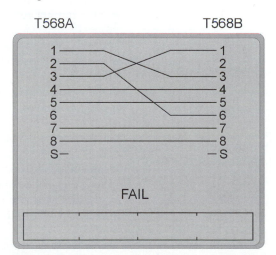

Figure 6-62: Crossed Pairs on Second Cable

Lab 24 Questions

1. Should a miswired cable be pulled immediately from a conduit?

2. Why does the control unit display the wire map information immediately in cases where pairs have been crossed, or miswired?

3. How widely do the test results vary between different pairs of the same cable?

4. What differences exist between the Level II tests conducted under the AUTOTEST menu with those conducted under the SINGLE TEST menu?

5. When viewing a plot for crosstalk, where does the cursor appear?

6. When the TDR Analyzer is activated, what does it mean when only one reflection is displayed on the Level II control unit?

CHAPTER SUMMARY

This chapter presented an overview of the current state of the cable-pulling industry, and various cabling parameters were discussed. Cable pulls covering greater and greater distances were attributed not only to the manufacture of stronger cable, but also to the emergence of technological improvements in the pulling lubricant industry. The importance of fire-retardant qualities in the manufacture of both cables and lubricants was discussed. Knowledge of fire suppression specifications by installers was stressed.

The role of conduit was described as was the benefit of preinstalling conduit groups before any major street or building construction is completed. The contrast between pulling cable and blowing cable through conduit was described, as were the differing criteria under which each installation method might best be utilized. The concept of a jam ratio was introduced as was the concept behind making calculations for the coefficient of friction.

You were introduced to the idea of tension limits, beyond which a data cable will be damaged or destroyed. The subject of multi-cable jamming was presented, along with a formula to calculate a three-cable jam ratio. Testing apparatus was described for the purpose of calculating COFs and measuring the effects of jamming. Important parameters included jam ratios, percentage of conduit fill, and the remaining conduit clearance following the pull.

The two basic configurations for three-cable pulls were discussed, immediately followed by Lab 19. During a pull, you received some experience with fish tape and were introduced to several guidelines on using it. You learned how to estimate pulling tension as the cable was pulled through the conduit and the importance of providing slack following the pull. Following the pull, you terminated the ends of the pulled cable with RJ45 plugs in the T568A configuration and conducted the appropriate tests on the cable using the Level II meters.

The technology of pulling lubricants was covered, and you were shown the COF formula used by cable manufacturers in the evaluation of various lubricants. The capability of pulling lubricants to reduce the negative aspects of friction, especially through multi-bend conduit, was documented.

During Lab 20, you gained experience in terminating CAT5 cable to a 110-type termination block by inserting the block into the existing run on the simulator grid panel. After removing the twine from the conduit path, you mounted the 110-type block between conduit 2 and conduit 3. After making a break in the cable run, you prepared the ends of the cable and punched the four twisted pairs down into the block. Then, you cut and terminated the conductors utilizing the blade on the impact tool. You used a 110-type connector block to join the second section of cable to the first section, on the 110 termination block, without having to use any type of jumper cables. You learned the correct way to orient the cable pairs by using the color code markings on the blocks. Once the 110-type termination block was properly inserted into the cable run, you performed the appropriate Level II testing, recorded the results, and compared those results with those for the initial cable pull.

For Lab 21, mounted an 89-series bracket, inserted a 66 M1-50 termination block into the bracket, and made a second break in the cable run. You then prepared the ends of the cable and punched them down on either side of the 66 M1-50 termination block. In order to complete the connections across this block, you used the 66-type connector clips and mounted them over the two middle quick-clips for each individual wire. You then performed the appropriate Level II tests on the cable run to ensure its integrity, and recorded the results. These test results were compared with those taken from the previous procedures, and any required actions were taken to maintain the certification of the run.

In Lab 22, two additional 110-type and 66-type blocks were mounted on the grid panel and inserted into the run. Level II testing was performed and recorded following the mounting and termination of each individual block. In order to maintain the certification of the entire cable run, the appropriate action was taken. In addition, a final Level II test was performed and documented once all of the corrections and adjustments on the run were completed.

The block terminations were altered in Lab 23 by first removing one, 66 M1-50 termination block (and its 89-series bracket) and two, 110-type termination blocks from the grid panel. Then, the remaining blocks were reconnected so that two, 66-type termination blocks and one, 110-type termination block made up the CAT5 UTP cable run. The reconfigured run was tested with the Level II meters, and the recorded results were compared with the previous lab procedures, especially with the results for three installed blocks from Lab 22. The blocks that were removed from the cable run during this procedure were stripped of wire fragments and connectors, and the various parts were returned to the designated storage area.

During Lab 24, the main sections of CAT5 cable were cut from the run and pulled out of their respective conduits. Then, the smaller section of cable running between the terminations blocks were severed. Following the removal of the remaining 66 M1-50 termination blocks (and their associated 89-series brackets), and the removal of the remaining 110-type termination block, the smaller sections of cable were pulled from their respective conduits. The wire fragments were removed and discarded, while the connector blocks/clips were removed from the blocks and returned to the storage area. The sections of conduit were also removed and stored. You used the remaining two sections of cable to create two terminations, one configured as T568A and the other as T568B. Using a 12- to 24-port patch panel and cables from previous lab procedures, you created a cable run with the first section of cable (configured as T568A) and ran the SINGLE TEST series with the Level II testers. After recording your test results, you substituted the second section of cable (configured as T568B) in the run. Then you ran the SINGLE TEST series again, and noted the differences. These results were also recorded and compared.

The following questions test your knowledge of the material presented in this chapter:

1. Name two recent developments primarily responsible for the ability of today's cable installers to pull longer lengths of cable than ever before.

2. What is the primary job of cable conduit?

3. State two benefits from being able to conduct longer cable pulls.

4. Why is it important for a cable installer to know the maximum stress a given cable can withstand?

5. Name two methods of cable blowing.

6. Why is it so important for cable installers to understand how to determine and use a jam ratio?

7. Why should multi-cable jamming be avoided if at all possible?

8. Name two possible configurations for a three-cable pull.

9. How does the weight correction factor apply to multi-cable pulls?

10. What is the main advantage to using pulling lubricants during a pull? What are the disadvantages?

CHAPTER 7

UNDERSTANDING BLUEPRINTS

OBJECTIVES

Upon completion of this chapter and its related lab procedures, you should be able to perform these tasks:

1. Characterize the various stages leading to the production of finished blueprints.

2. Differentiate between an architectural schematic drawing, and a set of preliminary drawings.

3. Establish why a cover sheet is such an important document.

4. Identify the items of information that would normally appear in the title block.

5. Describe the various types of lines and symbols that are typically found on an architectural drawing.

6. Specify the importance of the architect including building material symbols on a drawing.

7. Explain how topographic symbols help the contractor.

8. Contrast the main differences between an architect's scale and an engineer's scale.

9. Clarify the necessity for working drawings to include a detailed set of specifications.

Understanding Blueprints

INTRODUCTION

Technicians involved in the design and installation of LAN or telecommunications cabling must have a basic understanding about how to interpret architectural drawings (more commonly known as **blueprints**). This chapter is designed to provide the reader with a basic knowledge of blueprint reading.

The original process used to create blueprints was actually a photographic process using iron salts, and producing an image in Prussian blue. The process was very expensive and time-consuming, and resulted in the production of prints of architectural and other technical drawings having white images on blue backgrounds. These drawings were very difficult to read, and did not last for any appreciable amount of time. Fortunately, these types of blueprints are no longer manufactured, nor is the type of paper that was required to make them.

Originals, and even reproductions, of architectural drawings are done either by hand, or more often by using computer-based **Computer-Aided Design and Drafting (CADD)** programs. When a rendering is ready to be transferred to a hard copy, the original is printed using a plotter, on either vellum (a heavy off-white fine-quality paper) for normal use or on mylar (a plastic-coated and durable film), which can easily last for 20 or 30 years.

Copies of an original drawing are achieved by placing the translucent vellum against the chemically treated side of a diazo paper, and feeding the combination through a blueprint machine. As the vellum is exposed to light, the chemical side of the diazo paper is burned at areas where light can get through. The burned areas are chemically treated as the paper continues through the machine. The final appearance of the copy is that of blue lines against a white background, hence the name "blueline" to describe this type of copy. These types of architectural drawings are better-looking and longer-lasting than their earlier blueprint cousins. This printing technique, using blue images on white backgrounds, is commonly known as creating "blueline prints." These bluelines can lose their quality over long periods of time, and should therefore be protected from light and other destructive elements.

Bluelines cannot be reproduced from other bluelines. However, copies of blueline prints (called engineer's prints) can be made on large-format Xerox machines, if and when the blueline is all that is available. However, some details in the drawing are certain to be lost using this reproduction method. This problem is only compounded when a copy is made from a copy, etc. For blue-toned photographs produced by the blueprint process, the term "cyanotypes" is used. Modern language uses the term "blueprints" when referring to reproductions of technical drawings, or "cyanotypes" when referring to camera images or photograms.

It is most important for you, as a cabling installer, to understand that even though most engineers, designers, drafters, and architects use the symbols that have been adopted by the American National Standards Institute (ANSI) for use on plans or drawings, many designers and drafters modify these symbols to suit their own particular requirements. The symbols and abbreviations represented in this chapter may, or may not, represent or resemble the symbols that you will encounter when working with blueprints in the future.

blueprints

Computer-Aided
Design and Drafting
(CADD)

DRAWING STAGES

Finished blueprints are actually the result of several preliminary stages of illustrative work. These preliminary stages provide important opportunities for the various professionals involved with the work to preview the overall ramifications of any ideas that are incorporated in them, and to provide any necessary critical feedback.

Schematic Drawings

The first stage of drawings includes the basic scheme of operation, and normally will present a rough depiction of how the proposed project will work. These illustrations are referred to as **Schematic drawings**. Schematic drawings are of a theoretical nature, and are usually the first representation of the project owner's needs, requirements, and how he/she expects the various parts of the proposed project to function.

These schematic drawings are first submitted to the project owner for evaluation. Once the project owner has had an opportunity to review the schematics, he/she will be able to knowledgeably offer any necessary changes that may be required to keep the project on track. As you can imagine, this is a very critical first step.

It is important not to confuse these schematic drawings, as we mention them here, with electrical schematic drawings that are used to depict the very detailed electrical circuits of an appliance, machine, or electronic device.

Schematic drawings

Preliminary Drawings

Any changes that result from this initial evaluation are incorporated into the following stage of drawings, called **Preliminary drawings**. The preliminary drawings provide an even more refined and detailed view of how the proposed project will look and function.

Preliminary drawings

Working Drawings

Working drawings signify the final stage in the design process. The completed drawings become a "set" that incorporates all of the changes and refinements made by the designer, as he/she turns the preliminary drawings into working drawings.

Working drawings will include all the details that the contractor will need to prepare a precise cost estimate. The set of working drawings is organized in such a way that each particular field of work included in the proposed project is represented with its own drawing type.

Working drawings

Various drawing types include:

- **Architectural drawings**, which show a layout of the proposed project's floor plans, elevations, and details.

- **Structural drawings**, which characterize how various load-carrying systems will be built.

- **Mechanical/Electrical drawings**, which show the physical plant of the structure, such as lighting, power, plumbing, fire protection, and HVAC.

- **Site drawings**, which depict the relationship between the structure and the property it will occupy, including various site improvements, such as sanitary systems, utilities, and so on.

Written instructions, called **specifications**, are issued by the architect as part of the proposed project. For projects of limited detail, the specifications may be printed on the drawings. We will go into more detail on specifications later in this chapter.

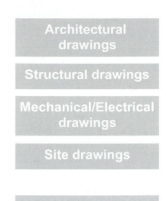

Cover Sheet

The very first sheet on any set of plans is the **Cover Sheet**. The cover sheet, shown in Figure 7-1, contains much of the information that the contractor would need in order to intelligently compete in the bidding for the proposed project.

Figure 7-1: Cover Sheet

The cover sheet provides the contractor with information that is crucial to the understanding of the drawings, and of the project as a whole. It generally lists much of the basic information, such as the name of the project, as well as its location, and the names of the architects, engineers, owners, and other notable people or firms that are involved in its design.

The cover sheet lists, in sequential order, the drawings that comprise the set. The drawing list is categorized by the number of drawings, and the title of the page on which it appears. Also listed are the specific requirements of the building codes that apply directly to the design of the proposed project. The required information includes the structure's total square footage, its use group designation, and its constructional type.

Another important element on the cover sheet is the listing of the abbreviations, or graphic symbols, that are used within the drawing set. Additionally, there is usually a section that contains notes specifically for the contractor, such as "All are to the face of the masonry" or "Dimensions shall be verified in the field." If there is not a separate set of specifications, the cover sheet may list the general technical specifications that will govern the quality of the materials used in the work.

Locus Plan

A **Locus Plan**, that displays the project's location with respect to preexisting, or well-known local landmarks and roadways, may be part of the cover sheet. An architectural rendering of the structure may also be included.

Title Block

Title Block

An example of a **Title Block** is shown in Figure 7-2, and is usually located in the lower, right-hand corner of the drawing. In a recent variation of this practice, some architectural firms are using customized layouts that reserve the entire right-hand portion of the sheet. Whichever variation is chosen, the title block should include the following information:

- The sheet or drawing number, identifying it as to which group and order it belongs.

- The drawing name or title, such as "Second Floor Plan".

- The date of completion of the drawing.

- The initials of the draftsperson.

- Any revisions to the final set of drawings.

The title block must also define whether the entire drawing is one scale, or, as is the practice in some cases, define the particulars as to how the scale may vary per individual detail.

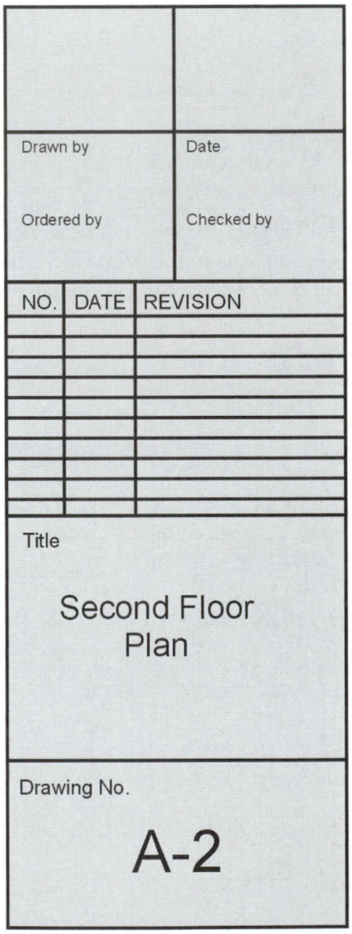

Figure 7-2: Title Block

Revisions

Working drawings are often revised at various times after their initial completion. Various recommendations for the correction, or clarification, of a particular component of the drawing may be needed, or major changes may require the redrafting of an entire sheet.

Small changes or clarifications are shown as revisions on the original sheet, and all revisions must be noted in the title block.

If there is insufficient space available within the title block to accommodate all of the notes, they should be placed as close to it as possible. The revision notes should be listed in order, by date and revision number, and the changes must be clearly marked by a **Revision Marker**, and circled with a scalloped line (that somewhat resembles a cloud). This revision marker is a triangle, with the number or letter designation of the revision located in the center of it, as shown in Figure 7-3.

Revision Marker

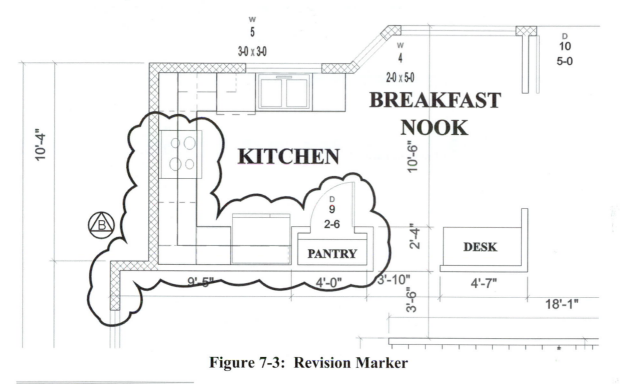

Figure 7-3: Revision Marker

Registration Stamp

Registration Stamp

Sets of plans for commercial projects require a **Registration Stamp** of the drafter or design professional, as shown in Figure 7-4. The stamp contains the architect's or engineer's name, and registration number. In addition, the signature of the individual is usually required over the stamp as well.

Lines

Because architectural drawings must provide a great deal of information in a relatively small amount of space, the use of text to provide this information would be extremely impractical. In order to alleviate this problem, various types of lines and symbols have been devised to provide the necessary information, without using a great deal of space, as shown in Figure 7-5.

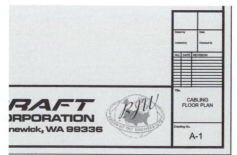

Figure 7-4: Registration Stamp

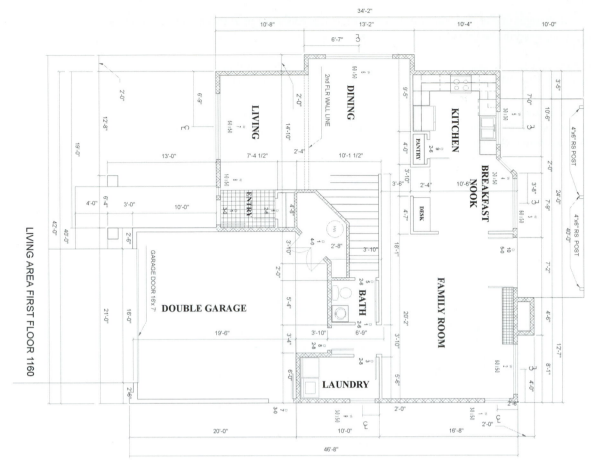

Figure 7-5: Lines and Symbols

The most commonly encountered line types in architectural drawings, as shown in Figure 7-6, are:

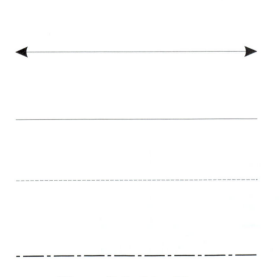

- **Main Object Lines**, which define the outline of the structure, or object. They are thick, unbroken lines that show the main outlines of the walls, floors, elevations, details, or sections.

- **Dimension Lines**, which provide the lengths of the main object lines. They are very light lines with triangles, resembling arrowheads, on each end. The number that appears in the center break of the dimension line represents the measurement of the specific main object line to which it refers.

- **Extension Lines**, which are used together with dimension lines, and are the light lines that extend beyond the main object lines. The arrowheads of the dimension lines usually reach and touch the extension lines.

- **Hidden Lines**, which are light dashes that indicate the outlines of an object normally hidden from view, either under or behind some other part of the structure. The dashes used in this line are usually of equal length.

- **Center Lines**, which are light lines with alternating long/short dashes, indicating the center of an object, and frequently labeled with the letter C superimposed over the letter L.

Figure 7-6: Line Types

Symbols

As already mentioned in the introduction of this chapter, the symbols used by one drafter or designer may not be exactly identical as those used by another. While most of the symbols are widely accepted and practiced, there will always be minor deviations, based on local conventions, such as the example shown in Figure 7-7. Although both of these symbols represent a single receptacle outlet, notice that one of them uses an additional line.

However, when you encounter any unfamiliar symbols and abbreviations, these will usually become clear as you study the drawings and check over the cover sheet of the working plans. If there is still some doubt, you will probably have to consult with the drafter or the designer of the proposed project.

Figure 7-7: Single Receptacle Outlet Symbol

Building Material Symbols

Building Material Symbols allow the drafter the opportunity to define the composition of the object being depicted. Various materials must be easily recognizable for the contractor to properly understand what the scope of the proposed project entails (concrete, brick, steel, framing wood, and so on). Table 7-1 presents a building material symbols listing.

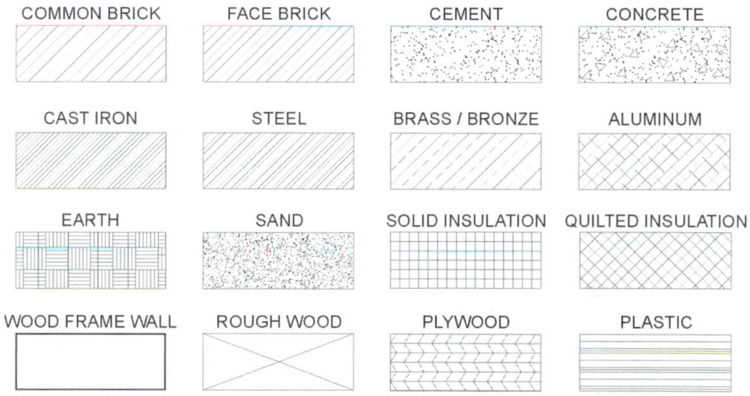

Topographic Symbols

Topographic Symbols provide the contractor with a standard format for recognizing the landscape, and objects, located around the outside perimeter of the proposed project. These types of symbols are extremely important to the communications contractor, if underground cables are part of the proposed project. If the graphic symbols call for a concrete or paved driveway as part of the proposed project, it would be much more practical to bury the cables before the driveway is actually constructed. Table 7-2 presents a listing of various topographic symbols.

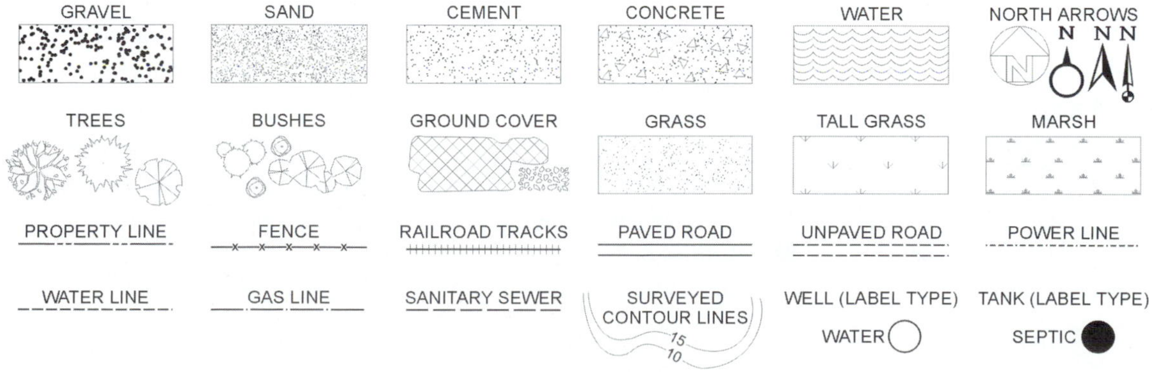

Table 7-1: Building Material Symbols

COMMON BRICK	FACE BRICK	CEMENT	CONCRETE
CAST IRON	STEEL	BRASS / BRONZE	ALUMINUM
EARTH	SAND	SOLID INSULATION	QUILTED INSULATION
WOOD FRAME WALL	ROUGH WOOD	PLYWOOD	PLASTIC

Table 7-2: Topographic Symbols

GRAVEL	SAND	CEMENT	CONCRETE	WATER	NORTH ARROWS
TREES	BUSHES	GROUND COVER	GRASS	TALL GRASS	MARSH
PROPERTY LINE	FENCE	RAILROAD TRACKS	PAVED ROAD	UNPAVED ROAD	POWER LINE
WATER LINE	GAS LINE	SANITARY SEWER	SURVEYED CONTOUR LINES	WELL (LABEL TYPE) WATER	TANK (LABEL TYPE) SEPTIC

Electrical Symbols

Construction plans include symbols showing the location and number of electrical outlets, lights, fans, and so on. Table 7-3 displays a listing of several standard **Electrical Symbols**.

Table 7-3: Electrical Symbols

SIMPLEX RECEPTACLE, MOUNT +18" AFF	FAN	CROSS-CONNECT JUMPER OR PATCH CORD
DUPLEX RECEPTACLE, MOUNT +18" AFF	PUSH BUTTON	ℓ_1 WORK AREA EQUIPMENT CABLE
QUADRAPLEX RECEPTACLE, MOUNT +18" AFF	BUZZER	ℓ_2 PATCH CORD OR JUMPER
EMERGENCY POWERED DUPLEX RECEPTACLE, MOUNT +18" AFF	BELL	ℓ_3 TELECOMMUNICATIONS CLOSET EQUIPMENT CABLE
EMERGENCY POWERED QUADRAPLEX RECEPTACLE, MOUNT +18" AFF	CH CHIMES	HUB DATA HUB / PATCH PANEL
GFCI PROTECTED RECEPTACLE, MOUNT +18" AFF	D MOTORIZED DAMPER	TELEPHONE TERMINAL CABINET
GFCI PROTECTED RECEPTACLE, MOUNT +48" AFF	LIGHTING FIXTURE	TELEPHONE OUTLET, MOUNT +18" AFF
DUPLEX RECEPTACLE, MOUNT +48" AFF	WALL MOUNTED LIGHTING FIXTURE	W TELEPHONE OUTLET, MOUNT +54" AFF
QUADRAPLEX RECEPTACLE, MOUNT +48" AFF	LIGHTING FIXTURE ON EMERGENCY POWER	DATA OUTLET, MOUNT +18" AFF
DUPLEX RECEPTACLE, MOUNT +7'–6" AFF	RECESSED LIGHTING FIXTURE	COMBO TELEPHONE/DATA OUTLET, MOUNT +18" AFF
DUPLEX RECEPTACLE, FLOOR MOUNTED	VP VAPOR PROOF LIGHTING FIXTURE	$\overset{WP}{S}, S, \underset{V}{S}$ SOUND SYSTEM; CEILING & WALL MOUNTED V - VOLUME CONTROL, WP - WEATHERPROOF
APPLIANCE RECEPTACLE	D LIGHTING DROP CORD	$\underset{E}{H}$ WALL MOUNTED SOUND SYSTEM /PAGING HORN, E - EXTERIOR, WATERPROOF
WIREMOLD SURFACE RACEWAY	S LIGHTING PULL SWITCH - CEILING	WALL CLOCK
E ELECTRIC CORD REEL	N HOUSE NUMBER LIGHTING	DOUBLE-FACE WALL CLOCK
J , J JUNCTION BOX	ELECTRICAL HEATER	D CABLE TV OUTLET
PANELBOARD, 277/480V	EMERGENCY KILL SWITCH	M^2 MICROPHONE OUTLET, 2 - NUMBER OF JACKS
PANELBOARD, 120/208V	MOTOR & CONNECTIONS	TV TV OUTLET
EMERGENCY PANELBOARD, 277/480V	G EMERGENCY GENERATOR	TP TEACHER'S PANEL
EMERGENCY PANELBOARD, 120/208V	S^a, S^a_3, S^a_4 SINGLE POLE, 3-WAY, & 4 WAY LIGHT SWITCH	P.B. FLUSH-MOUNTED PULLBOX
BRANCH CIRCUIT HOMERUN	$S^a_{K}, S^a_{K3}, S^a_{K4}$ SINGLE POLE, 3-WAY, & 4 WAY KEY OPERATED LIGHT SWITCH	F FIRE ALARM MANUAL PULL STATION
ISOLATED GROUNDED CONDUCTOR	S S TWO SINGLE POLE LIGHT SWITCHES	F FIRE ALARM AUDIO / VISUAL
GROUNDED CONDUCTOR	$S_3 S_3$ TWO 3-WAY LIGHT SWITCHES	A FIRE ALARM AUDIO ONLY
NEUTRAL CONDUCTOR		V FIRE ALARM VISUAL ONLY
PHASE CONDUCTOR	FLUORESCENT LIGHT FIXTURE	S FIRE ALARM SMOKE DETECTOR
DISCONNECT SWITCH	FLUORESCENT LIGHT FIXTURE CONNECTED TO EMERGENCY POWER SUPPLY	H FIRE ALARM HEAT DETECTOR
P MANUAL MOTOR STARTER WITH PILOT LIGHT	EXIT FIXTURE ON EMERGENCY POWER, CEILING MOUNTED	SD FIRE ALARM DUCT SMOKE DETECTOR
HOA MANUAL MOTOR STARTER WITH HAND-OFF-AUTOMATIC	EXIT FIXTURE ON EMERGENCY POWER, WALL MOUNTED	TS FIRE ALARM TAMPER SWITCH
COMBINATION MAGNETIC STARTER	S_E EMERGENCY LIGHTING SINGLE POLE SWITCH	F FIRE ALARM FLOW SWITCH
MAGNETIC STARTER MOTOR	T LINE VOLTAGE THERMOSTAT	P FIRE ALARM PRESSURE SWITCH
SECURITY SYSTEM DIRECTIONAL PASSIVE INFRA-RED MOTION DETECTOR	GROUNDING ROD	M_F FIRE ALARM MAGNETIC DOOR HOLDER
SECURITY SYSTEM OMNI-DIRECTIONAL INFRA-RED MOTION DETECTOR	K SECURITY SYSTEM REMOTE KEY SWITCH	G FIRE ALARM CONNECTED TO EMERGENCY GENERATOR
CR SECURITY SYSTEM CARD READER DEVICE		FP FIRE ALARM CONNECTED TO FIRE PUMP CONTROLLER
SECURITY SYSTEM CCTV CAMERA		

Plumbing Symbols

In the various trades, there are always symbols that are used to characterize trade-specific items. Table 7-4 displays a listing of **Plumbing Symbols** common to that particular trade.

Plumbing Symbols

Table 7-4: Plumbing Symbols

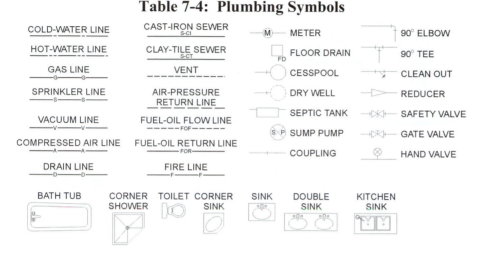

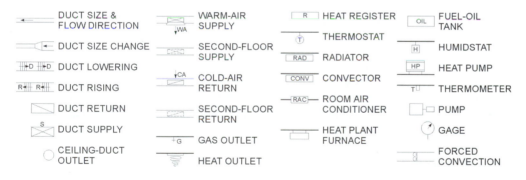

Climate-Control Symbols

Table 7-5 lists several common **Climate-Control Symbols**.

Climate-Control Symbols

Table 7-5: Climate-Control Symbols

Architectural Symbols

Architectural Symbols

Architectural Symbols make up the shorthand of the building industry. Table 7-6 lists several of the more common architectural symbols.

Table 7-6: Architectural Symbols

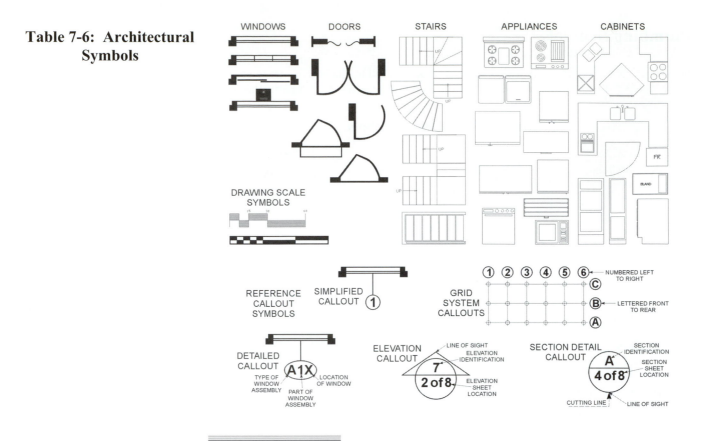

WINDOWS   DOORS   STAIRS   APPLIANCES   CABINETS

DRAWING SCALE SYMBOLS

REFERENCE CALLOUT SYMBOLS

SIMPLIFIED CALLOUT ①

GRID SYSTEM CALLOUTS

① ② ③ ④ ⑤ ⑥ — NUMBERED LEFT TO RIGHT
ⒸⒷⒶ — LETTERED FRONT TO REAR

DETAILED CALLOUT (A1X)
TYPE OF WINDOW ASSEMBLY
PART OF WINDOW ASSEMBLY
LOCATION OF WINDOW

ELEVATION CALLOUT
LINE OF SIGHT
ELEVATION IDENTIFICATION
7'
2 of 8
ELEVATION SHEET LOCATION

SECTION DETAIL CALLOUT
SECTION IDENTIFICATION
A'
4 of 8'
SECTION SHEET LOCATION
CUTTING LINE   LINE OF SIGHT

Drawing Scale

Depicting a building to actual size on a piece of paper would be out of the question. The renderings must be smaller than the actual-size buildings that they represent.

Architectural drawings keep their relationship to the actual size of the building by means of a **ratio**. A ratio is the relationship between two things in amount, size, or degree and is expressed as a proportion. Using an accepted ratio between full size, and what is seen on the drawings, is called **scale**, and there are two major types of scales used in reading plans.

Architect's Scale

The **Architect's Scale** may either be flat or three-sided. The three-sided version of the scale has ten separate scales, paired in five groups of two. They are:

- 1/8-inch and 1/4-inch.
- 1-inch and 1/2-inch.
- 3/4-inch and 3/8-inch.
- 3/16-inch and 3/32-inch.
- 1 and 1/2-inch and 3-inch.

One side of the scale is marked in inches, similar to a ruler. For example, when using a 1/4-inch architect's scale on a floor plan, each 1/4-inch of length on the drawing represents one foot. This also applies for the 1/8-inch scale, where each 1/8-inch line segment represents one foot of actual size. In fact, the most commonly used scales on floor plans and elevations are in 1/4-inch or 1/8-inch scale. Figure 7-8 illustrates an architect's scale.

Figure 7-8: Architect's Scale

Engineer's Scale

The **Engineer's Scale** is somewhat similar to the architect's scale. The main difference being the number of scales, and the project size that the scale is capable of depicting. Having six scales (10, 20, 30, 40, 50, and 60), they refer to how many feet per inch the drawing depicts. For example, the 10 scale refers to 10 feet per inch, while the 20 scale refers to 20 feet per inch, and so on.

Figure 7-9 illustrates an engineer's scale. Sometimes the architect may insert the letters NTS beside a specific detail in a drawing. The NTS is an abbreviation for "Not To Scale". This tells the contractor that the labeled detail is for illustration purposes only, and is not an accurately scaled part of the drawing.

Figure 7-9: Engineer's Scale

Specifications

Working drawings are always issued with a set of specifications. This holds true for even the simplest of projects, which will incorporate the specifications either placed directly on the drawings, or issued as a separate document. The specifications, more commonly called **specs**, should cover all of the work segments and items shown on the working drawings. It is not uncommon for a set of specifications to cover from fifteen to twenty pages.

For example, considering only the exterior siding to be used on a home or building, such details as material grades, types of fasteners to be used, size and color of individual pieces, manufacturer's name, brick style, mortar joint width, joint dressing style, mortar mix manufacturer, and so on, must all be spelled out, so that there will be no misunderstanding about what is expected.

For interior and exterior doors, details would have to include the exact locations for each one, the manufacturer and model for each, the dimensions, and the model/manufacturer of each handle or door knob to be used.

Manufacturer Instruction Sheets

AMP CONTROLLED DOCUMENTS

AMP

Assembly Procedures for
AMP* Four Position Universal
Data Connector Kits

Instruction Sheet
408-3195
05 OCT 94   Rev D

WALL MOUNT APPLICATION

Use Wall Outlet Kit 555268, 555450, 556373, 557397, or 557398 with cable listed in chart.

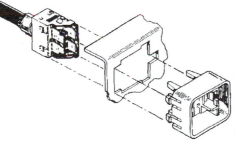

CABLE TYPE	ETL● VERIFIED TO FOLLOWING SPECIFICATIONS
1	4716748, 4716749, or 6339585
2●●	4716738 or 4716739
6	4716743
9	6339583, 74F8680, or 74F9730

MONTROSE† Part No. CBL 6748 and CBL 6749
are prepared the same as Type 2 Cable.

● ETL is a trademark of ETL Testing Laboratories, Inc.
●● Type 2 cable requires separation of voice and data pairs.
† MONTROSE Products Co., Auburn, MA.

PANEL MOUNT APPLICATION

Use Panel Connector Kit 555707 or 558212 with cable listed in chart.

PATCH CORD APPLICATION

Use Connector Kit 556898 or 558213 with cable listed in chart.

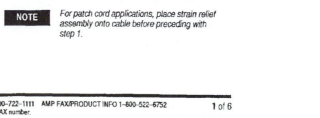

Figure 1   93-228, 93-229, 93-200

1. INTRODUCTION

This instruction sheet covers assembly procedures for the data connector kits listed in Figure 1.

NOTE *Dimensions are in metric units [with U.S. customary units in brackets].*

Reasons for reissue are provided in Section 3, REVISION SUMMARY.

2. ASSEMBLY PROCEDURES

Steps 1 through 10 are the same for panel mount, wall mount, and patch cord applications.

NOTE *For patch cord applications, place strain relief assembly onto cable before preceding with step 1.*

1 of 6

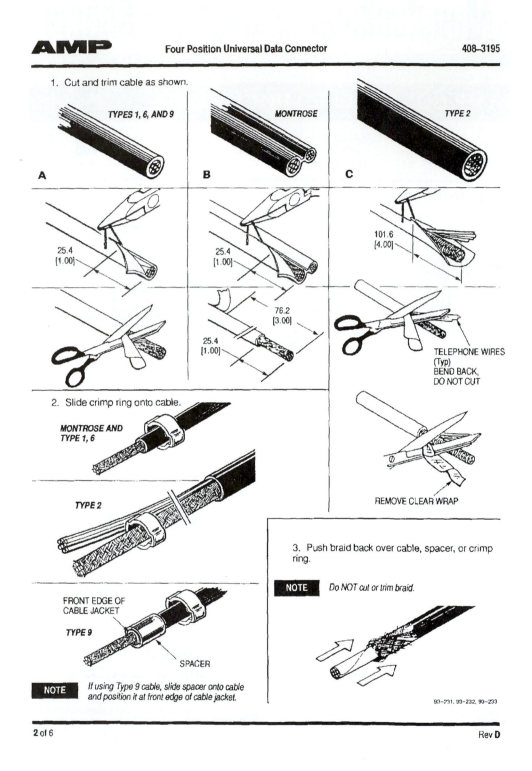

1. Cut and trim cable as shown.

A   TYPES 1, 6, AND 9

B   MONTROSE

C   TYPE 2

25.4 [1.00]

25.4 [1.00]

101.6 [4.00]

76.2 [3.00]

25.4 [1.00]

TELEPHONE WIRES (Typ) BEND BACK, DO NOT CUT

REMOVE CLEAR WRAP

2. Slide crimp ring onto cable.

MONTROSE AND TYPE 1, 6

TYPE 2

FRONT EDGE OF CABLE JACKET

TYPE 9

SPACER

NOTE   If using Type 9 cable, slide spacer onto cable and position it at front edge of cable jacket.

3. Push braid back over cable, spacer, or crimp ring.

NOTE   Do NOT cut or trim braid.

93–231, 93–232, 93–233

4. Remove foil and wire fillers. Trim wires as indicated.

6. Align stuffer cap with contacts. With your thumbs, gently seat stuffer cap over contacts.

NOTE If using Type 2 cable, slide spacer over foil. Remove 25.4 mm [1 in.] of foil.

VOICE PAIRS   FOIL   25.4 [1.00]

BRAID   CRIMP RING   SPACER   **TYPE 2**

15.8 [5/8]

R
O
G
B

5. Insert conductor ends into appropriate slots of stuffer cap, following the color code shown beneath the slots.

NOTE Type 9 cable application requires a Type 9 Stuffer Box 554821–2, which must be ordered separately.

NOTE Make sure conductor ends are flush with back of wire slots and press down into cap.

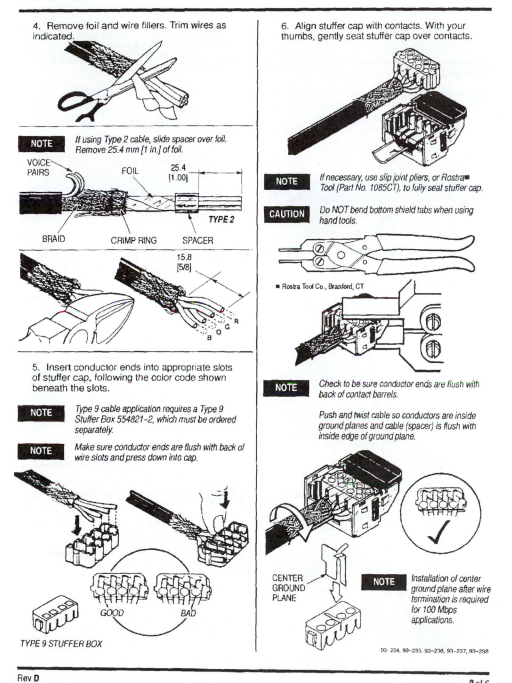

GOOD   BAD

TYPE 9 STUFFER BOX

NOTE If necessary, use slip joint pliers, or Rostra■ Tool (Part No. 1085CT), to fully seat stuffer cap.

CAUTION Do NOT bend bottom shield tabs when using hand tools.

■ Rostra Tool Co., Branford, CT

NOTE Check to be sure conductor ends are flush with back of contact barrels.

Push and twist cable so conductors are inside ground planes and cable (spacer) is flush with inside edge of ground plane.

CENTER GROUND PLANE

NOTE Installation of center ground plane after wire termination is required for 100 Mbps applications.

93–234, 93–235, 93–236, 93–237, 93–238

7. Insert upper ground plane at an angle, so that the two front tabs slide under the opening above the contacts.

CAUTION   *Do NOT allow individual wires to be cut or pinched between ground planes.*

NOTE   *Make sure holding tabs on side of upper ground plane go over outside of lower ground plane.*

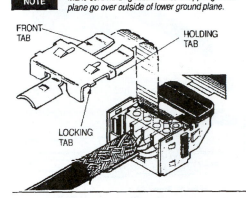

FRONT TAB   HOLDING TAB

LOCKING TAB

8. Snap upper ground plane in place. It is easier to engage locking tab on one side of uper ground plane and then tab on other side.

9. Push braid forward over top and bottom ground plane tabs. Slide crimp ring forward on cable and over top and bottom ground plane tabs and braid. Align crimp ring 2.38 ± 0.78 mm [3/32 ± 1/32 in.] from tip of ground plane tabs.

2.38 ± 0.78
[3/32 ± 1/32]

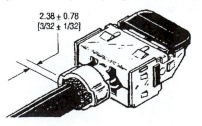

10. Crimp the crimp ring with end– or side–cutting pliers. Alternate crimping each side 2 or 3 times gently until crimp is secured evenly. Trim braid ends flush or within 1.57 mm [1/16 in.] max. of crimp ring front.

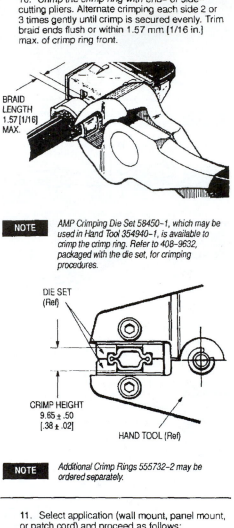

BRAID LENGTH 1.57 [1/16] MAX.

NOTE   *AMP Crimping Die Set 58450–1, which may be used in Hand Tool 354940–1, is available to crimp the crimp ring. Refer to 408–9632, packaged with the die set, for crimping procedures.*

DIE SET (Ref)

CRIMP HEIGHT
9.65 ± .50
[.38 ± .02]

HAND TOOL (Ref)

NOTE   *Additional Crimp Rings 555732–2 may be ordered separately.*

11. Select application (wall mount, panel mount, or patch cord) and proceed as follows:

93–239, 93–240, 93–241, 93–242, 93–243

WALL MOUNT APPLICATION INSTRUCTIONS

a. Align locking tabs and snap connector into wall outlet plate.

b. Fasten outlet plate to wall box with No. 6–32 screws provided.

NOTE

PUNCH HOLE FOR OPTIONAL DUST COVER. (554885–1 Ordered Separately)

CUT CABLE SHEATH 76.2 mm [3.00 IN.] (Optional—For Ground Clip On Support Bar)

TYPES 1, 6, AND 9 CABLE

AMP CABLE MANAGEMENT RING 556562 (Typ)

SUPPORT BAR 556742–1 (Typ)

PANEL MOUNT APPLICATION INSTRUCTIONS

THE FOLLOWING 482.6 mm [19 IN.] RACK MOUNTABLE PANELS WILL ACCEPT THE PANEL MOUNT CONNECTOR KITS	
555731–1	80 kits max. 400.05 [15 3/4] space
555706–1	64 kits max. 400.05 [15 3/4] space
556907–1	32 kits max. 222.25 [8 3/4] space
555573–1	20 kits max. 133.35 [5 1/4] space
556390–1	16 kits max. 138.35 [5 1/4] space
555571–1	16 kits max. 177.8 [7] space

a. Align connector with largest opening in panel cutout. Insert connector until front locking tabs are through the panel. Then, slide connector to the side.

b. Snap bezel onto front of panel over connector. Snap single latch side in first.

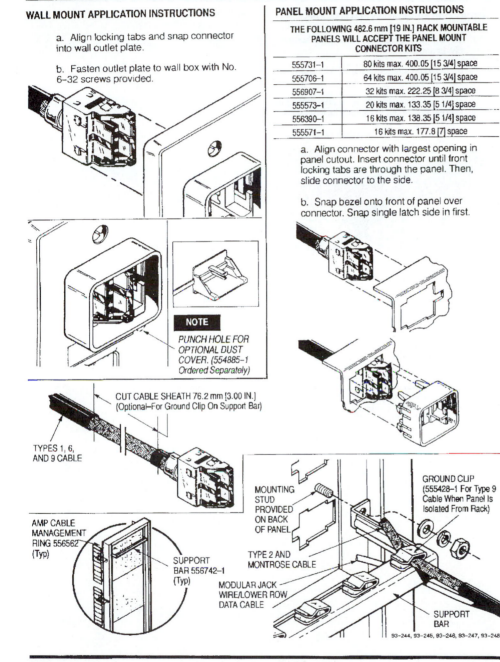

MOUNTING STUD PROVIDED ON BACK OF PANEL

GROUND CLIP (555428–1 For Type 9 Cable When Panel Is Isolated From Rack)

TYPE 2 AND MONTROSE CABLE

MODULAR JACK WIRE/LOWER ROW DATA CABLE

SUPPORT BAR

93–244, 93–245, 93–246, 93–247, 93–248

Rev **D**

PATCH CORD APPLICATION INSTRUCTIONS

a. Align connector with strain relief assembly as shown.

b. Insert connector until side tabs snap into strain relief.

c. Slide integral lock forward after mating to expose yellow dot to indicate locked position.

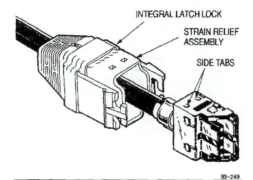

INTEGRAL LATCH LOCK

STRAIN RELIEF ASSEMBLY

SIDE TABS

93–249

3. REVISION SUMMARY

Revisions to this document include:

Per EC 0210-0769-94:

- Updated cable type specification data in Figure 1.
- Changed location of spacer on Type 2 cable.
- Changed hand crimping tool part number.
- Added illustration of Type 2 cable preparation.
- Changed illustration of center ground plane.

AMP

Communications Outlet
Dual Port Kits

Instruction Sheet
408–3199
12 MAR 98   Rev D

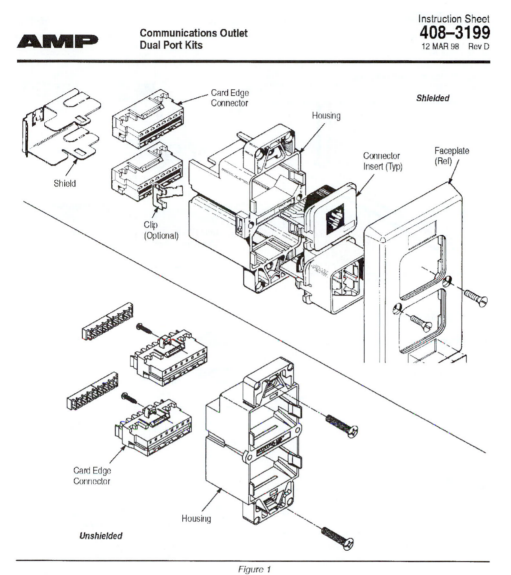

Figure 1

1. INTRODUCTION

This instruction sheet provides termination and installation procedures for AMP* Communications Outlet Dual Port Kits and the various connectors that can be used with these kits. Unshielded kits are available with 110Connect or AMP–BARREL* Card Edge Connectors and each has a different method of

termination. Shielded kits are available with AMP–BARREL Card Edge Connectors only.

NOTE *Dimensions on this sheet are in millimeters [with inches in brackets]. Figures are not drawn to scale.*

Reasons for reissue of this instruction sheet are provided in Section 10, REVISION SUMMARY.

AMP Incorporated, Harrisburg, PA 17105   TOOLING ASSISTANCE CENTER 1–800–722–1111   AMP FAX*/PRODUCT INFO 1–800–522–6752

This AMP controlled document is subject to change. For latest revision call the AMP FAX number.
©Copyright 1998 by AMP Incorporated. All Rights Reserved.
*Trademark

1 of 6

LOC B
Form 404–25 1/98

2. INSTALLATION

Dress cable as shown in Figure 2.

254 to 304.8
[10 to 12]

Figure 2

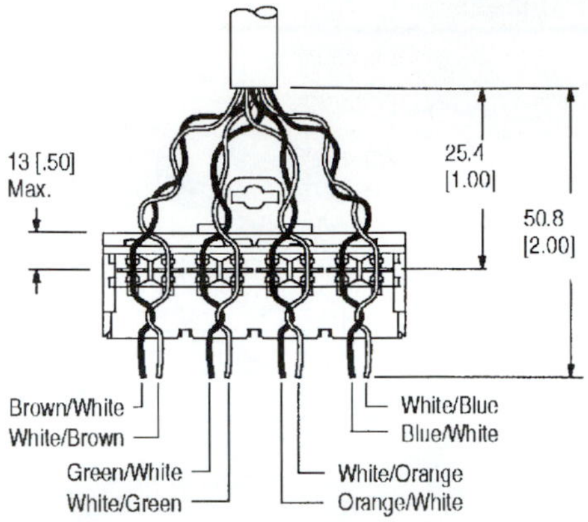

3. TERMINATION PROCEDURES

 The *110Connect card edge connector comes pre–installed in the dual port housing. The housing functions as a secure holding fixture while terminating.*

NOTE The *AMP–BARREL card edge connector must be terminated prior to installation into the dual port housing.*

3.1. 110Connect Card Edge Connector

NOTE The *110Connect connector is approved for use with 24 AWG solid wire with a maximum outside insulation diameter of 1.22 mm [.048 in.] (100 ohm UTP or STP).*

1. Prepare the cable as shown in Figure 3 by removing approximately 50.8 mm [2 in.] of the cable jacket.

2. Lace the twisted–pair conductors into the strain relief according to the appropriate color code. Lacing the center pairs first will help with proper wire dress.

CAUTION *Be sure to maintain the pair twists to within 13 mm [.50 in.] of the termination with Category 5 cables.*

● Trademark of Krone Inc.

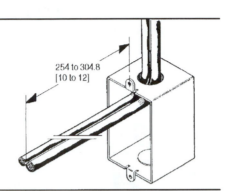

13 [.50]
Max.

25.4
[1.00]

50.8
[2.00]

Brown/White
White/Brown
Green/White
White/Green
White/Blue
Blue/White
White/Orange
Orange/White

Figure 3

3. Terminate conductors using AMP Impact Tool 569994–1, D814 impact tool, or KRONE● single wire tool. See Figure 4.

NOTE *Always use the low–impact setting of the impact tool. The cutoff blade may be used to trim conductors.*

NOTE *Stuffer caps are supplied in installation kits to provide additional strain relief. Stuffer caps are an option for panel application.*

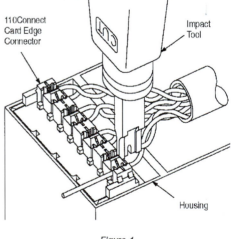

110Connect
Card Edge
Connector

Impact
Tool

Housing

Figure 4

3.2. AMP–BARREL Card Edge Connector

1. Trim the cable according to Figure 5.

 The AMP–BARREL connector is approved to terminate multi pair, copper conductors (22, 24, or 26 AWG shielded or unshielded, up to 8 pair) used in building communications wiring.

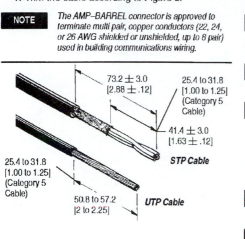
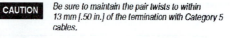

25.4 to 31.8 [1.00 to 1.25] (Category 5 Cable)

73.2 ± 3.0 [2.88 ± .12]

25.4 to 31.8 [1.00 to 1.25] (Category 5 Cable)

41.4 ± 3.0 [1.63 ± .12]

STP Cable

25.4 to 31.8 [1.00 to 1.25] (Category 5 Cable)

50.8 to 57.2 [2 to 2.25]

UTP Cable

 *The jacket on Category 5 cable should be stripped 25.4 to 31.8 [1 to 1.25].*

Figure 5

2. Insert conductor ends into appropriate slots of stuffer cap, following the color code shown beneath the slots. Make sure that the ends are flush with the back of the conductor slots. Maintain wire twist as far as possible on data cable. See Figure 6.

NOTE *A stuffer cap with a universal label is used for both the STP (150 ohm—red, green, orange, and black) and the UTP (100 ohm) cables.*

Good

Bad

Figure 6

CAUTION *Insert only one wire per slot.*

CAUTION *Be sure to maintain the pair twists to within 13 mm [.50 in.] of the termination with Category 5 cables.*

NOTE *Stuffer Punch Down Tool 556706–1 may be used to insert wires into stuffer caps.*

NOTE *Color code must be followed to ensure proper system operation.*

3. Align stuffer cap with contacts and press stuffer cap over contacts. Using slip joint pliers, fully seat the stuffer cap. See Figure 7.

CAUTION *Do NOT press on latch or mounting screw boss area in the middle of the connector.*

NOTE *Press on each side of latch or mounting screw boss until stuffer cap has bottomed with connector.*

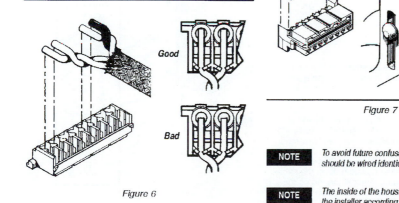

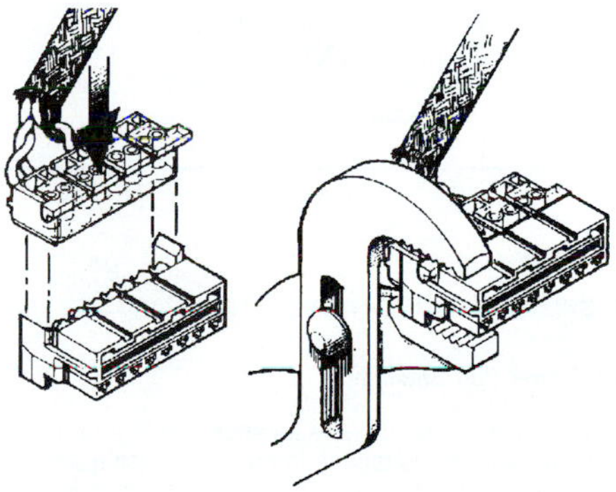

Figure 7

NOTE *To avoid future confusion, all outlets of a building should be wired identically.*

NOTE *The inside of the housing should be labeled by the installer according to the wiring junction.*

4. INSERTION PROCEDURE (Figure 8)

For shielded applications, snap a connector into the back of the housing. For unshielded applications, insert a connector into the back of the housing and secure it with a screw.

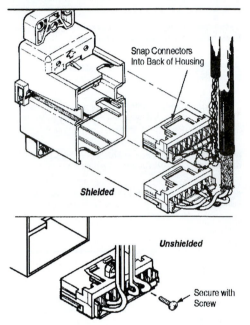

Shielded

Unshielded

Secure with Screw

Figure 8

5. SHIELDED APPLICATIONS

5.1. Lower Port Shielding

1. Slip shield over cable as shown in Figure 9. Insert shield into guides in back of housing and push shield onto housing until latches snap into place.

NOTE *To ensure optimum performance, be sure to follow national and local grounding ("earthing"), bonding, and EMC regulations and procedures*

2. Loop a cable tie around the shielded cable only. Partially tighten the tie and slide it down the cable, capturing the braid tightly against the tab on the shield. FULLY tighten and trim the cable tie using appropriate tool. See Figure 10.

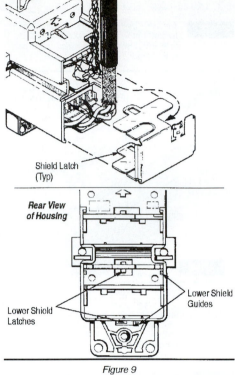

Snap Connectors Into Back of Housing

Shield Latch (Typ)

Rear View of Housing

Lower Shield Latches

Lower Shield Guides

Figure 9

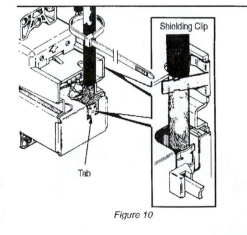

Shielding Clip

Tab

Figure 10

Rev **D**
Form 404–33 1/98

5.2. Upper Port Shielding (Figure 11)

This procedure is the same as lower port shielding (Paragraph 5.1.) except that the tab on the shield and the cable exit must be oriented downward. If both ports are to be shielded, remove knockouts from both shields and dress each cable through the other shield. Install upper and lower shields at the same time.

| NOTE | If required, mounting flanges may be trimmed for clearance in switch box. See Figure 11. |

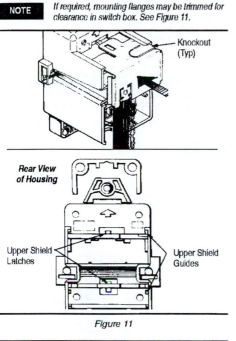

Figure 11

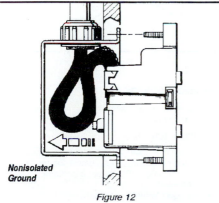

Nonisolated Ground

Figure 12

6. GROUNDED APPLICATIONS

6.1. Nonisolated (Normal) Ground (Figure 12)

If using a nonisolated ground system grounded at both the closet and at each AMP Communications Outlet (which will minimize EMI radiation), mount each housing as delivered in a grounded metal switch box.

Push housing and excess cable back into switch box and secure housing to box using screws captive to the housing.

6.2. Isolated Ground (Figure 13)

If using an isolated ground system grounded only at the communications closet (to minimize ground loops), install nylon shoulder bushings and nylon washers (supplied with AMP Communications Outlet Kit) as shown at each screw mount. First tape one turn of electrical tape around the body of the housing to insulate the housing from the metal switch box.

Using a screwdriver or similar tool, remove the thin plastic flash which secured each screw in the housing. Install a nylon shoulder bushing on the room side and a nylon washer between the housing and the switch box.

Re–install each screw to mount the box. Isolated grounding may also be achieved by using a plastic switch box. Push housing and excess cable back into switch box and secure housing to box.

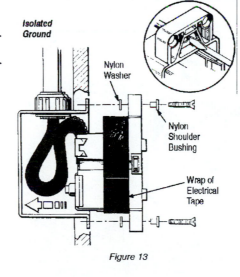

Figure 13

7. INSTALLATION OF INSERTS (Figure 14)

To install connector inserts, remove any protective dust covering from housing and push desired insert into appropriate port. Latches on insert must align with tabs on edge of housing. If a port is not used, install a blank insert into the port.

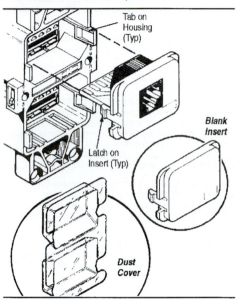

Figure 14

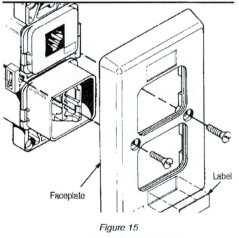

Figure 15

8. INSTALLATION OF FACEPLATES (Figure 15)

Make sure that tabs on edge of housing align properly with slots on back of faceplate. Label ports in the areas provided on the faceplate.

NOTE *The subassembly qualifies with Underwriters Laboratories Inc. listing when one or more module inserts and a faceplate are used. If only one insert is used, a blank insert must be provided for the unused portion.*

9. REMOVAL OF INSERTS (Figure 16)

1. Remove faceplate.

2. Insert a small screwdriver blade into gap on side of insert (between insert and tab on edge of housing) to disengage latch. As each latch is disengaged, rock insert to release that side.

3. Repeat procedure on other side; then pull insert out.

CAUTION *Do NOT try to pry insert out.*

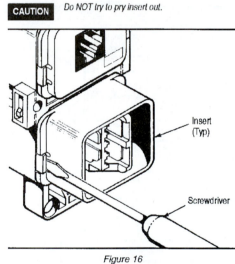

Figure 16

10. REVISION SUMMARY

Since the previous release of this sheet, the following changes were made:

Per EC 0990-1116-97

- Added illustration of unshielded kit in Figure 1
- Added procedures and illustrations for termination of 110Connect card edge connectors
- Changed format to standard layout

342 APPENDIX A

**Communications Outlet
Single Port Kits**

Instruction Sheet
408–3232
19 AUG 97   Rev B

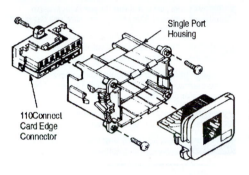

Single Port
Housing

110Connect
Card Edge
Connector

Figure 1

1. INTRODUCTION (Figure 1)

This instruction sheet provides termination and installation procedures for AMP* Communications Outlet Single Port Kits and the various connectors that can be used with these kits. There are 110Connect or AMP–BARREL* Card Edge connectors available and each has a different method of termination. Single port kits can be used in panel mount or furniture raceway applications. Optional faceplate kits and various adapter insert assemblies are available to meet either application.

NOTE *Dimensions on this sheet are in millimeters [with inches in brackets]. Figures are not drawn to scale.*

Reasons for reissue of this instruction sheet are provided in Section 9, REVISION SUMMARY.

2. TERMINATION PROCEDURES

NOTE *The 110Connect card edge connector comes pre–installed in a single port housing. The single port housing functions as a secure holding fixture while terminating.*

NOTE *The AMP–BARREL card edge connector must be terminated prior to installation into a single port housing.*

2.1. 110Connect Card Edge Connector

NOTE *The 110Connect connector is approved for use with 24 AWG solid wire with a maximum outside insulation diameter of 1.22 mm [.048 in.] (100 ohm UTP or STP).*

1. Prepare the cable as shown in Figure 2 by removing approximately 50.8 mm [2 in.] of the cable jacket.

2. Lace the twisted–pair conductors into the strain relief according to the appropriate color code. Lacing the center pairs first will help with proper wire dress. Figure 2 illustrates proper lacing for single port installation kits.

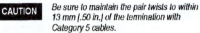
CAUTION *Be sure to maintain the pair twists to within 13 mm [.50 in.] of the termination with Category 5 cables.*

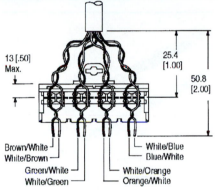

13 [.50]
Max.

25.4
[1.00]

50.8
[2.00]

Brown/White
White/Brown

White/Blue
Blue/White

Green/White
White/Green

White/Orange
Orange/White

Figure 2

3. Terminate conductors using AMP Impact Tool 569994–1, D814 impact tool, or KRONE● single wire tool. See Figure 3.

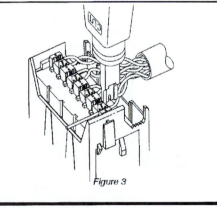

Figure 3

1 of 5

LOC B

NOTE	*Always use the low–impact setting of the impact tool. The cutoff blade may be used to trim conductors.*
NOTE	*Stuffer caps are supplied in installation kits to provide additional strain relief. Stuffer caps are an option for panel application.*

2.2. AMP–BARREL Card Edge Connector

1. Trim the cable according to Figure 4.

NOTE	*The AMP–BARREL connector is approved to terminate multi pair, copper conductors (22, 24, or 26 AWG shielded or unshielded, up to 8 pair) used in building communications wiring.*

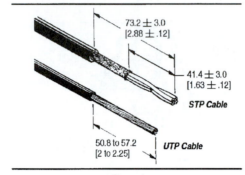

73.2 ± 3.0
[2.88 ± .12]

41.4 ± 3.0
[1.63 ± .12]

STP Cable

50.8 to 57.2
[2 to 2.25]

UTP Cable

Figure 4

2. Insert conductor ends into appropriate slots of stuffer cap, following the color code shown beneath the slots. Make sure that the ends are flush with the back of the conductor slots. Maintain wire twist as far as possible on data cable. See Figure 5.

NOTE	*A stuffer cap with a universal label is used for both the STP (150 ohm—red, green, orange, and black) and the UTP (100 ohm) cables.*

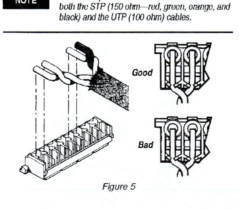

Good

Bad

Figure 5

CAUTION	*Insert only one wire per slot.*
CAUTION	*Be sure to maintain the pair twists to within 13 mm [.50 in.] of the termination with Category 5 cables.*
NOTE	*Stuffer Punch Down Tool 556706–1 may be used to insert wires into stuffer caps.*

3. Align stuffer cap with contacts and press stuffer cap over contacts. Using slip joint pliers, fully seat the stuffer cap. See Figure 6.

CAUTION	*Do NOT press on latch or mounting screw boss area in the middle of the connector.*
NOTE	*Press on each side of mounting screw boss until stuffer cap has bottomed with connector.*

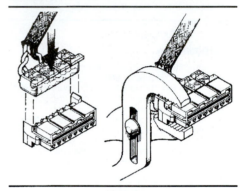

Figure 6

4. Insert connector into the back of the housing and secure with screw. See Figure 7.

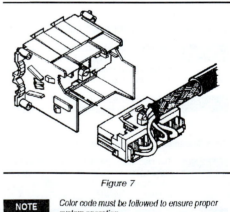

Figure 7

NOTE	*Color code must be followed to ensure proper system operation.*

5. Dress the cable to feed out either side of the housing. In shielded applications, use the clip to press the cable shield tightly against one of the three middle toothed ribs. See Figure 8.

NOTE *The center rib is used for small diameter cable, and the off-center ribs (depending on the cable exit) may be used for larger diameter cable (typically Type 1 shielded cable).*

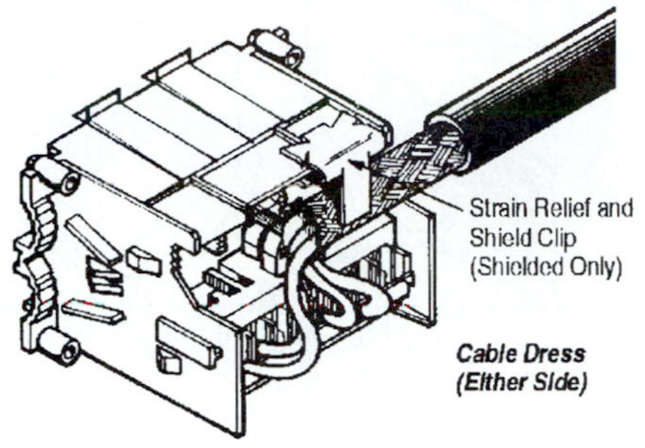

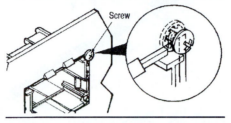

Strain Relief and Shield Clip (Shielded Only)

Cable Dress (Either Side)

Figure 8

3. PANEL MOUNT (Figure 9)

3.1. Unshielded Option

1. Snap housing into back of panel.

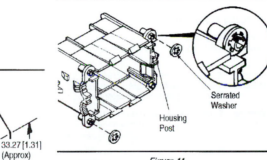

33.27 [1.31] (Approx)

Panel Cutout

47.75 [1.88] (Approx)

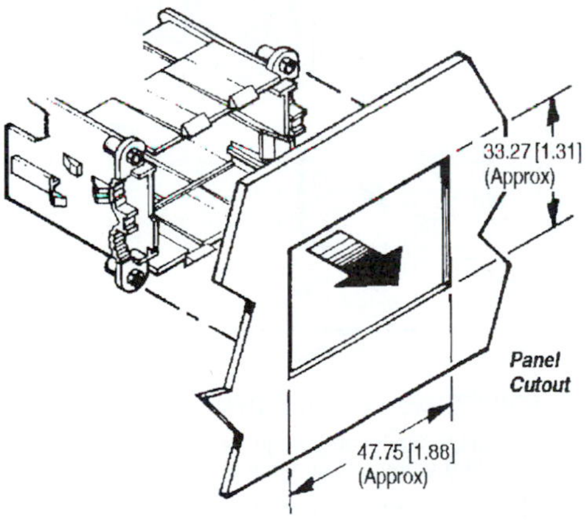

Figure 9

2. Install two No. 2 screws into two posts located diagonally apart. Tighten screws to clamp housing to panel. See Figure 10.

Screw

Figure 10

3.2. Shielded Option

Grounding Option

Inserting the housing into a metal panel will ground the housing; however, if an absolute electrical ground is required, proceed as follows:

1. Press the two No. 4 serrated washers onto two diagonal posts of the housing. See Figure 11.

Serrated Washer

Housing Post

Figure 11

2. Snap housing into BACK of panel.

3. Install the two No. 2 screws onto the two posts which hold the serrated washers. Tighten screws to clamp housing to panel. See Figure 10.

4. FURNITURE RACEWAY (Figure 12)

1. Pull cable through panel cutout.

2. Snap housing into BACK of bezel adapter.

3. If using furniture raceway faceplate kits, snap the bezel adapter and housing assembly into the raceway opening.

MANUFACTURER INSTRUCTION SHEETS   345

Existing Panel Cutout 68.6 [2.70] (Approx)

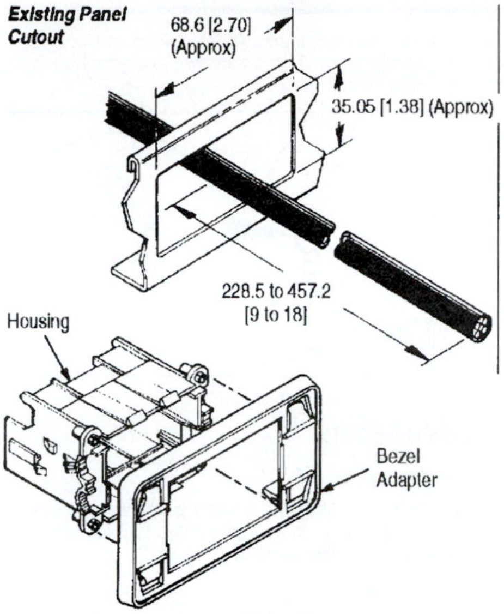

35.05 [1.38] (Approx)

Housing

228.5 to 457.2 [9 to 18]

Bezel Adapter

Figure 12

5. SHIELDED APPLICATIONS

1. Break away a portion of the tab on the side of the metal shield (side oriented with cable exit). See Figure 13.

NOTE *To minimize electromagnetic interference (EMI), remove a small portion of the tab at a time to provide a tight fit around the cable shield.*

NOTE *To ensure optimum performance, be sure to follow national and local grounding ("earthing"), bonding, and EMC regulations and procedures.*

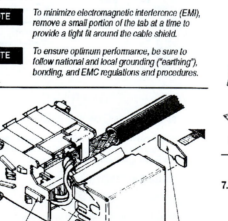

Guide

Latches

Remove to Fit Cable

Figure 13

2. Slide the metal shield onto the guides on the BACK of the housing. Push shield over housing until all four latches snap into place.

6. INSTALLATION OF INSERTS

1. Install connector inserts by pushing the insert into the appropriate port. Align insert latches with tabs on the edge of the housing. If a port is not used, install a blank insert into the port. See Figure 14.

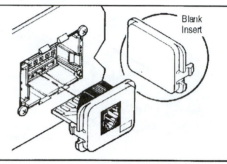

Blank Insert

Figure 14

2. If installing the faceplate kit, align and press the faceplate onto the FRONT of the housing. See Figure 15.

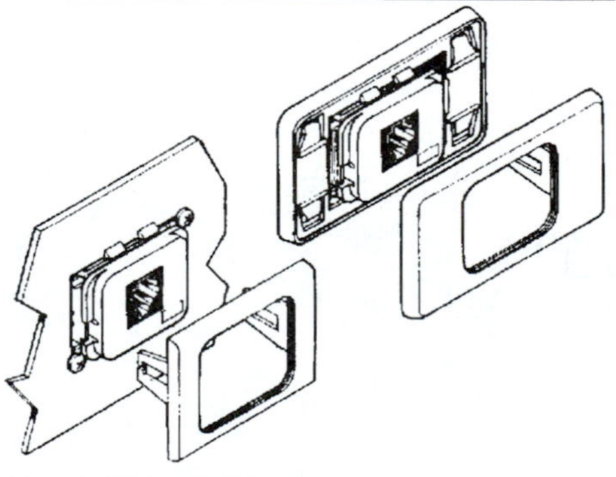

Figure 15

7. CONNECTOR INSERT REMOVAL

1. Pull the faceplate straight out or pry the faceplate at the slot on each side.

2. Insert a small screwdriver blade into the gap on the side of the insert (between the insert and the tab on the edge of the housing). As each latch disengages, rock the insert to release the side. See Figure 16.

3. Repeat procedure on the other side; then remove insert.

CAUTION   *Do NOT try to pry insert out.*

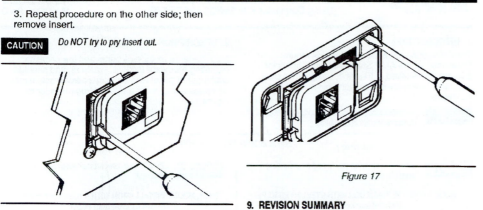

Figure 16

Figure 17

8. BEZEL ADAPTER REMOVAL

Release each corner latch with the blade of a small screwdriver while pulling outward lightly on the bezel adapter. See Figure 17.

9. REVISION SUMMARY

Revisions to this instruction sheet include:

Per EC 0990–0633–97:

- Product was redesigned

AMP

Rack Mount Installation of AMP* Four Position
Data Connector System Panel 557946–1
(Using Connector Kits 554000 and 557900)

Instruction Sheet
408–3321
15 MAR 94   Rev O

1. INTRODUCTION

This instruction sheet covers the rack mount
installation of the AMP Standard Data Connector
System Panel listed in Figure 1, using either steel
mounting hardware or insulating nylon mounting
hardware.

NOTE   *All dimensions are in metric units [with U.S.
customary units in brackets].*

2. DESCRIPTION

Each panel kit includes a panel and the necessary
mounting hardware (screws, washers, and shoulder
bushings). If panel is to be isolated from rack, it may
be necessary to order a Ground Lug Kit (refer to
Figure 1).

RACK MOUNT INSTALLATION

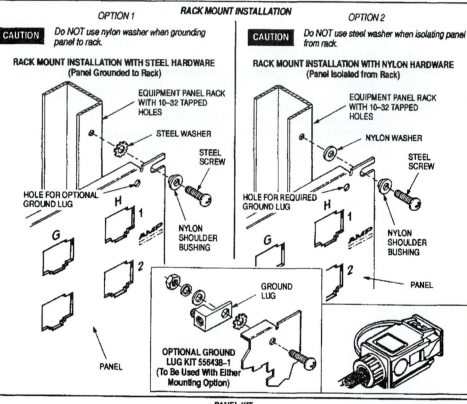

OPTION 1

CAUTION   *Do NOT use nylon washer when grounding
panel to rack.*

RACK MOUNT INSTALLATION WITH STEEL HARDWARE
(Panel Grounded to Rack)

OPTION 2

CAUTION   *Do NOT use steel washer when isolating panel
from rack.*

RACK MOUNT INSTALLATION WITH NYLON HARDWARE
(Panel Isolated from Rack)

PANEL KIT

PART NUMBER	DESCRIPTION	HARDWARE (Qty)				GROUND LUG HOLES PROVIDED PER PANEL
		STEEL SCREWS	STEEL WASHERS	NYLON WASHERS	NYLON SHOULDER BUSHINGS	
557946–1	64 Positions – 399.8 [15.74] Height	8	8	8	8	2

Figure 1

94–97

AMP Incorporated, Harrisburg, PA 17105   TECHNICAL ASSISTANCE CENTER 1–800–722–1111   AMP FAX/PRODUCT INFO 1–800–522–6752
This AMP controlled document is subject to change. For latest revision call the AMP FAX number.
©Copyright 1994 by AMP Incorporated. All Rights Reserved.
*Trademark

1 of 2

LOC B

The AMP Standard Data Connector Panel allows two options for panel-to-rack grounding:

(1) direct grounding of the panel to the rack using toothed steel washers between panel and rack, or

(2) use of an optional ground lug kit and ground wire (part no. 556438–1).

Panel-to-rack grounding options are shown in Figure 1. The ground lug kit may be used with either option, but it *must* be used to ground the panel if the panel is isolated from the rack. This option may be selected where powered equipment is installed in the same rack and there is concern over equipment ground noise being introduced onto the cable shield. The ground lug must be connected to a suitable building ground through a No. 6 AWG or larger

copper conductor. The ground lug setscrew must be torqued to 5–5.6 N•m [45–50 inch–pounds].

Any combination of these grounding options will meet a 200 milliohms maximum needed to satisfy the overall shield and ground path requirements of a 100 meter terminal–to–wiring closet link.

Connector or cable shield–to–panel grounding is shown in Figure 2.

 The optional ground lug kit will provide a redundant ground path of reduced resistance.

The bare metal shield of the panel–mounted standard data connector is exposed and inserted into the ground clip of the support bar which is attached to the panel.

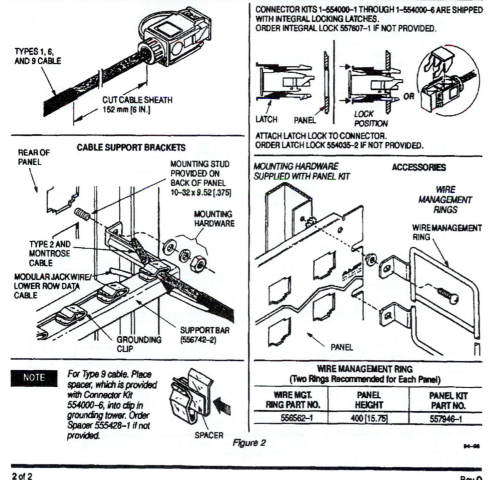

Figure 2

1. INTRODUCTION

These instructions cover the installation procedures for the AMP* 110Connect XC System connecting blocks and wiring blocks listed in Figure 1.

DESCRIPTION	PART NUMBER
3–Pair Connecting Block	558400–1
4–Pair Connecting Block	558401–1
5–Pair Connecting Block	558402–1
50–Pair Wiring Block with Mounting Legs	558841–1
100–Pair Wiring Block with Mounting Legs	558842–1
300–Pair Wiring Block with Mounting Legs	558843–1
50–Pair Wiring Block	558839–1
100–Pair Wiring Block	558840–1

Figure 1

NOTE *All dimensions are in millimeters [with inch equivalents in brackets].*

Reasons for reissue are provided in Section **4**, REVISION SUMMARY.

2. INSTALLATION PROCEDURES

2.1. 25–Pair Cable

1. Strip outer cable jacket back a minimum of 254 mm [10 in.].

2. Route cables into wiring block as shown in Figure 2. Lace each pair into position in index strip slots while maintaining correct tip (white) and ring (colored) sequence. Complete lacing of index strip and inspect for accuracy (use black markings as reference).

3. Use AMP Impact Tool 569994–1, AT&T● 788J1 Impact Tool (with cutoff blade), or an equivalent to seat each conductor and shear off conductor end (see Figure 3).

4. After all index strips have been wired and trimmed, orient first connecting block at left end of top row and seat using an AT&T 788J1 Impact Tool or equivalent. See Figure 3. Working from left to right, orient and seat each additional block in row. Complete remaining rows following same procedure.

NOTE *Care must be taken to align the connecting block with the blue marking on the left.*

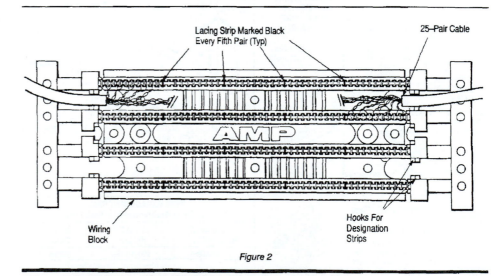

Lacing Strip Marked Black Every Fifth Pair (Typ)

25–Pair Cable

Wiring Block

Hooks For Designation Strips

Figure 2

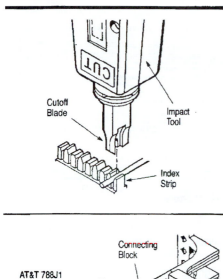

Cutoff
Blade

Impact
Tool

Index
Strip

AT&T 788J1
Impact
Tool

Connecting
Block

Blue

Index
Strip

Figure 3

2.2. Data Applications

For installations utilizing 4–pair cables and 4–pair connecting blocks, the following procedures should be followed:

1. Position the 4–pair cables against the back panel and determine which cables will be routed to the left side and which cables to the right side of the wiring block. Separate each group into bundles of 6 cables and tie wrap loosely.

2. Mount wiring block in desired location on back panel. If more than one block will be required and cables feed from the ceiling, begin at top of panel and work downward. If cables feed up from the floor, begin at bottom of panel and work upward.

3. Select the 2 cable bundles that will be terminated in each half of this first wiring block. Position bundles under block at right and left ends and feed individual cable ends out between mounting legs below appropriate wiring channel. The cables on the right end will be terminated in the right half of the wiring block and the cables on the left end will be terminated in the left half. See Figure 4.

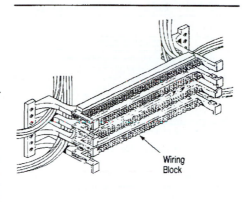

Wiring
Block

Figure 4

4. Following the sequence shown in Figure 5, route each cable around end of wiring block and into the channel between the index strips. Cut cable to appropriate length and strip back only enough cable jacket to allow lacing of pairs.

5. Place each pair into selected position on index strip and pull down into the slot on each side of a high tooth. Allow no more than 13 mm [0.5 in.] untwist in each pair.

6. Complete routing of pairs to index strip on each side of channel and inspect for accuracy.

7. Use Impact Tool 569994–1, AT&T 788J1 Impact Tool (with cutoff blade), or an equivalent to seat each conductor and shear off conductor end. (see Figure 3).

8. After all index strips have been wired and trimmed, orient first connecting block at left end of top row using an AT&T 788J1 Impact Tool or equivalent (see Figure 3). Working from left to right, orient and seat each additional block in row. Complete remaining rows following same procedure.

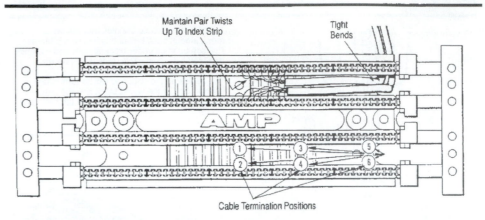

Maintain Pair Twists
Up To Index Strip

Tight
Bends

Cable Termination Positions

Figure 5

2.3. Workmanship Guidelines (To Help Ensure Performance of Installation)

1. Strip only enough cable jacket to allow termination of pairs.

2. Place tension on twisted pairs when bending and routing in order to maintain twist and close conductor spacing.

3. Maintain pair twist up to point of termination in index strip. Allow no more than 13 mm [0.5 in.] to untwist.

3. TERMINATING CROSS–CONNECT WIRES

NOTE *Only 22 through 26 AWG insulated wire should be terminated to the connecting block. (Maximum diameter over insulation is 1.3 mm [.050 in.].)*

1. Cross–connect wires are terminated in the top of the connecting block using Impact Tool

569994–1 or KRONE■ Universal Wire Insertion Tool. See Figure 6. Use low impact setting.

CAUTION *Do NOT use 788J1 five pair Impact Tool to terminate cross–connect wires in connecting blocks.*

2. The wires should be completely inserted to the bottom of the wire slot and the excess wire should be removed by the cutting edge of the impact tool.

4. REVISION SUMMARY

Revisions to this document include:

Per EC 0990–0954–96:

- Changed part number of AMP impact tool
- Updated illustration of wiring block in Figures 2 and 5 to show black marking on lacing strips (every fifth pair)

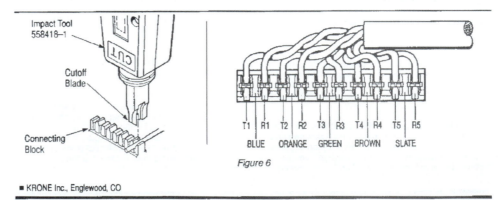

Impact Tool
558418–1

Cutoff
Blade

Connecting
Block

T1 R1 T2 R2 T3 R3 T4 R4 T5 R5

BLUE   ORANGE   GREEN   BROWN   SLATE

Figure 6

■ KRONE Inc., Englewood, CO

AMP

Four Position EMC
Data Connector Kit 558585

Instruction Sheet
408-3344
17 MAR 99   Rev B

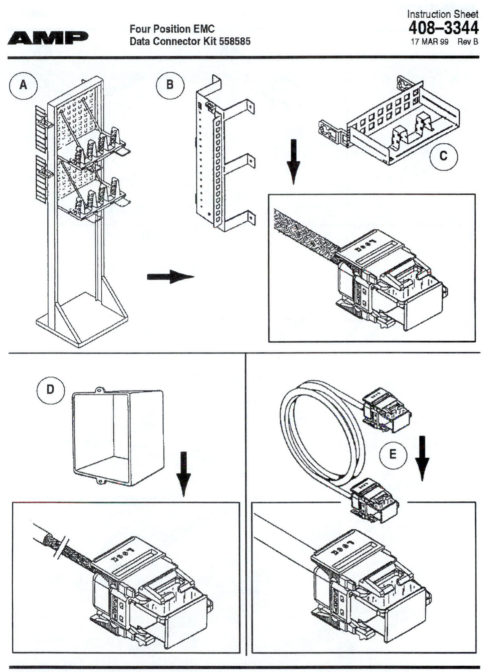

AMP Incorporated, Harrisburg, PA 17105   TOOLING ASSISTANCE CENTER 1-800-722-1111   AMP FAX*/PRODUCT INFO 1-800-522-6752
This AMP controlled document is subject to change. For latest revision call the AMP FAX number.
©Copyright 1999 by AMP Incorporated. All Rights Reserved.
*Trademark

1 of 4

LOC B

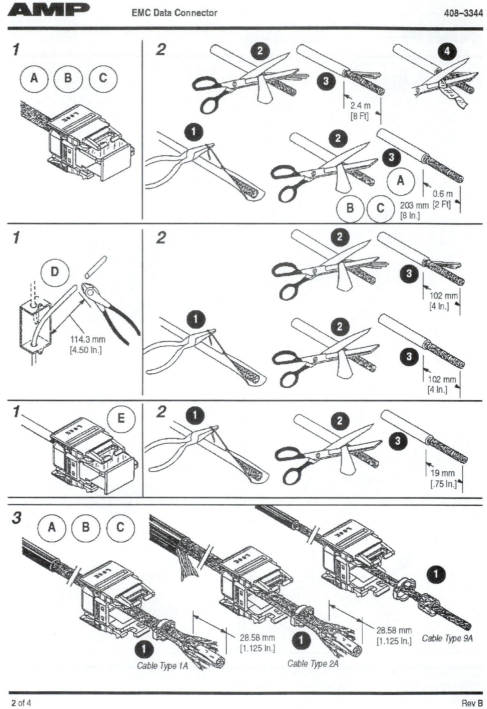

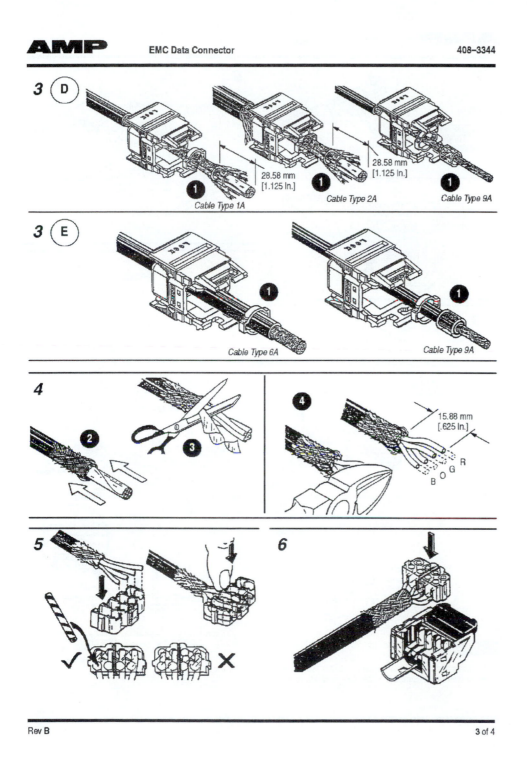

3 D

28.58 mm [1.125 In.]

Cable Type 1A

28.58 mm [1.125 In.]

Cable Type 2A

28.58 mm [1.125 In.]

Cable Type 9A

3 E

Cable Type 6A

Cable Type 9A

4

4   15.88 mm [.625 In.]

R
O G
B

5

6

MANUFACTURER INSTRUCTION SHEETS   355

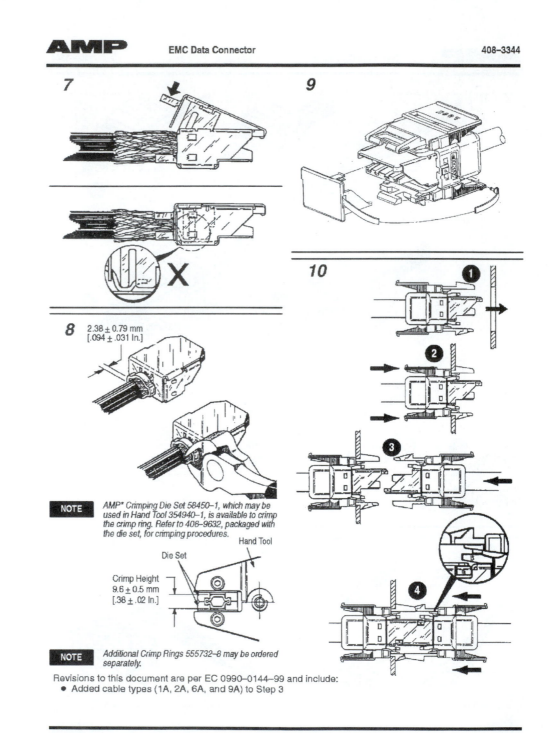

7

8   2.38 ± 0.79 mm
[.094 ± .031 In.]

9

10

NOTE   *AMP* Crimping Die Set 58450–1, which may be used in Hand Tool 354940–1, is available to crimp the crimp ring. Refer to 408–9632, packaged with the die set, for crimping procedures.*

Hand Tool

Die Set

Crimp Height
9.6 ± 0.5 mm
[.38 ± .02 In.]

NOTE   *Additional Crimp Rings 555732–8 may be ordered separately.*

Revisions to this document are per EC 0990–0144–99 and include:
- Added cable types (1A, 2A, 6A, and 9A) to Step 3

Shielded and Grounded
110Connect Panel Mount Jacks
(Category 3 and 5)

Instruction Sheet
408–3354
10 JUN 98   Rev D

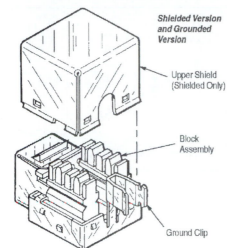

*Shielded Version
and Grounded
Version*

Upper Shield
(Shielded Only)

Block
Assembly

Ground Clip

Figure 1

1. INTRODUCTION

This instruction sheet provides termination
procedures for shielded and grounded
AMP* 110Connect Panel Mount Jacks.

 *All dimensions are in millimeters [with inches in
brackets]. Illustrations are for reference only and
are not drawn to scale.*

Reasons for reissue of this sheet are provided in
Section 4, REVISION SUMMARY.

2. DESCRIPTION

The jacks accept four pair twisted–pair cable with
22 to 26 AWG solid conductors or 20 to 26 AWG
stranded conductors. The maximum conductor
insulator diameter is 1.27 mm [.050 in.].

3. TERMINATION

 *Wire termination requires the use of AMP Impact
Tool 569994–1, KRONE● Universal Wire
Insertion Tool, or stuffer cap.*

1. Strip cable jacket back 50.8 mm [2.00 in.] as
shown in Figure 2.

2. Fold metal foil back over cable jacket. Trim foil,
leaving 9.52 mm [.375 in.] of foil over jacket.

3. Remove clear wrapping from twisted–pair wires.

 Do NOT cut the drain wire.

*For Shielded Cable
with Insulation Less
than 6.1 [.24] Diameter*

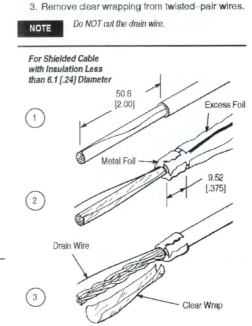

50.8
[2.00]

Excess Foil

Metal Foil

9.52
[.375]

Drain Wire

Clear Wrap

NOTE: Not to Scale

*Wire Dress for Shielded Cable with
Insulation More than 6.1 [.24] Diameter and
with Foil on Outside of Foil Lamination*

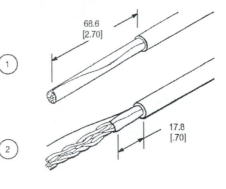

68.6
[2.70]

17.8
[.70]

Figure 2

1 of 3

LOC B
Form 404–25 1/98

4. Align cable in ground clip — with open–side of foil facing out and end of cable jacket flush with top of clip. See Figure 3.

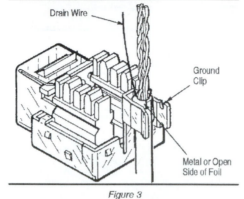

Drain Wire

Ground Clip

Metal or Open Side of Foil

Figure 3

5. Using pliers, pinch tabs on ground clip around cable foil until tabs are parallel. See Figure 4.

Drain Wire

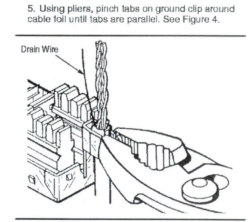

Figure 4

6. Grasp drain wire with needle–nose pliers. Bend wire around clip tabs in a figure–8 pattern, using slots provided in tabs. Use all of drain wire, and end with wire between tabs. Wrap 1 1/2 turns of 12.7–mm [.50–in.] wide vinyl electrical tape around ground clip. See Figure 5.

CAUTION *Exercise caution when installing heat shrink tubing. Excessive heat could damage the connector and wiring.*

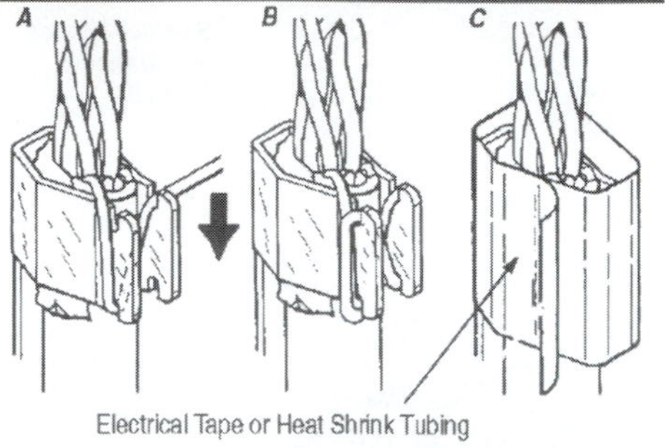

A   B   C

Electrical Tape or Heat Shrink Tubing

Figure 5

7. Bend conductors back at right angle to clip. Do NOT untwist pairs. Bend ground clip forward between blocks. See Figure 6.

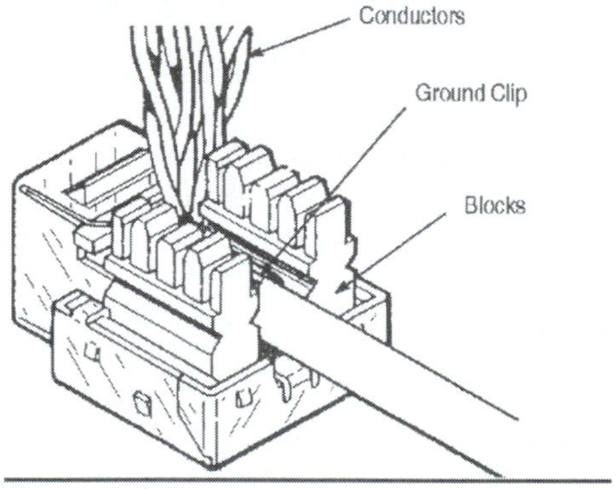

Conductors

Ground Clip

Blocks

Figure 6

8. Pull orange pair and brown pair through openings on opposite sides. Bend wires around blocks and align them with appropriate color code slots (use minimum slack while maintaining twist). See Figure 7.

9. Use tool to punch wires down into slot and cut off excess wire. Remove loose wire pieces.

CAUTION *Set impact setting for low–impact. Orient tool so cutoff blade is on inside of block.*

358   APPENDIX A

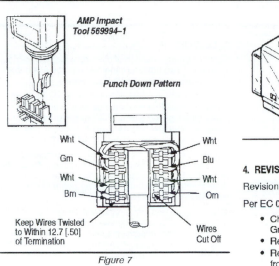

AMP Impact Tool 569994-1

Punch Down Pattern

Wht
Grn
Wht
Brn

Wht
Blu
Wht
Orn

Keep Wires Twisted to Within 12.7 [.50] of Termination

Wires Cut Off

Figure 7

10. Repeat Steps 8 and 9 with blue pair and green pair.

11. Press conductors flat against blocks. For shielded applications, slide the top shield over the housing until it snaps in place. See Figure 8.

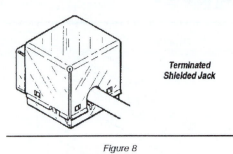

Terminated Shielded Jack

Figure 8

4. REVISION SUMMARY

Revisions to this document include:

Per EC 0990–0611–98:

- Changed title to include "Shielded and Grounded".
- Removed unshielded illustration from Figure 1.
- Removed table of description and part numbers from Figure 1.
- Changed introduction paragraph to include shielded and grounded.
- Removed first sentence from description.
- Removed paragraph in termination section.
- Removed references to unshielded versions.
- Removed unshielded option from Figure 7.

AMP

Coaxial Active Tap Assemblies
228752–[ ] and 415195–[ ]

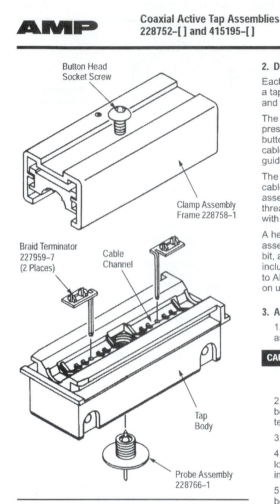

Button Head
Socket Screw

Clamp Assembly
Frame 228758–1

Braid Terminator
227959–7
(2 Places)

Cable
Channel

Tap
Body

Probe Assembly
228766–1

Figure 1

1. INTRODUCTION

This instruction sheet covers the installation of AMP*
Coaxial Active Tap Assemblies 228752–[ ] and
415195–[ ], designed for low–profile application.
Read these instructions thoroughly before starting
assembly.

Reasons for reissue of this instruction sheet are
provided in Section 4, REVISION SUMMARY.

2. DESCRIPTION (Figure 1)

Each assembly consists of a clamp assembly frame,
a tap body, two braid terminators, a probe assembly,
and a protective dust cover (not shown).

The frame, which slides onto the tap body, features a
pressure block that holds the cable in place and a
button head socket screw. The tap body features a
cable channel that retains the cable and an internal
guide slot to accept a printed circuit (pc) board.

The braid terminations are designed to penetrate the
cable jacket and provide a positive ground. The probe
assembly features a spring–loaded contact, and
threads into the tap body, ensuring permanent contact
with the center conductor.

A hex wrench and combination application tool
assembly, which includes a socket wrench and drill
bit, are required to install the taps. These items are
included in AMP Installation Tool Kit 228917–1. Refer
to AMP Instruction Sheet 408–9014 for instructions
on using the tool kit.

3. ASSEMBLY PROCEDURE

1. Using the installation tool kit, remove the probe
assembly from the tap body. See Figure 2.

> **CAUTION** *The probe assembly is installed in the tap body to protect the probe tip during shipment; however, the probe assembly must be removed **before** the cable is installed.*

2. Slide the clamp assembly frame from the tap
body. Check to make sure that the braid
terminators are in the tap body. See Figure 1.

3. Determine tapping location on the cable.

4. Using the hex wrench supplied with the tool kit,
loosen the button head socket screw. Insert cable
into cable channel of tap body.

5. Slide the clamp assembly frame onto the tap
body.

6. Thread the button head socket screw into the
frame until the pressure block bottoms on the track
and holds the cable securely. See Figure 3.

> **CAUTION** *To avoid damage to the frame, do NOT overtighten the button head socket screw.*

7. Using the application tool supplied with the tool
kit, drill a hole through the cable to the center
conductor by rotating the coring tool end. Refer to
Figure 3.

> **NOTE** *The coring tool end is designed with a stop to prevent overdrilling.*

LOC B

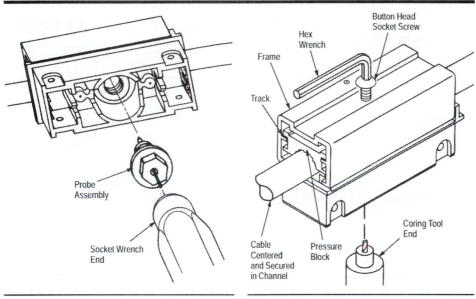

Figure 2

Figure 3

8. Inspect the drilled hole to make sure no particles of the cable shield or other debris remains in the hole.

9. Using the socket wrench end of the application tool, thread the probe assembly into the tap body until the probe assembly bottoms. See Figure 2.

CAUTION *Hand-tighten the probe assembly only. Do NOT overtighten; otherwise damage to the probe assembly could result.*

10. Align the probe and braid terminator posts with the contacts on the pc board, then place tap onto pc board.

11. Secure the tap body to the pc board with screws and nuts, rivets, or other suitable hardware. See Figure 4.

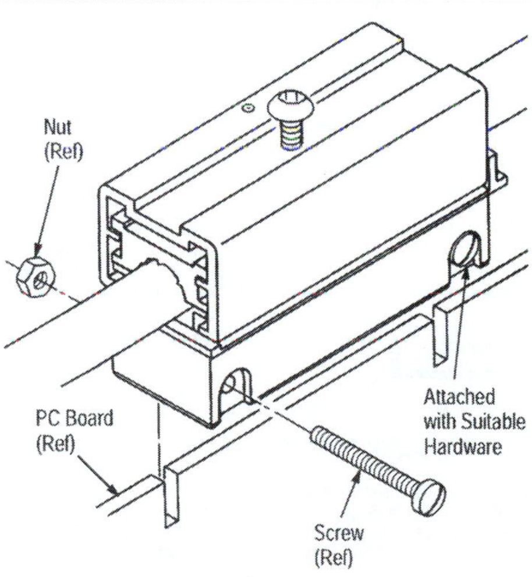

4. REVISION SUMMARY

Revisions to this instruction sheet include:

Per EC 0990-0239-96:

- Changed "kit" to "assembly"
- Added Assembly 415195-[ ]
- Added Step 1 and CAUTION to Section 3
- Modified Steps 2 and 4 in Section 3

Figure 4

AMP* INSTALLATION
TOOL KIT 228917-1
(For Low-Profile Coaxial Active Tap)

IS 9014

RELEASED
7-24-91

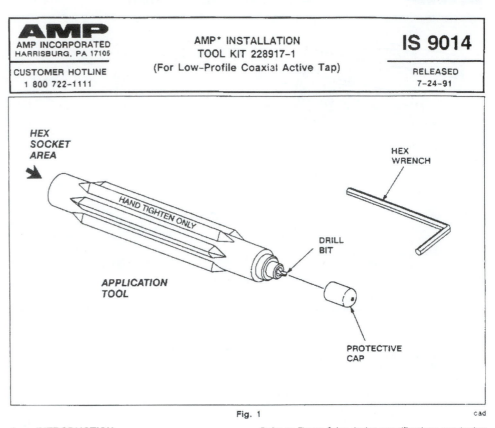

Fig. 1

cad

1. INTRODUCTION

This Instruction Sheet (IS) provides information concerning AMP Installation Tool Kit 228917-1 which is used to install AMP Low-Profile Coaxial Active Taps. Read these instructions thoroughly before using the tool kit.

 NOTE — *All dimensions on this sheet are in inches.*

2. DESCRIPTION (Figure 2)

Each tool kit features a hex wrench and an application tool (with a drill bit and a protective cap). Refer to the instructions packaged with the tap (IS 6814) for information concerning tap installation.

 NOTE — *When using the application tool, follow the inscription on the tool handle* — **HAND TIGHTEN ONLY.**

Refer to Figure 2 for design specifications required to ensure continuous use and reliability of the tools.

3. MAINTENANCE AND INSPECTION

To ensure proper operation, the tool should conform to the dimensions in Figure 2. It is recommended that the tool be inspected immediately on its arrival and at regularly scheduled intervals. The bit is not replaceable and the tool is not repairable. The tool should be replaced when worn or damaged. It is recommended that the tool be stored in a clean, dry place and cleaned with a soft, lint-free cloth. Replacement parts (Figure 2) and additional tools can be purchased from:

CUSTOMER SERVICE (38-35)
AMP INCORPORATED
P.O. BOX 3608
HARRISBURG, PA 17105-3608

1 OF 2

AMP INSTALLATION KIT 228917-1

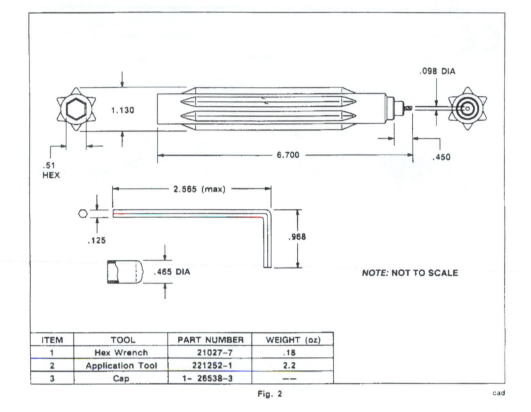

.098 DIA

1.130

.51
HEX

6.700

.450

2.565 (max)

.125

.968

.465 DIA

NOTE: NOT TO SCALE

ITEM	TOOL	PART NUMBER	WEIGHT (oz)
1	Hex Wrench	21027-7	.18
2	Application Tool	221252-1	2.2
3	Cap	1- 26538-3	--

Fig. 2

cad

AMP

LAN–LINE* Thinnet Tap System for LAN–LINE
Tap Assemblies 222503–[ ] and 414913–[ ],
and Drop Cable Assembly 222675–[ ]

Instruction Sheet
408–9365
25 MAR 96   Rev F

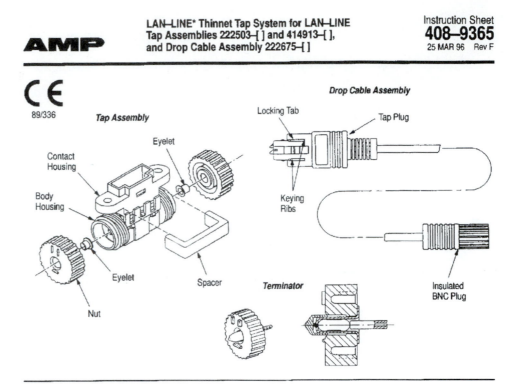

$\mathsf{C}\,\mathsf{E}$
89/336

Tap Assembly

Contact
Housing

Eyelet

Body
Housing

Nut

Eyelet

Spacer

Drop Cable Assembly

Locking Tab

Tap Plug

Keying
Ribs

Insulated
BNC Plug

Terminator

1. INTRODUCTION

This instruction sheet describes the installation and use of AMP* LAN–LINE Tap Assemblies 222503–[ ] and 414913–[ ], which is used in Ethernet network ("Thinnet") applications. Drop Cable Assembly 222675–[ ], Terminator 222504–[ ], and Wall Cover Kit 222754–1 are also covered. The cable assembly connects a device (such as a computer terminal) to the tap. The 50Ω terminator can be installed in the tap assembly for a permanent termination of a network line.

 AMP LAN–LINE Thinnet tapping system is for use in IEEE 802.3/ISO 8802–3, 10Base2 networks.

Reasons for reissue of this sheet are provided in Section 4, REVISION SUMMARY.

2. DESCRIPTION

The tap assembly includes a body housing, a contact housing, two nuts, and a plastic spacer. The center contact of the coaxial network cable connects to the

tap through insulation–displacement type contacts inside the body housing. The cable braided shield makes contact with a metal disc, also inside the body housing.

The tap can be mounted in a wall–mounting plate, or it can be used in free–hanging applications.

The drop–cable assembly is provided as a finished unit. The assembly is made of twin coaxial RG–58–type cable, a tap plug (with keying ribs and a locking tab), and an insulated BNC plug connector that attaches to the device being added to the network.

Terminators are permanently installed in the tap when the tap is located on the end of a network cable, thus defining the end of the network segment.

3. INSTALLATION

3.1. Installing the Tap Assembly

CAUTION *AMP Thinnet Tap is designed for only one application cycle. Once removed from a cable or terminator, the tap cannot be re–used.*

AMP Incorporated, Harrisburg, PA 17105   TOOLING ASSISTANCE CENTER 1–800–722–1111   AMP FAX*/PRODUCT INFO 1–800–522–6752
This AMP controlled document is subject to change. For latest revision call the AMP FAX number.
©Copyright 1996 by AMP Incorporated. All Rights Reserved.
*Trademark

1 of 4

LOC B

CAUTION *Make sure that you have located the proper cable before applying the tap assembly. The tap assembly must be installed to IEEE 802.3/ISO 8802–3, 10Base2 compliant cable.*

1. Turn the system power OFF.

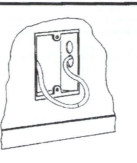

2. Locate the network bus cable and determine the desired location on the cable. Place a mark on the cable at the approximate center of the tap location.

3. Clean–cut the cable (90°) at the mark, then strip 12.7+0.002 mm [.500+.050 in.] of outer jacket from each cable end. If required, strip–length markings are provided on each nut.

4. Carefully slide each wire nut, threaded end outward, onto the appropriate cable end.

5. Flare the braid away from the cable dielectric. **Remove ALL foil shield, making sure that no dielectric is removed.**

6. Slide a metal eyelet, flanged end outward, onto each cable end. While holding the eyelet, slide the nut up to the eyelet and *press the eyelet into the hole in the nut.* Make sure that the cable braid is captured between the eyelet and the nut.

7. Insert each cable end into the respective side of the body housing, making sure to push the cables in as far as possible. The end of the eyelet flange should bottom on the body housing's ground contact, and *no strands from the braid should enter the tap hole.* If using a terminator, insert it into the body housing at this time.

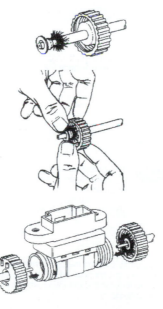

8. Partially screw the nuts onto the body housing. DO NOT tighten the nuts onto the plastic spacer.

9. Remove and discard the plastic spacer, then hand–tighten the nuts.

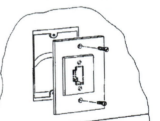

10. Support the face of the contact housing on a firm surface, and push the body housing downward. The two latches on either side of the body housing will engage the slots in the contact housing with a "click."

NOTE	*Check that the two housings are fully engaged. If not, repeat Step 10.*

3.2. Using the Wall Plate

If desired, the tap may be mounted with the optional wall plate. The two screws supplied with the tap are used to mount the tap onto the wall plate. Two longer screws are supplied in the wall plate kit to fasten the wall plate onto a box. To mount the tap with the wall plate:

1. Position the completed tap assembly onto the plate, making sure to align the respective mounting holes.

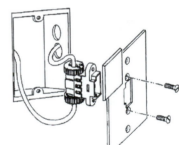

2. Using the two short screws, attach the tap assembly to the wall plate.

CAUTION	*Be sure to use only the screws supplied with the tap when mounting the assembly. Using the longer screws, supplied with the wall plate, may cause the tap housing and the body housing to partially separate, resulting in network failure.*

3. Carefully position the tap assembly into the wall box, making sure to avoid pinching or kinking the bus cable. Align the mounting screws in the wall plate and wall box.

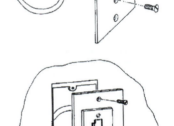

4. Insert the two long screws through the wall plate and into the wall box. Tighten the screws securely.

If desired, the tap assembly can be installed on a panel. Refer to this diagram for the correct cutout dimensions:

3.3. Using the Drop Cable Assembly

To install the drop cable assembly:

1. Align the keying ribs on the drop cable plug with the keyed area of the tap contact housing.

2. Push the plug into the tap until it bottoms completely (the locking tab on the plug should click), then pull back lightly on the plug. If the plug disengages from the tap assembly, repeat the insertion.

To disconnect the drop cable assembly:

1. Lightly push the plug into the tap assembly.

2. While maintaining pressure on the plug, press the plug's locking tab inward.

3. Pull back lightly on the plug. If the plug does not easily disengage from the tap assembly, repeat the disconnection procedure.

When the tap is not in use, the protective dust cover should be rotated to cover the tap cavity.

3.4. Qualification

AMP LAN–LINE Thinnet tapping system is designed to comply and verify with EN55022 Class B and EN50082–1 provided that the driving devices also comply with these standards.

4. REVISION SUMMARY

Since the previous release of this instruction sheet, the following changes and additions were made:

Per EC 0990–0240–96:

- Modified title of instruction sheet
- Added Drop Cable Assembly 222675–[ ] to title
- Added CE logo to page 1
- Added NOTE to Section 1
- Added information to CAUTION in Section 3
- Added Paragraph 3.4

NOTE: Wall Cover Kit 222754–1 includes two 6–32 X 1/2 in. oval head screws and two 6–32 X 3/8 in. oval head screws.

3.86 ± .076 mm [.152 ± .003 in.] Dia, 2 Holes

A   24.89 mm [.980 in.]
B   17.01 mm [.670 in.]
C   11.55 mm [.455 in.]
D   9.22 mm [.363 in.]
E   35.56 mm [1.400 in.]

ETA International Certifications

INTRODUCTION

Evolving Technologies Association International Incorporated (ETA International) is a corporation dedicated to providing technicians with international certifications. By teaming with Wrightco Technologies, Incorporated, an internationally recognized leader in technology training, ETA International is involved with providing an ongoing training component, along with a dedication to curriculum development necessary to ensure current and up-to-date technical training.

ETA's training philosophy is geared towards ensuring that only quality international certifications are awarded. These certifications can be achieved through the application of approved technical education in combination with hands-on testing of the specific technologies and standards.

ETA International certifications signify that the professional technician has successfully demonstrated an understanding of the technical knowledge and a possession of the hands-on proficiencies required in his/her chosen field of technology. These internationally accepted certifications signify the depth of knowledge and the level of expertise, therefore, placing those ETA Certified Technicians among the worldwide elite. Accordingly, ETA certifications are nonvendor-specific, thus providing the assurance of transferability of technical skills among a wide variety of vendor products.

Finally, ETA's professional exams are continually reviewed and upgraded by various international certification committees to ensure that the testing corresponds to relevant and up-to-date industry standards

INDUSTRY RECOGNITION

The recognition of value in the certification process by both business and employee associations has resulted in the growth of education and testing facilities worldwide.

Businesses

Businesses now recognize the benefits of hiring ETA Certified Technicians, and many of them proudly promote the fact that they have hired employees that are certified by the leader in quality technology certifications. Many times it is mandated that certified technicians be employed prior to the awarding of major contracts.

Governmental agencies are also realizing the cost effectiveness of requiring certified technicians to be employed on projects that are funded wholly, or partially, by public funds.

Employees

Employees are finding that an ETA Certification can result in quicker promotions and higher pay than what can be expected by employees without ETA Certifications. As you might expect, persons entering the technology workforce possessing ETA Certifications are usually hired first, and for a greater starting salary, than those without certification.

ETA Certifications Available

ETA International currently offers the technical certifications listed below.

- Enhanced Data Cabling Installer
- Fiber Optics Cable Installer
- Electronic Security Alarm Installer
- Telecommunications Technician
- Wireless Communications Technician
- Network Administration Technician

TAKING AN ETA CERTIFICATION EXAM

The following information is provided for those wishing to sit for one of the ETA Certification examinations listed above.

Registration

You may register for an exam by contacting ETA International at 877-512-3382. Alternately, you may register by completing the necessary forms at our web site **www.etainternational.org**, or you may e-mail us at **cert@etainternational.org**.

An example application form is shown on the following page in Figure B-1.

ETA International

ETA CERTIFICATION INFORMATION ___/___/___

Examination Date

Name:_____ MI:____ Social Security #:_____-_____-_____
Address:_____ City:_____
State:_____ Zip:_____ Phone:(____)-_____-_____

---

1. Which certification course are you interested in?

    _____ETA Fiber Optic Installer Certification
    _____ETA Enhanced Data Cabling Installer Certification
    _____ETA Electronic Security Alarm Installer Certification
    _____ETA Telecommunication Technician Certification
    _____ETA Wireless Communications Technician Certification
    _____ETA Network Administrator Technician Certification

2. Do you have any other ETA certifications?   (   )Yes (   )No

    If "yes", which certificate do you have?_____

3. Are you color blind?   (   )Yes (   )No

4. Employment:

    Present employer _____

    Address _____ City:_____ State:_____

    Length of employment_____ Employer's Phone:_____

    Past employer _____

    Address _____ City:_____ State:_____

    Length of employment_____ Employer's Phone:_____

    Past employer _____

    Address _____ City:_____ State:_____

    Length of employment_____ Employer's Phone:_____

    (If additional space is needed, please use the back of this form.)

I certify to the best of my knowledge the above information is correct

Signature_____ Print_____

FOR OFFICE USE ONLY   ☐ Retake   Grade_____

Payment Received $_____   Received by:_____
   Ck. Number: _____

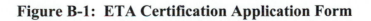

Figure B-1: ETA Certification Application Form

Certification Exam Fees

- Enhanced Data Cabling Installer - $155.00

- Fiber Optics Cable Installer - $150.00

- Electronic Security Alarm Installer - $150.00

- Telecommunications Technician - $155.00

- Wireless Communications Technician - (Available Soon)

- Network Administration Technician - (Available Soon)

Payment for the Certification Exam

You must pay for each exam on or before sitting for the exam, and the exam fees may be paid anytime prior to actually taking it. For methods of payment please contact ETA International at 877-512-3382.

Cancellation

You may cancel, or reschedule, a Certification Exam no later than 48 hours prior to the actually scheduled exam's date and time.

Testing Procedures

Please arrive at the exam site at least 15 minutes prior to the scheduled exam. Bring with you the following items:

- A current photo ID.

- The exam fee, or a receipt of payment of the exam fee.

- A calculator, pen or a pencil.

- A copy of any previous ETA certifications.

- Your tools for the hands-on portion of the certification.

Grading of the Certification Exams

Upon completion of the Certification Exam, your Test Administrator will forward the exam to the international headquarters office, where the exam will be graded. Upon achieving a passing grade of 80%, or better, your certification document will be issued within 30 days.

Certification Exam Retakes

Should you not pass the certification exam at your initial sitting, you are permitted one free retake. This free retake must be completed within thirty days from the grading of your first attempt. During the retake, it is advisable to take the full allotted time to review your exam prior to turning it in to the test administrator. Should you fail to pass the exam with the first free retake, you must wait a minimum of 90 days to sit for an additional retake. Keep in mind that a fee will be charged for any additional retakes.

Those who do not pass their certification exam within one year of their first sitting will be required to begin the entire process again, beginning with a new application for certification, and paying a new fee according to the current standard fee schedule.

ETA International Certified Training

ETA International licenses a selected group of high-quality training institutes to provide a variety of ETA-approved training programs leading to the International Certifications. For information about the approved educational providers, or to apply for becoming an approved educational provider, please contact the ETA International office for additional details.

Glossary

A

Adapter: A device that enables any or all of the following: a.) Different sizes or types of plugs to mate with one another or to fit into a telecommunications outlet/connector, b.) The rearrangement of leads, c.) Large cables with numerous wires to fan out into smaller groups of wires, d.) Interconnection between cables.

Administration: The method for labeling, identification, documentation, and usage needed to implement moves, additions, and changes for the telecommunications infrastructure.

Attenuation: The reduction in the strength of a signal.

American National Standard Institute (ANSI): A private, non-profit, non-governmental national organization which serves as the primary coordinator of standards within the United States in relation to ISO standards.

American Wire Gauge (AWG): Standard measuring gauge for nonferrous conductors. Gauge is a measure of the diameter of the conductor. The higher the AWG number, the thinner the wire.

Analog: Voice or Data signals, which are continuous variables from positive amplitude to negative amplitude.

Anomaly: A location on a network cable, where the impedance of the cable changes abruptly.

B

Backbone: A facility (for example, pathway, cable, or conductors) between telecommunications closets, floor distribution terminals, the entrance facilities, and the equipment rooms within or between buildings.

Bandwidth: The information capacity of a signal, measured in Hz for analog and bps for digital carriers.

Baseband: Uses the entire bandwidth (all of the frequencies) of a medium to send a single signal. Compare to Broadband.

Bonding: The permanent joining of metallic parts to form an electrically conductive path that will assure electrical continuity and the capacity to conduct safely any current likely to be imposed on it.

Bridged Tap: The multiple appearances of the same cable pair at several distribution points.

Broadband: Uses a large range of frequencies over a single medium to transmit various simultaneous signals, using a different frequency as a transmission channel. Compare to Baseband.

C

Cable: An assembly of one or more conductors or optical fibers within an enveloping sheath, constructed so as to permit the use of the conductors singly or in groups.

Cable Sheath: A covering over the conductor assembly that may include one or more metallic members, strength members, or jackets.

Central Office (CO): It is the location that houses a switch to serve local telephone subscribers.

Channel: The end-to-end transmission path between two points at which application-specific equipment is connected.

Characteristic Impedance: Total opposition (DC resistance and AC resistance) to the flow of AC that a network cable would have if the cable were infinitely long.

Closet, Telecommunications: An enclosed space for housing telecommunications equipment, cable terminations, and crossconnect cabling. The closet is the recognized location of the crossconnect between the backbone and horizontal facilities.

Coaxial Cable: A transmission cable consisting of two conductors insulated from one another, and enclosed in a polyethylene jacket.

Computer & Communications Industry Association: A trade organization consisting of data communications, computer, and specialized common carrier services companies.

Conduit: A raceway of circular cross-section of the type permitted under the appropriate electrical code.

Connecting Hardware: A device providing mechanical cable terminations.

Crossconnect: A facility enabling the termination of cable elements and their interconnection, and/or crossconnection, primarily by means of a patch cord or jumper.

Crossconnection: A connection scheme between cabling runs, subsystems, and equipment using patch cords or jumpers that attach to connecting hardware on each end.

Crosstalk: Unwanted signal transfer between adjacent cable pairs.

Customer Provided Equipment: Refers to telephone equipment such as key systems, PBXs, answering machines, fax machines, for example, that are located at the customer's premise.

D

Daisy Chain: In telecommunications, a wiring method where each telephone jack in a premises is wired in series from the previous jack. Daisy chain is not the TIA/EIA preferred wiring method.

DB An abbreviation of decibel. Unit is used to denote the loss or gain of signal strength.

Decibel (dB): A measure unit of signal strength, as a relation between a transmitted signal and a standard signal source.

Demarcation Point (DEMARC): A point where the operational control or ownership changes.

Digital: Voice or data signals that are encoded into a binary state, such as 0's or 1's that represent either on or off.

Distribution Frame: A structure with terminations for connecting the permanent cabling of a facility in such a manner that interconnection or crossconnections may be readily made.

Duct: A single enclosed raceway for wires or cables; a single enclosed raceway for wires or cables usually used in soil or concrete; an enclosure in which air is moved.

E

Entrance Facility, Telecommunications: An entrance to a building for both public and private network service cables including the entrance point at the building wall and continuing to the entrance room or space.

Entrance Point, Telecommunications: The point of emergence of telecommunications conductors through an exterior wall, a concrete floor slab, or from a rigid metal conduit or intermediate metal conduit.

Entrance Room or Space, Telecommunications: A space in which the joining of inter- or intra-building telecom backbone facilities takes place.

Equipment Cable (Cord): A cable or cable assembly used to connect telecommunications equipment to horizontal or backbone cabling.

Equipment Room, Telecommunications: A centralized space for telecommunications equipment that serves the occupants of the building. An equipment room is considered distinct from a telecommunications closet because of the nature or complexity of the equipment.

Event: Any significant occurrence within a system or network that requires attention. A significant occurrence isolated while testing a system or network.

F

Far-End Crosstalk: Interference of two signals traveling in the same direction in a cable, where the stronger signal overrides the weaker signal.

Fiber Optics: A transmission medium that uses glass strands to transmit high data rates through optical signals.

Frequency Division Multiplexing (FDM): A method of transmitting simultaneous signals over a single medium by using different frequencies for each signal.

G

Giga: A prefix referring to 1 billion or 10^9.

Ground: A conducting connection, whether intentional or accidental, between an electrical circuit or equipment and the earth, or to some conducting body that serves in place of the earth.

H

Hertz (Hz): A unit of frequency measurement. The unit is used to express the varying amplitudes of a wave in relation to time. One hertz is equal to one cycle per second.

Heterodyne: To create a new frequency by the interfering with, or the mixing together of, two or more separate frequencies.

Horizontal Cabling: The cabling between and including the telecommunications outlet/connector and the horizontal crossconnect.

Horizontal Crossconnect: A crossconnect of horizontal cabling to other cabling; for example, horizontal, backbone, or equipment.

Hub: A high-speed switch at the center of a network that provides a connection among multiple computers in a star-configured network.

Hybrid Cable: An assembly of two or more cables (of the same or different types or categories) covered by one overall sheath.

I

Impedance: Opposition to the flow of AC signals. Impedance varies with the frequency of the applied signal.

Inductance: The property of a medium that opposes changes in current. It is usually an undesirable characteristic because it causes signal attenuation.

Institute of Electrical Engineers (IEEE): Organization of engineers and electronic professionals that developed the IEEE 802.x standards for the physical and data-link layers of the OSI reference model used for various network configurations.

Insulation Displacement Connector (IDC): A type of wire connection device in which the wire is "punched down" into a double metal holder and as it is, the metal holders displace the insulation from the wire, thus causing the electrical connection to be made.

Interconnection: A connection scheme that provides for the direct connection of a cable to another cable or to an equipment cable without a patch cord or jumper.

Interface: A boundary or connection between two adjacent objects, materials or protocols.

Intermediate Crossconnect: A crossconnect between first level and second level backbone cabling.

Intertone Distortion: The misinterpretation by the receiving circuitry as to which key is being pressed.

J

Jumper: An assembly of twisted pairs without connectors, used to join telecommunications circuits/links at the crossconnect.

L

Link: A transmission path between two points, not including terminal equipment, work area cables, and equipment cables.

Local Area Network (LAN): Nodes networked together in a geographical confined area, such as in a building, college campus, office building, etc.

Logical Topology: The way a network operates, as compared to a physical topology.

Loop: A pair of wires originating at the central office, and terminating at a telephone, or a telephone system at the premise.

M

Main Crossconnect: A crossconnect for first level backbone cables, entrance cables, and equipment cables.

Media, Telecommunications: Wire, cable, or conductors used for telecommunications.

Medium: The physical component, such as a copper wire, optical fiber, or open air waves that carries a transmission signal and/or an electrical current.

Modular Jack: A telecommunications female connector.

Modular Plug: A telecommunications male connector for wires or cords.

Multiplexing: The combination of multiple signals/channels onto a single medium sharing the same bandwidth.

N

Near-End Crosstalk: Interference of two signals traveling in opposite directions in the same cable where the stronger signal overrides the weaker signal.

Node: A device that interfaces to a network through a medium and is capable of communicating with other network devices. For example, computer, printer, server, and repeaters are all called nodes.

Noise: Unwanted random signals that may degrade or distort data signals.

Nominal Velocity of Propagation (NVP): The speed of a signal through a cable, expressed as a percentage of the speed of light. Typically the speed of a signal through a cable is between 60% and 80% of the speed of light.

O

Open: A disruption in an electrical signal that occurs when a continuous conductor has been cut or when two conductors are not in contact with each other.

Outlet Box, Telecommunications: A nonmetallic or metallic box mounted within a wall, floor, or ceiling and used to hold telecommunications outlet/connectors or transition devices.

Outlet/Connector, Telecommunications: A connecting device in the work area on which horizontal cable terminates.

P

Passive: Pertaining to a specific topology that does not actively move the data signal along the network.

Patch Cord: The length of cable with connectors on both ends used to join telecommunications circuits/links at the crossconnect.

Patch Panel: A crossconnect system of connectors that facilitates administration.

Pathway: A facility for the placement of telecommunications cable.

Physical Topology: The physical installation of a network cabling system, as compared to a logical topology.

Plenum: Air ducts and open spaces (over suspended ceilings or under raised flooring) used to distribute environmental air (heat or air conditioning) in order to meet specific fire regulations.

Plenum Cables: Cables certified by UL to be used in plenum areas. Plenum cables are fire retardant and do not emit toxic fumes when burned.

POTS: Plain Old Telephone Service. A regular residential analog telephone.

Private Branch Exchange (PBX): A smaller version of a telephone company's central office switching center that is owned by the customer.

Propagation Delay: The time for an electrical signal to travel the length of a cable.

Pull Tension: The pulling force that can be applied to a cable without affecting specified characteristics for the cable.

Punch-down tool: An impact tool used to terminate copper conductors to a specialized IDC type connector.

Q

Quad Wire: A cable that contains four insulated copper conductors that are not twisted together.

R

Raceway: Any channel designed for holding wires or cable.

Return Loss: The loss of a signal's strength in a cable due to signal reflections.

Reversed Pair: An error in a network' cabling where the pins on a cable are reversed between conductors of the cable.

S

Server: A computer that provides shared resources to other nodes within the network.

Shield (Screen): A metallic layer placed around a conductor or group of conductors, either as the metallic sheath of the cable itself, or as the metallic layer inside a nonmetallic sheath.

Short: A disruption in an electrical signal that occurs when any two conductors are in contact with each other.

Single Mode (SM): A type of fiber optic cable. Data transmission in which modal dispersion is eliminated by radiating only one wavelength of laser light through a narrow core. The cable is limited to a diameter of 8.7/125 microns.

Splice: A joining of conductors generally from separate sheaths.

Star Topology: A networking topology in which each telecommunications outlet/connector is directly cabled to the distribution point.

Switch: A data transmission device that passes data only to a segment of the network containing the destination host.

T

Telecommunications: Any transmission, emission, or reception of signs, signals, writings, images, and sounds, that is information of any nature by cable, radio, optical, or other electromagnetic systems.

Telephony: The technology and the manufacture of telephone equipment that is capable of electrically transmitting sound between distant stations.

Time Division Multiplexing (TDM): Uses all of the frequencies of a medium to send simultaneous transmission signals by using individual time slots for each transmission signal.

Tip/Ring: Terminology originally used by the telephone company to describe the polarity of wire pairs. The Tip was wired to the positive side of the battery, and the Ring was wired to the battery's negative side.

Trunk: A single cable also referred to as backbone or main.

Tone Generator: Test equipment used to generate an audible for locating wire pairs in UTP cable.

Topology: The physical or logical arrangement of a telecommunications system.

Transfer Impedance: The ratio of the induced voltage of the conductors enclosed by the shield to the shield of the cable, connector, or cable assembly.

Transition Point: A location in the horizontal cabling where flat undercarpet cable connects to round cable.

Twisted Pair: A cable pair made of two conductors that are twisted together to minimize the crosstalk to other pairs. The more twists per foot, the greater the reduction in crosstalk.

U

Unshielded Twisted Pair: A twisted pair cable with no shielding within its jacket.

W

Work Area (Workstation): A building space where the occupants interact with telecommunications terminal equipment.

Work Area Cable (Cord): A cable assembly connecting the telecommunications outlet/connector to the terminal equipment.

Acronyms

A

AAL	ATM Adaptation Layer
ACR	Attenuation-to-Crosstalk Ratio
ANSI	American National Standards Institute
ASTM	American Society for Testing and Measuring
ATM	Asynchronous Transfer Mode
AUI	Attachment Unit Interface
AWG	American Wire Gauge

B

BER	Bit Error Rate
BICSI	Building Industry Consulting Service International
BISDN	Broadband Integrated Services Digital Network
BNC	British National Connector/Bayonet-Neil-Concelman
BSR	Board of Standards Review

C

CCIA	Computer and Communications Industry Association
CENELEC	European Committee for Electrotechnical Standardization
CMP	Communications Plenum Cable
CN	Communications General Purpose Cable
CO	Central Office
CP	Consolidation Point
CPE	Customer Provided Equipment or Customer Premise Equipment
CSA	Canadian Standards Association
CSMA/CD	Carrier Sense Multiple Access/Collision Detection

D

dB	Decibel

E

EF	Entrance Facilities
EIA	Electronic Industries Alliance or Electronic Industry Association
ELFEXT	Equal-Level Far-End Crosstalk
EMI	Electromagnetic Interference
EP	Entrance Point
ER	Equipment Room

F

FCC	Federal Communications Commission
FD	Floor Distributor
FDDI	Fiber Distributed Data Interface
FDR	Full-Duplex Repeater
FEXT	Far-End Crosstalk
FIPS PUB	Federal Information Processing Standard Publication
FTTD	Fiber To The Desktop

H

HC	Horizontal Crossconnect
HVAC	Heating, Ventilation, and Air Conditioning
Hz	Hertz

I

IAEI	International Association of Electrical Inspectors
IC	Intermediate Crossconnect
ICEA	Insulated Cable Engineers Association
IDC	Insulation Displacement Connector/International Data Corporation
IEC	International Electrotechnical Commission
IEEE	The Institute of Electrical and Electronics Engineers
ISDN	Integrated Services Digital Network
ISO	International Organization for Standardization

ITU-TSS	International Telecommunications Union-Telecommunications Standardization Section		**S**	
			SRL	Structural Return Loss
			STP	Shielded Twisted Pair
L				
			T	
LAN	Local Area Network			
LANE	LAN Emulation		TBB	Telecommunications Bonding Backbone
Lbf	Pound force		TC	Telecommunications Closet
			TDM	Time-Division Multiplexing
			TDR	Time Domain Reflectometry
M			TDX	Time Domain Crosstalk
			TGB	Telecommunications Grounding Busbar
Mp/s	Megabits per second		TIA	Telecommunications Industry Association
Mbps	Megabits per second		TO	Telecommunications Outlet
MC	Main Crossconnect		TP	Twisted Pair/Transition Point
MCM	Multiple Circular Mils		TP-PMD	Twisted Pair-Physical Media Dependent
MDF	Main Distribution Frame		TSB	Technical Systems Bulletin/Telecommunications Systems Bulletin
MDI	Medium-Dependent Interface			
MHz	Megahertz		TSSC	Telecommunications Standards Subcommittee
MMO	Modular Multimedia Outlet			
MMTA	MultiMedia Telecommunications Association			
MSAU	Multi-Station Attachment Unit			
MUTOA	Multi-User Telecommunications Outlet Assembly		**U**	
			UDC	Universal Data Connector
			UL	Underwriters Laboratories
N			USOC	Universal Service Order Code
			USTA	United States Telephone Association
NEC	National Electric Code		USTSA	United States Telephone Suppliers Association
NEMA	National Electrical Manufacturers Association		UTP	Unshielded Twisted Pair
NEXT	Near-End Crosstalk			
NFPA	National Fire Protection Association			
NIC	Network Interface Card		**W**	
NIR	Near-end Crosstalk-to-Insertion Loss Ratio			
NVP	Nominal Velocity of Propagation		WA	Work Area
			WAN	Wide Area Network
P			**X**	
PAR	Project Authorization Request			
PBX	Private Branch Exchange		X	Crossconnect
PCM	Pulse-Coded Modulation			
PDAM	Proposed Draft Amendment			
PIN	Project Initiation Notice			
POTS	Plain Old Telephone System			
PVC	Polyvinyl Chloride			
R				
RFI	Radio-Frequency Interference			
RH	Relative Humidity			
Rms	Root mean square			

Index

RC Jet-Cobra Race Car SE-1030

Leave the competition in the dust.....build this 1/18 scale, 2 wheel drive Baja racer from Marcraft. Select one of two forward speeds, punch the turbo power and go for it. The Jet-Cobra features a 5-function pistol grip controller that incorporates a turbo-boost circuit. Independent front and rear suspension provide excellent handling on sharp, high-speed corners. You'll find state-of-the-art IC technology coupled with fundamental transistor circuitry to demonstrate important electronics components such as: RF signal transmission and reception, digital information encoding and decoding, and motor control theory. A 72-page manual provides a thorough understanding of the electronics principles. As low as $36.95.

MARCRAFT Electronics Kits

PC Technician's Tool Kit IC-345

This professional technician's kit contains 29 of the most popular PC service tools to cover most PC service applications. Zipper case is constructed out of durable vinyl with room for an optional DMM and optional CD-ROM service disks. Includes: case (with zipper) has external slash pocket for extra storage (13½" L x 9¾" H x 2 3/8" W), slotted 3/16" screwdriver, IC Inserter, slotted 1/8" screwdriver, 3-prong parts Retriever, Phillips # 1 screwdriver, self-locking tweezers, Phillips # 0 screwdriver, tweezers, precision 4 pc (2 slotted / 2 Phillips) screwdriver set, inspection mirror, driver handle, penlight, # 2 Phillips / slotted ¼" screwdriver bits, 6" adjustable wrench, # T-10 Torx / T-15 Torx screwdriver bits, 4 ½" mini-diagonal, ¼" nut driver, 5" mini-long nose, 3/16" nut driver, anti-static wrist strap, 5" hemostat, part storage tube, IC Extractor. (Note: Meter not included) As low as $36.95

Sonic Rover SE-1029

The Rover is an exciting hands-on electronics project, providing an easy way to learn basic transistor, amplifier and switching circuitry. It teaches the fundamentals of sound detection and amplification, as well as switching circuits, DC motors and gear ratios. The front-mounted microphone sensor can be activated by a touch or noise. When the sensor encounters an object, or hears a loud voice command, it will automatically stop, back up, turn to the left 90 degrees, and then resume its forward motion. A dual-colored LED switches between Green and Red to indicate forward and reverse motion. The 48-page manual details breadboarding explorations and circuit construction for this 9-transistor project. As low as $13.95.

Power Supply SE-1014

project that teaches half-wave, full-wave, and full-wave, and ll-wave bridge rectification. When finished, it's a usable power pply featuring four d.c. output voltage selections. An isolation ansformer is included and is enclosed in a durable case for fety. The manual has 32 pages of information. As low as $9.95

Analog Multimeter SE-1028

The Analog Multimeter teaches the importance of electronic "basics". This project functions as an AC/DC meter and is designed to cover all aspects of meter theory including diode rectification and protection…and how resistors are used to limit current and drop voltage. This kit is first breadboarded and tested in a series of informative circuits and explorations, which are detailed in the 48-page instruction manual. As low as $26.95.

ORDERING INFORMATION

School Purchase orders: Terms are net 30 days.
Direct Student orders: Must be accompanied by check, money order, credit card or shipped C.O.D. Shipping Charge: $5.00 per kit to cover shipping and handling charges, for C.O.D. orders add $5.50.

QUANTITY DISCOUNT PRICE LIST

Model	Description	1-4	5-9	10-99	100+
E-1014	Power Supply	12.95	11.95	10.95	9.95
E-1028	Analog Multimeter	32.95	30.95	28.95	26.95
E-1029	Sonic Rover	17.95	16.95	14.95	13.95
E-1030	R/C Race3 Car	42.95	40.95	37.95	36.95
C-345	Deluxe Tech Tool Kit	39.95	38.95	37.95	36.95

Order Toll Free
1-800-441-6006

MARCRAFT

Marcraft International Corporation
100 N. Morain - 302, Kennewick, WA 99336

Your IT Training Provider
(800) 441-6006

A+ Certification

This book provides you with training necessary for the A+ Certification testing program that certifies the competency of entry-level (6 months experience) computer service technicians. The A+ test contains situational, traditional, and identification types of questions. All of the questions are multiple choice with only one correct answer for each question. The test covers a broad range of hardware and software technologies, but is not bound to any vendor-specific products.

The program is backed by major computer hardware and software vendors, distributors, and resellers. A+ certification signifies that the certified individual possesses the knowledge and skills essential for a successful entry-level (6 months experience) computer service technician, as defined by experts from companies across the industry.

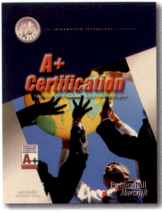

Network+ Certification

Network+ is a CompTIA vendor-neutral certification that measures the technical knowledge of networking professionals with 18-24 months of experience in the IT industry. The test is administered by NCS/VUE and Prometric™. Discount exam vouchers can be purchased from Marcraft.

Earning the Network+ certification indicates that the candidate possesses the knowledge needed to configure and install the TCP/IP client. This exam covers a wide range of vendor and product neutral networking technologies that can also serve as a prerequisite for vendor-specific IT certifications. Network+ has been accepted by the **leading networking vendors** and included in many of their training curricula. The skills and knowledge measured by the certification examination are derived from industry-wide job task analyses and validated through an industry wide survey. The objectives for the certification examination are divided in two distinct groups, Knowledge of Networking Technology and Knowledge of Networking Practices.

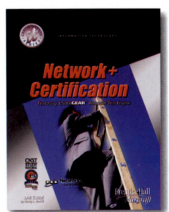

i-Net+ Certification

The i-Net+ certification program is designed specifically for any individual interested in demonstrating baseline technical knowledge that would allow him or her to pursue a variety of Internet-related careers. i-Net+ is a vendor-neutral, entry-level Internet certification program that tests baseline technical knowledge of Internet, Intranet and Extranet technologies, independent of specific Internet-related career roles. Learning objectives and domains examined include Internet basics, Internet clients, development, networking, security and business concepts.

Certification not only helps individuals enter the Internet industry, but also helps managers determine a prospective employee's knowledge and skill level.

Linux+ Certification

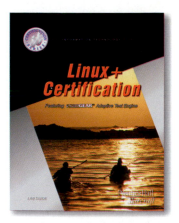

The Linux+ certification measures vendor-neutral Linux knowledge and skills for an individual with at least 6 months practical experience. Linux+ Potential Job Roles: Entry Level Helpdesk, Technical Sales/Marketing, Entry Level Service Technician, Technical Writers, Resellers, Application Developers, Application Customer Service Reps.

Linux+ Exam Objectives Outline: User administration, Connecting to the network, Package Management, Security Concept, Shell Scripting, Networking, Apache web server application, Drivers (installation, updating, removing), Kernel (what it does, why to rebuild), Basic printing, Basic troubleshooting.

Server+ Certification

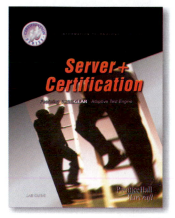

Server+ certification deals with advanced hardware issues such as RAID, SCSI, multiple CPUs, SANs and more. This is vendor-neutral with a broad range of support, including core support by 3Com, Adaptec, Compaq, Hewlett-Packard, IBM, Intel, EDS Innovations Canada, Innovative Productivity, and Marcraft.

This book focuses on complex activities and solving complex problems to ensure servers are functional and applications are available. It provides an in-depth understanding of the planning, installing, configuring, and maintaining servers, including knowledge of server-level hardware implementations, data storage subsystems, data recovery, and I/O subsystems.

Data Cabling Installer Certification

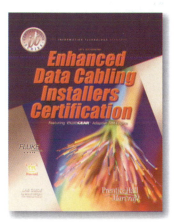

The Data Cabling Installer Certification provides the IT industry with an introductory, vendor-neutral certification for skilled personnel that install Category 5 copper data cabling.

The Marcraft *Enhanced Data Cabling Installer Certification Training Guide* provides students with the knowledge and skills required to pass the Data Cabling Installer Certification exam and become a certified cable installer. The DCIC is recognized nationwide and is the hiring criterion used by major communication companies. Therefore, becoming a certified data cable installer will enhance your job opportunities and career advancement potential.

Fiber Optic Cabling Certification

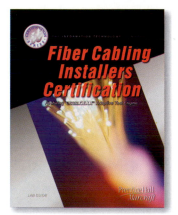

There is a growing demand for qualified cable installers who understand and can implement fiber optic technologies. These technologies cover terminology, techniques, tools and other products in the fiber optic industry. This text/lab book covers basics of fiber optic design discipline, installations, pulling and prepping cables, terminations, testing and safety considerations. Labs will cover ST-compatible and SC connector types, both multimedia and single mode cables and connectors. Learn about insertion loss, optical time domain reflectometry, and reflectance. Cover mechanical and fusion splices and troubleshooting cable systems. This Text/Lab covers the theory and hands-on skills needed to prepare you for fiber optic entry-level certification.

You've earned it!

You've studied hard and now it's time to take the certification exam.

Call Marcraft toll free to take advantage of your **EDUCATIONAL DISCOUNT COUPON.** See coupon for details.

Call toll free **1-800-441-6006**

MARCRAFT

E-mail: discountvoucher@mic-inc.com